# CAREER OPPORTUNITIES in BIOTECHNOLOGY and DRUG DEVELOPMENT

## ALSO FROM COLD SPRING HARBOR LABORATORY PRESS

### Handbooks

*An Illustrated Chinese-English Guide for Biomedical Scientists*

*At the Bench: A Laboratory Navigator,* Updated Edition

*At the Helm: A Laboratory Navigator*

*Experimental Design for Biologists*

*Lab Dynamics: Management Skills for Scientists*

*Lab Math: A Handbook of Measurements, Calculations, and Other Quantitative Skills for Use at the Bench*

*Lab Ref, Volume 1: A Handbook of Recipes, Reagents, and Other Reference Tools for Use at the Bench*

*Lab Ref, Volume 2: A Handbook of Recipes, Reagents, and Other Reference Tools for Use at the Bench*

*Safety Sense: A Laboratory Guide,* 2nd Edition

### Histories and General Science Interest

*Abraham Lincoln's DNA and Other Adventures in Genetics*

*Is It in Your Genes? The Influence of Genes on Common Disorders and Diseases That Affect You and Your Family*

*Pandora's Baby: How the First Test Tube Babies Sparked the Reproductive Revolution*

*The Strongest Boy in the World: How Genetic Information Is Reshaping Our Lives*

*Times of Triumph, Times of Doubt: Science and the Battle for Public Trust*

*The Unfit: A History of a Bad Idea*

*The Writing Life of James D. Watson*

# CAREER OPPORTUNITIES in BIOTECHNOLOGY and DRUG DEVELOPMENT

Toby Freedman

COLD SPRING HARBOR LABORATORY PRESS
Cold Spring Harbor, New York • www.cshlpress.org

**CAREER OPPORTUNITIES IN BIOTECHNOLOGY AND DRUG DEVELOPMENT**


Printed in the United States of America

| | |
|---|---|
| **Publisher** | John Inglis |
| **Acquisition Editor** | David Crotty |
| **Development Director** | Jan Argentine |
| **Developmental Editors** | Tracy Kuhlman, Siân Curtis, Michael Zierler, and Rachel Lagodka |
| **Project Coordinator** | Mary Cozza |
| **Production Editor** | Pat Barker |
| **Desktop Editor** | Lauren Heller |
| **Production Manager** | Denise Weiss |
| **Marketing Manager** | Ingrid Benirschke |
| **Sales Account Manager** | Elizabeth Powers |
| **Cover Designer** | Ed Atkeson |

Library of Congress Cataloging-in-Publication Data

Freedman, Toby.
Career opportunities in biotechnology and drug development / Toby Freedman.
p. cm.
Includes bibliographical references and index.
ISBN 978-0-879698-80-5 (pbk.: alk. paper)
1. Biotechnology--Vocational guidance. 2. Drug development--Vocational guidance. 3. Life sciences--Vocational guidance. I. Title.

TP248.2.F737 2007
660.6023--dc22

2007022863



For a complete catalog of all Cold Spring Harbor Laboratory Press publications, visit our website at www.cshlpress.org.

All World Wide Web addresses are accurate to the best of our knowledge at the time of printing.

# Contents

# Preface

I WAS INSPIRED TO WRITE THIS BOOK BECAUSE I WANTED to assist talented people in their search for satisfying employment in the life sciences industry. Since academia generally does not prepare individuals for careers in industry, this book aims to serve as a resource for making the transition—a transition that I myself have made. Given the complexity and large variety of careers available, my intent is to help job seekers make meaningful and informed decisions about their careers and to provide enough information to enable them to find the job that best utilizes their natural capabilities, educational backgrounds, and interests.

The material for this book was generated from interviews with more than 200 biotechnology and pharmaceutical executives, many of whom work at the vice president level. These contributors graciously and generously shared information about their career paths and the fields that they had chosen—information that some wished had been available before they finally found their career niche.

While you are reading, keep in mind that although every effort has been made to be comprehensive and accurate, the biotechnology and drug development industry is large, and details will vary from company to company. In addition, the perceptions of each interviewee will differ, based on his or her personal experiences. To obtain a refined, well-rounded perspective containing alternative viewpoints, an average of ten people were interviewed for each chapter.

This book describes in detail a wide variety of careers in biotechnology and drug development. Each of the main chapters presents a different career area, with descriptions of the different types of jobs and their day-to-day responsibilities, insights into the pros and cons of the job and what it takes to be successful, practical tips for obtaining employment, and much more. A resource guide is included at the end of each chapter. In addition, several introductory chapters provide guidance about resumes, job searches, and informational interviewing. The highlighted "points of interest" that appear alongside the text throughout the chapters were derived from the interviews. The book is written in a format that allows easy scanning, so that you can quickly find relevant information about the particular jobs in which you are interested.

I wish each of you great success in your career. May it be enjoyable and productive, because we vitally need more individuals to help discover and develop new products for the many serious and chronic illnesses that still remain without adequate medical treatment options.

If you would like to provide feedback for use in future editions of this book, please feel free to contact me or visit my Web site at www.careersbiotech.com/.

# Acknowledgments

THE CONTENT OF THIS BOOK IS BASED ON INTERVIEWS granted by more than 200 industry executives. Most of the interviews lasted an hour or two (sometimes more!), and I am very grateful for the generosity of those who participated in this comprehensive project. Many people reviewed drafts of the chapters to which they had contributed, often in addition to being interviewed.

I acknowledge with thanks and deep gratitude the following individuals for their contribution to this project, in addition to those who chose to remain anonymous: Mircea Achiriloaie, Angelie Agarwal, Priya Akkihal, Betsy Alberty, Detlef Albrecht, Linda Anderson, Paul Anderson, Shari Annes, Jack Anthony, Ximena Ares, Paul Armel, James Audet, Laurie Averill, Greg Baigent, James Barrett, Steven Barriere, Michael Biros, Debbie Jo Blank, Leonard Blum, Robert Blum, Lawrence Bock, Bonnie Bowers, Tanya Boyaniwsky, Erin Brubaker, Katherine Call, Joseph Carlino, Casey Case, Mary Cassoni, Lois Chandler, John Choi, Shelley Chu, Paul Clarkson, Suzanne Coberly, Derek Cole, Barbara Coleman Preston, Rebecca Coleman, Thomas Coll, Anthony Czarnik, Deborah Dauber, Christophe Degois, David DeNola, Tina Doede, Joe Don Heath, Meredith Dow, Ramesh Durvasula, Richard Eglen, Ronald Ellis, Klaus Esser, Douglas Fambrough, Chris Fibiger, John Fiddes, Alvan Fisher, Lawrence Fisher, Michael Flashner, Matt Foehr, Jason French, Vikki Friedman, Gina Fusaro, Nancy Gadol, Bruno Gagnon, Chris Garabedian, Sabine Geisel, Jack Geltosky, Karen Georgiou, Krishna Ghosh, Martin Giedlin, Richard Gill, Jane Green, Bill Guyer, Ann Hanham, Judith Hasko, Paul Hawkins, Diane Heditsian, Steven Highlander, Tamar Howson, James Huang, Annabella Illescas, Karin Immergluck, Nancy Isaac, Kent Iverson, Bahija Jallal, Stuart Johnston, Claudia Julina, Michael Kalchman, Deborah Kallick, Mark Karvosky, Tariq Kassum, Daniel Kates, Douglas Kawahara, Brian Kearney, Ravi Kiron, Gautam Kollu, Anne Kopf-Sill, Richard Kornfeld, Thane Kreiner, Diane Krueger, Steven Kuwahara, Daniel Lang, John Leung, Michael Levy, Deborah Lidgate, Jason Lilly, Bill Lindstaedt, Otis Littlefield, Anna Longwell, Michael Louie, Heath Lukatch, Carol Marzetta, Phil McHale, Paula Mendenhall, Angela Miller, Krys Miller, Madeline Miller, Nancy Mize, Sandhya Mohan, Randall Moreadith, Sriram Naganathan, Carol Nast, Richard Newmark, Mika Newton, James Nickas, Larry Norder, Scott Ogg, Julia Owens, Nandan Oza, Barbara Paley, Eric Peacock, Michael Penn, Matthew Perry, Matthew Plunkett, Renee Polizotto, Eric Poma, Olga Potapova, Mark Powell, Michael Powell, Nancy Pryer, Yolanda Puga, Doug Rabert, Rebecca Redman, Carolina Reyes, Dorian Rinella, Nadine Ritter, Charlotte Rogers, Peggy Rogers, Ellen Rose, Daniel Rosenblum, Philip Ross, Ruedi Sandmeier, Molly Schmid,

William Schmidt, Birthe Schnegelsberg, James Schwartz, Sushma Selvarajan, Peter Shabe, Pratik Shah, Andy Shaw, Laura Spiegelman, Olena Stadnyuk, Aron Stein, Alexander Stepanov, Frank Stephenson, Michelle Stoddard, Helen Street, Anantha Sudhakar, Milla Sukonik, Karen Talmadge, Alan Taylor, Robin Taylor, Klaus Theobald, Silke Thode, Elizabeth Tillson, Chris Van Dyke, Kimberly Vanover, Keith Vendola, Martin Verhoef, Peter Virsik, David Walsey, Dara Wambach, Hong Wan, Michael Warner, Robin Wasserman, Anna Waters, Darin Weber, Ken Weber, Jennifer Wee, David Weitz, Phyllis Whiteley, Mary Wieder, Oriana Wiklund, Michael Williams, Eric Witt, Jason Wood, Chris Wubbolt, Robert Yip, Angie You, Peter Young, and Evgeny Zaytsev. In addition, Francesca Freedman, David Grosof, and Jurgen Weber generously provided useful discussions about the book.

## SPECIAL THANKS

Some individuals deserve special thanks for their contributions. First, I would like to express my deep gratitude to Molly Schmid for reviewing and adding her scientific and managerial wisdom to many of the career chapters. I would also like to thank Joseph Carlino, Betsy Alberty, Angelie Agarwal, and Bill Lindstaedt for reviewing several chapters. Chris Garabedian, Anthony Czarnik, and Pratik Shah deserve special recognition for several significant and beneficial discussions.

I owe a great deal of thanks to my editor, Tracy Kuhlman, for her ability to simplify complex technical material. Not only is she a very competent editor, but being a scientist herself, she provided a great deal of intellectual input into the chapters. I also thank the staff at Cold Spring Harbor Laboratory Press, including John Inglis, David Crotty, Jan Argentine, Mary Cozza, Patricia Barker, Lauren Heller, and Denise Weiss for their support and for being accommodating.

I am very grateful to my husband, Peter Symonds, for his editorial input and for providing a nonbiotechnology viewpoint. And last, but most importantly, this book would not have been possible without the input of my father, William Freedman. He tirelessly edited and reviewed each chapter—some of them as many as four or five times—cheerfully and without complaint.

# 1

# The Pros and Cons of Working in Industry

## Why Make the Transition?

MANY PEOPLE ARE ATTRACTED TO THE BIOTECHNOLOGY INDUSTRY because it gives them an opportunity to contribute to the development of drugs that benefit mankind. Although the work can be challenging, it can also be fun and rewarding. You can make the most of your talents and continue to grow while exploring industry's many interesting vocational areas. The science is exciting, and the people are bright, passionate, and highly motivated. You don't have to rely on grants, the compensation is good, and your efforts could have a worldwide effect. The industry is growing, and it serves a basic need in modern society—human health.

*The future of the biotechnology industry holds immense promise.*

### WHY DO SO MANY PEOPLE ENJOY WORKING IN BIOTECHNOLOGY AND DRUG DEVELOPMENT?

The following is a list of reasons why people in biotechnology and drug development enjoy their jobs, based on discussions with over 200 industry professionals who have successfully (and without regret!) made the transition from academia or medical practice.

*If you are doing what you love, then it's not really "work."*

#### *An Opportunity to Make a Positive Contribution to Science and Society*

The satisfaction that comes from developing drugs that promise to benefit mankind is a primary reason that people enjoy their careers in biotechnology and drug development. Industry provides the daily sense that your work is something really worthwhile.

*From a researcher in industry: "I would not have been able to develop drugs and get them into the market had I stayed in academia."*

### A Collegial Environment

Being a part of a team can be very appealing. It is motivating to work with other intelligent people who are working toward a mutually common goal, and a strong sense of camaraderie often develops as a result of sharing the successes and failures of a team effort.

#### The Benefits of Working in Teams

The biotechnology industry is populated with bright, motivated individuals, and when they act as a team, they can accomplish far more than what can be achieved by a solo investigator. Because the people employed in industry's many disciplines are poised and ready to work together in a coordinated fashion, it is much easier to tackle complex projects in a company than in academia.

#### A Focus on Productivity

In academia, especially for graduate students and postdoctoral fellows, much time can pass without a clear sense of progress. In industry, you will have clearly defined individual goals each year, and the focus is on timelines and productivity. Industry is more goal-driven and focused on quickly arriving at productive decisions. Projects don't languish, and those that fail are quickly terminated.

#### Your Efforts Are More Applied

In academia, the end product is knowledge. Industry, however, allows you not only to learn and contribute to science, but also to make products. You will be able to see the application of your endeavors and add value to the company by developing products with commercial and social utility.

*The academic currency is based on publishing papers. In industry, it is based on developing products.*

#### Career Diversity and Personal Growth

Over time, people's interests change and their attitudes evolve. One of the major advantages of working in industry is that there is an abundance of alternative career areas to explore. You can evaluate your own capabilities and seek out new areas that best suit your interests, personality attributes, and skill sets. As your career develops, you will be exposed to a broad range of disciplines, and there will be opportunities to learn new skills and technologies. One position can lead to another, and you may eventually reach areas that, based solely on your academic training, would previously have been inaccessible.

#### A Stimulating Environment

Because biopharma companies are at the forefront of new drug development, there are a myriad of opportunities for learning about the latest advances in technology and science. This, in combination with the daily variety of on-the-job surprises and challenges, keeps the work environment stimulating and the quality of science high.

*The challenges associated with creating new treatments for diseases are boundless. It is difficult to be bored in industry.*

#### *Natural Selection for Those with Exceptionally Good Managerial Skills*

In academia, as long as you publish and receive grants, you can continue in leadership roles regardless of your ability as a manager. In industry, however, difficult personalities have limited options and seldom are given managerial authority.

*Good managers are appreciated and well rewarded in biotechnology.*

Industry places a high value on creating a thriving working environment where employees can excel and be productive, so there is a strong emphasis on developing and retaining high-quality managers. Companies even invest time and money in the training of supervisors. As a result, these employees develop improved managerial skills and gain incremental line-management responsibilities, resulting in better-run organizations. To learn more about management, see Chapters 7 and 22.

#### *Connections to Academia*

Some industry projects involve much collaboration with experts from academia, and many employees remain connected to universities. Companies often encourage scientists and clinicians to continue publishing papers and to make presentations at meetings. This exposure enhances both the individual's and the company's reputation. Often, employees can retain adjunct faculty positions and doctors can continue to treat patients. Careers in discovery and preclinical research, clinical development, medical affairs, technical applications and support, sales (particularly field application scientists), business development, law, and venture capital remain closely associated with academia.

#### *Immense Resources*

Large companies have seemingly unlimited resources and an infrastructure that would be difficult to obtain in academia. It is easier to conduct "big science," research on a grand scale, simply because of the magnitude and expense involved. People in industry generally have access to the latest scientific technology, and there is less reliance on grants.

#### *It's Not a One-Way Door Anymore*

Generally speaking, entry into a career in industry is no longer a one-way door, although there is still some career risk. If you are in basic research, academic positions will continue to be available throughout your career as long as you are productive and continue to publish. For clinical practice, however, depending on the specialty and circumstances, your previous training may become obsolete after a limited time, and a return to active practice may require additional training.

#### *Financial Rewards*

Few become extraordinarily wealthy in the biotechnology industry, but it does pay well in the form of salary and stock options. If you are interested in earning a lot of money, you should consider reading the chapters on health care finance and executive leadership. Additional chapters to review are those on law, business development, management consulting, sales, marketing, and clinical development.

### Career Development and Mentoring

Managers are responsible for the training, mentoring, and career development of employees. Objective and constructive feedback gained from performance reviews can help you define and focus on areas in which improvement is needed.

### Weekends and Vacations

In some biotechnology careers, employees work from 9 to 5. Others demand longer hours, but most allow weekends off. Keep in mind that people do work intensely, but when you go home, depending on your position, you may be able to leave your work at the office or lab. If a 9-to-5 job appeals to you, consider a career in technical support, medical information, quality, market research, or drug safety.

### Fewer Geographic Limitations

Many positions allow employees to work from a home office, alleviating the need to relocate for a new job and making it easier to sustain two-career families. If working from home appeals to you, look for a position in sales, as a medical science liaison, as a patent agent, in life science information management, or as a field application specialist.

## WHY YOU MIGHT THINK TWICE ABOUT JOINING INDUSTRY

*Although there are many advantages of working in biotechnology and drug development, there are also drawbacks. Before you commit to a job in industry, you should be aware of the issues described in the list below.*

### Less Job Security

*You are hired at will—and you can be fired at any time.*

There is little job security in biotechnology and pharmaceutical companies; this is probably the biggest drawback to working in this industry. Technologies can quickly become obsolete and projects can fail. Companies implement cost-cutting measures and there is a growing trend to outsource work overseas. Even the large pharmaceutical companies have mergers and acquisitions, reorganizations, and shifts in strategy. All of these events can lead to layoffs.

*The biotechnology industry has a history of booms and busts; employment rates rise and fall in synchrony with market cycles.*

Small biotechnology companies experience the highest risk, because their financial positions are the most precarious. They develop early discovery platforms with unknown technical issues and tend to have sparse product pipelines and limited resources. If one project fails, the company may be forced to fold, whereas a project failure in a large company may have little effect.

Careers in discovery research, in particular, offer less job security than other careers. This is because during difficult financial times, companies focus their efforts and resources on products closest to generating sales, typically those products in the clinical or product development stages. When planning your career, an initial foray into basic research may be the easiest entrance, but you might want to obtain new skills along the way to facilitate lateral moves into other positions with greater job security, such as in clinical, medical, or regulatory affairs.

Fortunately, layoffs usually do not happen precipitously. Typically, there are warning signs that a department may close—mergers and acquisitions often take years to complete, and it takes many months for the executive team to decide which departments will stay and which will go. Most people do eventually find employment, because hiring companies recognize them as innocent victims of circumstances. Large companies can afford compensation packages for employees who are let go, and they also employ outplacement firms to assist candidates in their career development and job search efforts.

Keep in mind that there are things you can do to minimize the impact of job instability. For example, if you work in a major biopharma hub, such as San Francisco or Boston, chances are good that you will not need to relocate. You can learn new skills to make yourself more marketable. You can work at large companies with robust pipelines; if a project fails, there will be many others to work on. Most importantly, be flexible. If you are asked to work in a therapeutic area about which you know little, try to expand your knowledge base and learn new disciplines.

#### *Lack of Control over Decision Making*

In many vocational areas, the results of your efforts will likely become part of a higher-level, broader picture, where upper management makes strategic decisions. Even if the science is good, your project can be suddenly terminated, based not only on critical safety data, but also on factors such as internal politics and the market. It is a commonly experienced frustration to have spent a significant amount of time, energy, and enthusiasm on a project only to have it terminated because of business-related factors beyond your control.

#### *Stress and Deadlines*

Depending on your job, industry can be quite stressful. Deadlines can be constant. If you thrive in a fast-paced work environment and can handle pressure well, you might enjoy a career in management consulting, health care finance, regulatory affairs, clinical development, business development, or law (not surprisingly, the jobs that pay the most).

#### *Less Autonomy*

In academia and the medical profession, there is considerable autonomy, as long as your research meshes with government funding. You have much more control over your time and the direction of your efforts. These freedoms are not as easy to obtain in industry, especially in lower-level positions, and you might have to adjust your working style. You may, for example, be required to work in therapeutic areas in which you have limited knowledge; decisions about your work may be made without your knowledge or input; and you will likely be expected to adhere to tight timelines. Few positions in industry allow independent lone rangers to flourish.

It takes time to adjust to these aspects of work and to learn to be flexible. It takes self-discipline to learn how to work in a team where decisions are made by consensus.

### *Less Freedom to Explore*

Industry is product-focused and, due to time constraints, your freedom to fully explore new questions and concepts may be limited. You may have to concentrate on only the most urgent tasks. This is probably one of the biggest challenges for new employees from academia.

### *Less Public Recognition for Your Work*

Many companies encourage employees to publish in order to establish a reputation for scientific excellence. Patent considerations and outside competition, however, may prevent you from discussing data or publishing papers.

### *The Team Is What Matters*

Your technical skills are important, but so is your ability to work well in a team (see "Team Player" in Chapter 2).

### *Professional Risk*

If you become removed from the practice of science or medicine for too long, it can be difficult to return to academia or clinical practice. You take a substantial risk when you embark on an industry career.

### *Many More Rules and Regulations*

Technology and drug development are heavily regulated by the FDA and other governmental organizations, so there are many rules to follow. There also may be accompanying paperwork, depending on your position. If you would rather not work in a process-oriented environment, consider a career in discovery research, sales, marketing, or venture capital.

### *Bureaucracy*

When you are part of a large organization, initiatives can move slowly. It can be frustrating to see committee decisions delay processes. Smaller companies tend to be more agile, but they have less infrastructure to help you get your job done. Additionally, the higher up you go, the more meetings there tend to be. This additional burden on your time makes it difficult to get your work done.

### *Travel*

Some positions require extensive travel, which makes it difficult to raise a family or enjoy time off. On the other hand, travel allows you to visit places where you would not otherwise have gone. For those who seek the excitement of travel, check out careers in management consulting, health care finance, clinical development, sales, marketing, business development, project management, or technical applications (field application scientists).

If you do not want to travel, consider law, regulatory affairs, quality, operations, manufacturing, and life science information technology.

#### Less Exposure to Patients

Some companies allow physicians to spend only a limited time attending to patients. Those with medical practice experience may miss this part of their previous careers.

## RECOMMENDED BOOKS AND MAGAZINES

#### Books about Biotechnology

Abate T. 2004. *The biotech investor: How to profit from the coming boom in biotechnology.* Owl Books, New York.

Alberts B., Johnson A., Lewis J., Raff M., Roberts K., and Walter P. 2002. *The molecular biology of the cell,* 4th edition. Garland Science, New York.
*This is the "bible" in cell and molecular biology; it is used extensively as a textbook for college science classes.*

Bazell R. 1998. *Her-2: The making of herceptin, a revolutionary treatment for breast cancer.* Random House, New York.

Ridley M. 2006. *Genome: The autobiography of a species in 23 chapters (PS).* Harper Perennial, New York.

Robbins-Roth C. 2001. *From alchemy to IPO: The business of biotechnology.* Perseus Publishing, Cambridge, Massachusetts.
*This book provides a general historical perspective of the biotechnology industry.*

Tagliferro L. and Bloom M.V. 1999. *The complete idiot's guide to decoding your genes.* Alpha Books, New York.

Werth B. 1994. *The billon dollar molecule: One company's quest for the perfect drug.* Simon & Schuster, New York.

#### Magazines about Biotechnology

*The Scientist, Magazine of the Life Sciences* (www.thescientist.com) is a monthly trade journal that covers the life sciences including research, technology, and business.

*Nature Biotechnology* (www.nature.com/nbt/index.html)

*Science* (www.sciencemag.org)

*Genetic Engineering & Biotechnology News* (www.genengnews.com)

#### Free Biotechnology On-line News Services

BioSpace's GenePool (www.biospace.com)

FierceBioResearch (www.fiercebioresearch.com)

FierceBiotech (www.fiercebiotech.com)

Biotechnology Industry Organization (www.bio.org)

Signals (www.signalsmag.com)

#### Biotechnology News

If you can afford it, BioWorld Online (www.bioworld.com) and BioCentury (www.biocentury.com) are probably two of the best detailed news sources about the biotechnology industry. Another excellent journal is Windhover's *In Vivo* Magazine (www.windhover.com).

#### For Those Interested in Bio-IT

*Bio-IT World Magazine* (www.bioitworld.com)

FierceCIO (www.fiercecio.com)

Scores of other organizations offer free news services for particular areas of interest.

#### Career Books

Anderson N. 2004. *Work with passion: How to do what you love for a living,* 3rd edition. New World Library, Novato, California.

Johnson S. 1998. *Who moved my cheese? An amazing way to deal with change in your work and in your life.* G.P. Putnam's Sons, New York.

Nelson Bolles B. 2006. *What color is your parachute? 2007. A practical manual for job-hunters and career-changers.* Ten Speed Press, Berkeley, California.

#### Biotechnology-specific Career Books

Robbins-Roth C. 2005. *Alternative careers in science: Leaving the ivory tower,* 2nd edition. Elsevier, Amsterdam.

2004. *Making the right moves: A practical guide to scientific management for postdocs and new faculty.* Howard Hughes Medical Institute (Chevy Chase, Maryland) and Burroughs Wellcome Fund (Research Triangle Park, North Carolina) (www.hhmi.org/resources/lab management/moves.html)
*This free book is based on courses held by the Burroughs Wellcome Fund and Howard Hughes Medical Institute.*

Moore F. and Penn M. 2006. *Finding your north.* PotentSci, Emeryville, California. (www.findingyournorth.com)
*This book is especially useful for those who are considering or are in the process of obtaining a scientific or medical degree. It was written by scientists and medical doctors about the personal difficulties they encountered in advanced education.*

#### Books about Management in the Life Sciences

Cohen C. and Cohen S. 2005. *Lab dynamics: Management skills for scientists.* Cold Spring Harbor Laboratory Press, Cold Spring Harbor, New York.

Sapienza A. 2004. *Managing scientists: Leadership strategies in scientific research.* Wiley-Liss, Hoboken, New Jersey.

#### Fun-to-read General Business Books

Collins J. 2001. *Good to great: Why some companies make the leap...and others don't.* HarperCollins, New York.

Gladwell M. 2000. *The tipping point: How little things can make a big difference.* Little, Brown and Company, New York.

Ries A. and Trout J. 2001. *Positioning: The battle for your mind,* 3rd edition. McGraw-Hill, New York.

*Other books by Al Ries and Jack Trout are also recommended.*

### About Working with People

Bolton R. and Bolton D.G. 1996. *People styles at work: Making bad relationships good and good relationships better.* Ridge Associates, New York.

### For Women

Mendell A. 1996. *How men think.* Ballantine Books, New York.

Klaus P. 2003. *Brag! The art of tooting your own horn without blowing it.* Warner Books, New York.

2

# How to Excel in Industry

## What to Expect and What Is Expected of You

*The standards used to measure performance in industry are very different from the ones used in academia!*

AS YOU READ ABOUT THE VARIOUS CAREERS within biotechnology and drug development, you will begin to notice commonalities among the skills and personality traits that contribute to individual success. Companies are only as good as their people, so it is extremely important to find hard-working, intelligent employees who interact well with others and understand the needs and goals of the company. These are the types of people who thrive in an industrial environment.

The following characteristics were mentioned repeatedly by the more than 200 professionals interviewed for the content of this book. Elements of these points may be used as criteria in performance appraisals, so keep them in mind as you consider whether or not industry is the right place for you.

- ***Interpersonal skills.*** Most employees work in a team environment, so the ability to develop rapport and get along with other people is essential. This includes being aware of other people's feelings, being attentive to the temperament of others, and having a positive, can-do attitude (see below, Interpersonal Skills).
- ***Communication skills.*** For most positions, excellent written and verbal communication skills are essential. The team environment of industry requires that you be explicit, objective, and understandable. This requires being able to speak up at meetings, effectively state your opinions, and logically defend arguments without offending others (see below, Interpersonal Skills).
- ***Being a team player.*** Most companies operate in a multidisciplinary and matrixed environment, in which coworkers depend on each other and work as a team (see below, Team Player). For an example of a matrixed organization, see Chapter 7, Figure 7-1.
- ***Diplomacy/influencing skills.*** For managerial positions, it is important to learn how to manage others in order to promote smoothly running, effective teams. Most people in management positions will eventually encounter difficult situations that call for tactful diplomacy.

- ***The ability to multitask and work productively in a fast-paced environment.*** Because of the fast-paced nature of the biotechnology industry, just about every job requires the ability to work on multiple projects and perform several functions.

- ***Being adaptable and flexible.*** Almost nothing is permanent in industry. Projects begin and end, managers come and go, and companies change their priorities. It's a dynamic work environment, and it requires the ability to quickly adapt to new situations. It is important to be flexible and able to consider alternative possibilities and objectively evaluate and accept other people's viewpoints.

*Projects, management, departments, and even companies come and go.*

- ***The ability to think strategically.*** Every career described in this book benefits from the ability to think ahead when planning. To make the best choices, you need to understand complex scenarios and anticipate consequences.

- ***Creative problem-solving skills.*** Obstacles are expected in biotechnology and drug development, so it is helpful if you are a quick thinker who can efficiently evaluate and respond to difficult situations. During such times, it is important to remain focused on solutions instead of problems.

*Nobody likes to work with whiny complainers!*

- ***The ability to understand the customer's point of view.*** There is a customer in every job; even if you are working at the bench, your boss, your peers, or the person down the line from you in product development can be considered customers. It is important to be able to objectively view processes and products from the customer's perspective and to be empathetic with their needs. A consistent theme in these careers, particularly for service functions, is that "happy clients" are the epitome of success.

- ***Being able to see "the forest through the trees."*** It is important to keep the big picture in mind while also recognizing that excellent performance often requires paying attention to the smallest of details. While working on projects, it is beneficial to keep in mind the company's goals, the patent landscape, the competition, the company's financial situation, and a host of other factors.

- ***Being analytical.*** It is to your advantage if you can solve problems by pulling concepts apart and logically reassembling them.

## AND JUST WHAT ARE GOOD "INTERPERSONAL SKILLS," ANYWAY?

Just about everybody includes "interpersonal skills" in their resumes, but what does that actually mean? The professionals interviewed responded to this question with the following points:

- ***Having a consistently positive attitude.*** Interpersonal skills include having a positive, can-do attitude. Obstacles are not seen as sources for complaint but are challenges to be enthusiastically overcome. Having a keen sense of humor is a definite plus.

*People enjoy working with people who enjoy working.*

- *The ability to build rapport and engender trust.* This includes sincerely caring about coworkers and being credible, responsive, and effective. Also mentioned were being empathetic, listening carefully to other people's problems, and offering helpful suggestions; understanding and accepting other people's different viewpoints, being perceptive in analyzing other people's emotions, and being able to fulfill promises.

*It is not merely WHAT you do but HOW you do it that often matters most.*

- *Communicating well with others.* Communication skills are part of having good interpersonal skills. This involves communicating clearly and succinctly; being able to speak openly in a nonconfrontational manner; and being able to work smoothly with people who have different personality types and different backgrounds.

## WHAT DOES IT MEAN TO BE A "TEAM PLAYER?"

A team-player attitude is important when working in a matrixed environment with coworkers from multiple disciplines. Performance evaluations often rate an employee's ability to project a team player mentality. Being a team player includes the following:

*Recognizing that the primary goal is the success of the team.* Most projects in this industry are far too extensive and complicated to be accomplished by a single person; they require the efforts of many people working together. As a team player, it is important to recognize that you are a part of something much larger than yourself. Your colleagues should be able to rely on you to perform high-quality work that promotes the success of the entire program.

*Projects require the efforts of many people working harmoniously together.*

*Interacting well with coworkers.* What you say and do affects other people's feelings and productivity. Being a team player includes being genuine and considerate when you interact with colleagues, being able to negotiate and compromise, and being able to recognize and take responsibility for difficult problems.

*A willingness to share knowledge.* It is important not only to be productive, but also to share your knowledge. For the overall good of the team and the project, you may need to be willing to let go of your own desire for credit and accolades.

*Taking initiative.* For the overall benefit of the program and team, you may need to take the initiative and delve into difficult problems even if they are not your direct responsibility.

*Active participation in team meetings.* Valued team members are active participants in meetings. They listen, understand, ask questions, and try to contribute to discussions. They tend to understand what the team is doing and how its work fits into bigger projects, and they recognize other people's contributions.

*Team players enjoy synergistically collaborating with their colleagues to accomplish more than they could have done independently.*

*Understanding the dynamics of team development.* Every team undergoes a series of defined stages which ultimately lead to a functioning group. It is important to be aware that these dynamics take place and to recognize that a certain amount of conflict is vital for the development of a high-performing team.

# 3

# So You Want a Job in Biotechnology and Drug Development...

## Finding Your Way In

IF YOU ARE CURRENTLY LOOKING FOR A NEW POSITION, finding that right job can be a long, intensive, and sometimes grueling process—in fact, looking for a job can be a full-time job! Depending on the economy and your background, expect to spend at least four to six months finding employment.

*Looking for a job can be a full-time job in itself!*

Seeking employment in industry for the first time can be especially difficult, because you are unproven AND competing against other bright people with prior industry experience. A good approach is to become acquainted with your industry cohorts in your area of scientific expertise and contact them early in the job search process. **Above all else, do not take it personally if you are not immediately successful—do not despair! It is tough out there!** If you are persistent, you will eventually find the job you want.

## HINTS AND TIPS FOR YOUR JOB SEARCH

Included below are some useful pointers that might assist you in your job-seeking efforts. Be sure to also review the recommendations for writing an industry-targeted resume (see Chapter 4), which is your most important marketing tool.

*The most effective approach to finding a job is to use every approach!*

### *Network as much as possible*

Expand your personal network and become acquainted with as many people as possible. A survey conducted by the Science Advisory Board (www.scienceboard.net) revealed that networking is by far the most successful means of finding employment.

*Sometimes it's not what you know but who you know that matters.*

- Attend and speak at as many professional meetings as possible.
- Join local biotechnology associations and attend the meetings regularly.

*Networking increases your net worth!*

- Apply to societies that are in your area of expertise; for example, the American Chemical Society (ACS), the Drug Information Association (DIA), the American Society for Microbiology (ASM), the American Society for Cell Biology (ASCB). Most associations have career Web sites that post job opportunities.
- Join networking societies, including university affiliations and groups centered around extracurricular interests.
- Attend career fairs and visit company representatives to discuss your background and job qualifications.

### *Create business cards for networking purposes*

If you are currently employed, offer your home number and E-mail address to ensure that companies and acquaintances can contact you if you leave your current job. If you do not have a personal E-mail address, set up a Yahoo! or gmail.com E-mail account; they are free. You can purchase inexpensive ($10), professional business cards at on-line companies such as www.vistaprint.com or print them on your own laser printer on special business card stationery found in most office supply stores. Keep the card simple, such as:

*It's a simple matter of statistics: the more you apply, the greater your chances of finding a job.*

**Toby Beth Freedman, Ph.D.**
Author, *Career Opportunities in Biotechnology and Drug Development*

Personal E-mail Address
Phone Number

### *Use the Internet to find a job*

An excellent career site is www.biospace.com, and there are many others (see the list at the end of this chapter). Apply as frequently and to as many positions as possible. Believe it or not, this method does work, and plenty of people find employment this way.

Several of these Web sites, including www.biospace.com, allow you to post your resume on-line for free. Recruiters and human resources personnel use these sites to find people with backgrounds suitable for available positions.

#### *Identify and apply to companies that are likely to need expertise in your specialty*

Keep in mind that your specialty may be in a therapeutic area or have a technical focus. If you know people in those companies, send them your resume directly and inform them that you are looking for employment. An alternative is to send an unsolicited resume to the director or vice president—found on the company's Web site or a meeting announcement. While networking at meetings, keep an eye out for name tags that indicate which people are from companies of interest to you. Introduce yourself, briefly state your interest in employment at the company, and ask for a referral to an appropriate person within the organization, preferably the hiring manager.

- To identify companies in your area of expertise, conduct a Google search (www.google.com) for conferences in your specialty. The conference programs will list the major companies and speakers in the industry.
- If you want to identify companies in your local area, search www.biospace.com.
- You can also search the NIH CRISP database for companies that have received federal grants. Go to www.crisp.cit.nih.gov, click on CRISP query form, select SBIR/STTR under activity, and submit your query. You will find company names and E-mail addresses of patent holders.
- Many public and university libraries provide access on-line or on disk to CorpTech (www.corptech.com), BioScan: Biotechnology Industry Database, Hoovers (www.hoovers.com), or Recombinant Capital (www.recap.com), which allow job seekers to create "most desirable companies" lists by searching the databases for companies according to location or research focus.

#### *If you want to apply to companies that are hot, follow the news*

Subscribe to free, daily E-mails from organizations such as www.biospace.com, www.fiercebiotech.com, or BIO's SmartBrief newsletter, found at www.bio.org. Another excellent source of information is www.signalsmag.com. These Web sites will help you find out which companies are performing well and which have had layoffs, which companies have had recent, large cash infusions from venture capital, and more.

#### *Read books about how to interview well*

Many are available, and they are quick reads. Interviewing information is also available on www.ScienceCareers.com. Practice interviewing, and hone your skills. Take advantage of the career center at your local university to practice mock interviews. Don't be afraid to request informational interviews with people you have met if you are seriously interested in a company.

#### *Interview as much as possible*

Companies interview many people to be confident they have selected the best prospective employee. You should follow the same strategy. The more companies you interview, the more likely you will find your ideal opportunity.

*Always follow your gut instinct*

After an interview, even if everything appears logical on the surface, you should listen carefully to that nagging gut feeling. If your instinct about a position is negative, it is time for some due diligence. Conduct your own research about the company—ask previous employees about the work environment, corporate culture, employee turnover, or the chances of product development success.

*Choose your mentor, not the company*

An extremely important element of a prospective job is your future boss. Be sure that he or she is a person with whom you can work well. During your interview, ask potential colleagues questions such as: What is your boss like to work with? Is your boss fair? What do you like most and least about your boss?

*It's far better to join a mediocre company with a great boss than a great company with a crappy boss.*

*Make sure that you research the companies as much as they research you*

Remember that the interview is not only a chance for a company to learn about you, but it is also your opportunity to learn about your potential work environment.

*Be nice to human resources and other administrative personnel*

It does not matter how little they may understand science or how insignificant their role might appear to you (it is not!)—they can make life difficult or they can perform miracles for you.

*The best time to negotiate your salary is when the job is first offered, not after you are hired*

Do not undervalue yourself! Your salary is set once you are hired, and it sometimes can be difficult to receive substantial raises. If you start out with a low base salary, it may take longer to increase your income, and it will be tremendously disheartening when other, less qualified candidates are hired at higher salaries.

Do your homework beforehand and be prepared to reveal your salary expectations, even on the first interview (i.e., provide an acceptable range—this is not the time to negotiate or discuss compensation details!). It shows that you are genuinely interested in the job. The company may want to know your salary expectations early in the process, just to determine whether you are in the right ballpark. Be sure to ask the human resources person what the company philosophy is regarding salary during your interview. Consult Web sites that offer salary surveys (see recommended Web sites at the end of this chapter) or discuss salary ranges with your friends, favorite recruiters, or human resources professionals who have access to the Radford Survey.

*When presented with a job offer, always negotiate for something*

Negotiations are not limited to salary. Other terms include sign-on bonus, stock options, vacation time, and relocation packages. Negotiating shows the employer that you want the company to respect your credentials and worth.

*Tips for applying to large companies*

- Whenever you send your resume electronically to a company's Web site, be sure to target a specific position by including a job requisition number. Otherwise, your resume may be shunted off to a database version of electronic "nowhere land." Always include an objective so that human resources can better find the right job for you.
- Use a simple Microsoft Word document format for your resume, and do not put your address in a header, footer, or special text box. Special formatting can cause valuable information to be lost or misplaced when the resume is processed by the job application database. Keep in mind that symbols, pictures, and special characters do not often survive importation into databases.
- It is okay to send your resume to multiple job postings at a large company, as long as you include the requisition number. Be strategic: If all of your applications go to the same hiring manager, you run the risk of appearing desperate! Instead, try to choose job postings in different departments.
- If you remain interested in working for a particular company but your resume or contact information has changed, send an updated resume.
- The best way to apply for a position is to be referred by an employee. If you do not have a personal referral, it is better to apply directly for a position on a company Web site than to go through a job-posting Web site such as www.biospace.com. The information on the company Web site is usually more current.
- And finally, the most important rule: **Network, Network, Network!** Most companies offer their employees cash incentives for finding new hires.

*Enjoy the process*

Most people detest the idea of looking for employment, but it can be rather fun. Try to think of job finding and interviewing as an exciting way to make new acquaintances and learn new things. If you present a positive attitude and enjoy the process, your shining optimism could well win the approval of—and employment by—the hiring managers.

## WORKING WITH RECRUITERS

As tempting as it is to try to rely on others, *you* need to find the best position for *yourself*. Recruiters may be helpful, but ultimately they are not working for you. Their client is the hiring company, not the candidate. Recruiters can assist you if they happen to have a requisition for a position that fits your background (this is rare!).

*Contingency recruiters* are paid only if the candidate is hired (often a shotgun approach), whereas *retained recruiters* are paid for the service of finding and recruiting, regardless of who identified the hired candidate (a more targeted approach).

The good recruiters are easy to spot; they can speak your scientific "language," know the companies in the market well, can gauge personalities, and advise on matches with the corporate culture of an organization. They help prepare you for interviews and offer feedback afterward. The best recruiters will provide long-term career advice and offer their personal opinions on whether an opportunity is right for you. They will view your relationship with them as long term, and not simply reduced to one particular placement.

*Be careful who has your resume and where it goes!*

Take note of the following advice when working with recruiters:

- Treat recruiters well—most take notes during your conversation, and your positive or negative tone will be noted forever in their database.
- Try to develop rapport with recruiters that you respect. Help them with their searches by providing recommendations for candidates whom you know. If you are helpful, they will be more likely to call you when they are asked to fill that great opportunity that fits your experience.
- ALWAYS ask recruiters to submit your resume only to companies for positions that you have pre-approved. Once you release your resume to recruiters or a database such as Monster or BioSpace, you lose control over its circulation, and some unethical recruiters may send it, unsolicited, to companies without your knowledge or approval. As a consequence, your resume might inadvertently be sent multiple times to the same companies.
- Work with recruiters who specialize in the life sciences and your area and level of expertise. Some conduct searches specifically for positions in research science, manufacturing and operations, marketing and sales, etc. Recruiters also specialize in levels of experience ranging from research technicians to executives. "Executive search" generally means people with 10–15 years of industry experience.
- It is generally a good idea to send your resume unsolicited to recruiters. Most recruiters will retain your resume and not respond unless there is an opportunity for which you are qualified.
- Perform your own due diligence when considering an opportunity—do not rely only on a recruiter's perception.
- Do not work with recruiters who want to charge you money for their assistance. These people represent employment agencies, and they are generally not trained to be helpful with scientists' career issues.

## WHAT TO LOOK FOR IN A START-UP COMPANY

*Think like a venture capitalist.*

Established, large companies generally provide better job security and have more prestigious reputations, and it is advisable to work for larger companies before joining smaller ones. Start-up companies, on the other hand, can be quite exciting, but the failure rate of such businesses is high. Before joining a start-up, carefully select your most promising opportunities to increase your chances of gaining financial wealth and building an impressive resume. When evaluating companies for employment possibilities, think like a venture capitalist. What do venture capitalists look for?

### *Management*

Venture capitalists invest in management teams with a track record of success. When assessing management teams, which are listed on a company's Web site, look for the following:

- A history of success
- Relevant industry experience
- A balanced team (expertise is spread over various areas)
- A group of people surrounding the founders (inventors) who have extensive operational experience and a reputation for getting things done

### *Technology and Market Potential*

Technology is perhaps as important as management. Consider the competitive landscape and market opportunity. Ask yourself if the technology is credible, promising, and based on solid scientific ground.

What to look for:

- Strong scientific rationale with proof of principle
- Large market potential and awareness of the competitive landscape
- Strong intellectual property (IP) position
- A straightforward product development and regulatory path
- Widely applicable, revolutionary, or innovative technology
- A balanced portfolio with multiple products at various stages of development

What you can do to assess the technology:

- Search for related articles on the Internet (e.g., www.google.com).
- Conduct a Boolean patent search at the U.S. Patent and Trademark Office Web site (www.uspto.gov) to identify and evaluate patents held by the company and its competitors. You can also conduct patent searches on the World Intellectual Property Digital Library (WIPO; www.wipo.org).
- Survey your friends about the technology and the company's reputation.
- Contact experts in the field.
- Review previous conferences on the subject and learn who the invited speakers were.

### *Investors and Financing*

*You'd better be certain that the company can be financed for both the short and long term.*

A less obvious but equally effective way to assess the potential of a company is to evaluate its investors. If the company is backed by top-class investors, other high-quality investors will be more likely to participate in serial financing rounds. People often assume that top-notch investors have conducted extensive due diligence before investing, so companies in which they invest *must* be good. On the other hand, if the initial investors have a tarnished reputation, why would top-rate investors want to risk joining them in subsequent rounds?

Because significant investors usually earn a board seat (so that they can watch how their money is being managed), you can assess investors' merit by reviewing the board members' backgrounds on the company's Web site. Look for investors with extensive experience funding successful biotechnology companies and preferably with previous experience in the industry. If the members of the board are industry outsiders or have invested in unsuccessful companies, you have good reason to be more cautious.

There are many ways to evaluate public companies. Some indicators are technology position, management, number of products on the market, and stock performance. You can read about the stock performance of public companies at Yahoo's finance Web site (http://finance.yahoo.com). In addition, you can visit the U.S. Securities and Exchange Commission (SEC) Web site (www.sec.gov; follow the EDGAR link), enter the company's ticker symbol, and read the relevant documentation. Available information includes how many shares the CEO and senior management team own, how much they are selling, the salaries of the executive team, how well the company is performing, and more.

### *Corporate Culture*

*Each company has its own unique corporate culture.*

Corporate culture (a company's management style) eventually will have a significant impact on your work environment, so it is important to assess early in the interviewing process whether you will be able to excel in that environment.

Evaluating the nature of a corporate culture before joining a company can save you from the aggravation of working in a difficult environment or having a short stay in a company. In companies with a productive corporate culture, people generally work harmoniously together. In other companies, people may be contentious, and there is often a high level of personnel turnover as a result.

After an interview, try to consider the little things that might define the culture. Think about the following...

- How pleasantly did the receptionist greet you at the door?
- Look for small details, such as the cleanliness of bathrooms—these places tend to reflect a sense of pride (or not).
- How were you treated by employees?
- Were the employees you met passionate, enthusiastic, or skeptical? What was their mood?

- If you asked different employees an important question and received consistent answers, chances are that there is good company-wide communication. If you did not receive the same answer from everyone, it may indicate a lack of effective communication in the company.
- Watch for the "revolving door" phenomenon—a constant turnover of employees signals an unstable and unpleasant work environment.
- During an interview, most employees will usually only say good things about the company. The people who can speak most candidly about a company are usually previous employees. Keep in mind that the people you talk with may have views that have been skewed by personal experience. A former employee who left the company unwillingly, for example, may offer only negative comments about the place.

*The single best way to receive an honest and candid assessment of a company is to speak to ex-employees.*

> ***The rumor mill***
>
> Ask your friends about companies and hiring managers. If you have a limited network, there is a biotechnology rumor mill at www.biofind.com. You can read anonymous, candid comments about companies posted by former and present employees.

## CAREERS IN BIOTECHNOLOGY AND DRUG DEVELOPMENT FOR NON-SCIENTISTS

Entering the complex and sophisticated biotechnology and pharmaceutical industries can be a daunting task for those without science backgrounds.

There are several functional areas in biotechnology that do not require a science background, such as human resources, finance, engineering, facilities, and some information technology (IT) fields. There are other areas where a science degree is advantageous but not necessary: legal careers, marketing, finance, supply chain management, corporate affairs, sales, project management, and more.

There are many people without science backgrounds who have been quite successful in biotechnology. Here are some ways they made the leap:

- Network! Most people in biotechnology without science backgrounds got there by networking.
- It is advantageous to have a skill that is unique for a biotechnology company; e.g., you have had experience helping a company go public.
- Expect the possibility of a downgrade in your title, level of responsibility, or pay when you enter a new industry. There is a price to pay for apprenticeship!
- Consider alternatives to biotechnology companies. Pharmaceutical companies and contract research organizations offer many different kinds of jobs. Also consider medical

device and bio-IT companies. Medical device companies, in particular, offer highly attractive careers with the same altruistic appeals as biotechnology. In addition, product development takes less time, offering better industry dynamics. Bio-IT is a growing industry in which many companies need computer scientists and engineers.

## RECOMMENDED RESOURCES

### Web Sites that Offer Job Postings

***Biotechnology and Drug Development Job Sites***

BioSpace (www.biospace.com)

Craigslist (www.craigslist.org)

***Scientific Organizations***

American Association of Pharmaceutical Scientists (www.aapspharmaceutica.com)

American Chemical Society (ACS; acswebcontent.acs.org/home.html)

The American Society for Cell Biology (ASCB; www.ascb.org)

American Society for Microbiology (ASM; www.asm.org)

*Chemical & Engineering News*-Chemjobs (www.cen-chemjobs.org)

Drug Information Association (DIA; www.diahome.org)

Federation of American Societies for Experimental Biology (FASEB; www.faseb.org)

International Scientific Products Exchange (www.ispex.ca/companies/employment.html)

ISPE (www.ispe.org)

Thomson CenterWatch (www.centerwatch.com/jobwatch)

U.S. Food and Drug Administration (www.fda.gov)

### Salary Surveys

The American Association for the Advancement of Science, (AAAS; www.sciencemag.org) offers job satisfaction and salary surveys of US-based life scientists.

The American Chemical Society (www.acs.org) publishes *Chemical & Engineering News*, a free weekly magazine which publishes yearly salary surveys; you can apply for the newsletter at www.chemistry.org.

*R&D* magazine (www.rdmag.com) offers job satisfaction and salary surveys.

The Radford Survey is expensive, but most human resources personnel have access.

Salary.com (www.salary.com) may be over-inflated and applies to all industries.

*The Scientist* (www.the-scientist.com), a free science magazine, occasionally publishes salary surveys.

VentureOne (www.ventureone.com) provides a compensation database called CompensationPro, which lists compensation data at various levels, from VPs to technicians, for privately held companies in every industry (you can try the demo for free).

# 4

# The Biotechnology Industry Resume

## Putting Your Best Foot Forward

*The goal of a resume is to illustrate what you can do for a company.*

*Like a thesis, a resume is a work in progress.*

YOUR RESUME IS YOUR PRIMARY MARKETING TOOL and is usually your first introduction to the hiring managers in a company. What human resources (HR) personnel and hiring managers initially look for in a resume are technical skill sets that match the position requirements and a track record of success. You want to set the best possible first impression, so pay particular attention to detail, avoid errors, and be aware of the aesthetic appearance of your resume.

There are many ways in which to draft your resume, and you will undoubtedly receive differing opinions on this. Consider what is written here as a guideline, and incorporate other people's editorial opinions as you see fit. There are many resources available for help, some of which have been listed at the end of this chapter.

1. Scientists tend to be modest about their achievements, but this is one occasion when it is better to show off your greatest accomplishments. Think about how to WOW your audience. What are your greatest WOW factors? Highlight these strengths and achievements and emphasize them early in your resume—don't hide them in the text.

2. Use the order of the sections in your resume to highlight your most significant accomplishments first. It is acceptable to be sparingly redundant. For example:

   - If you attended a top-tier school, list your education first (unless you have been out of school for more than five years).
   - If you published a paper in a high-profile journal, include it under accomplishments or experience—in addition to your publications section.
   - If your work resulted in a patent, put it in the first paragraph under core expertise/accomplishments.

- If you received a prestigious fellowship or award, include a section about awards before education or under experience.
- If you have had research experience with a high-profile principal investigator, list it with your advisor's name in the education section and put that section near the top of your resume.

3. Within the experience section, industry resumes are arranged chronologically beginning with the most recent experience.

4. Include not only what you DID, but also what you ACHIEVED. An achievement is something that was accomplished because of effort or skill, and it often has special merit or significance. For example:

*It's not only about what you DID, it's about what you ACHIEVED.*

   - What you did is "climbed a mountain," but "climbed Mt. Everest" is an achievement.
   - What you did is "studied gene X," but being "the first to identify and characterize gene X, which has been shown to be important in small cell carcinoma" is an achievement.

   If you have trouble discerning this subtle difference, ask yourself these questions about your experiences: "What was the result of this action?" and "Why is it significant?"

5. If you state that you have expertise in a particular area, you need to back it up with facts that validate your claim. Use action words such as "identified," "discovered," and "determined." Consider the following examples when describing your expertise:
   - "great in sales" versus "increased sales by 20% and brought in $1M in revenues"
   - "great molecular biologist" versus "discovered and identified the first gene involved in breast cancer"

6. Don't give yourself credit where credit is not due. Inconsistencies may surface during the interviews or referencing and it's just not worth risking embarrassment or losing a potential employment opportunity.

*Above all else... never exaggerate! It may come back to bite you!*

7. It is difficult to be objective and to know how much or how little information to include in a resume. The best way to build your resume is to consult with your friends or career counselors to help select and highlight your most significant accomplishments. Contrary to popular belief, it is acceptable to have multiple pages. There are, however, dissenting opinions on this topic.

8. State an objective in your resume and cover letter. Make it easy for HR to know where to forward your resume by including the job requisition number and a description of the position that you seek. Examples of objectives are listed in the resume template.

9. Most people tailor their resume to each position for which they apply. It is common to craft several types of resumes to suit different positions.

10. Feel free to apply for positions in which you are interested, not merely the jobs for which you think you are qualified. You won't get what you want if you don't ask for

it. If you are applying for a position for which you have no prior experience, be sure to highlight those areas in your background that might justify why you should be considered for such a position.

11. It is common courtesy to write cover letters. Be sure to include the requisition number of the job for which you are applying and relay to the reader why you are interested in that position or company.

12. Include references in a resume only if you have informed the people that you are looking for a job and have asked permission to list them.

13. E-mail your resume as an attachment compatible with both Macintosh and PC formats. For best results, use Microsoft Word. The hiring manager will be more likely to retain a copy of your resume if it is sent as an attachment. If it is embedded in the text of an E-mail, it might be lost.

14. Use the file name "Last name, First name, date.doc" instead of "resume.doc" for your resume. HR professionals receive hundreds of resumes, and you don't want your file to be lost on their computers amid all of the other "resume.doc" files.

***More resume tips***

Be concise. HR personnel will likely spend no more than 30 seconds scanning your resume, and hiring managers are exceptionally busy people, too. They may not read fine print or long paragraphs. Use bullets whenever possible.

Hiring managers do pay attention to punctuation and grammar. You may not get past HR if there are grammatical or punctuation errors. Use the past tense consistently, even when describing your current position of employment. Have your friends read and re-read your resume for typos.

Avoid jargon, abbreviations, acronyms, etc. For example, use "biotechnology" instead of "biotech" and "postdoctoral fellow" instead of "postdoc."

*Spellcheck, Spellchick, Spellheck!!!*

Try to use the same typeface throughout the resume—do not make it difficult to review. Use italics or boldface when appropriate, but sparingly. Do not use fancy formatting with lines and shading or double columns. Single spacing is hip, but not everyone agrees. You want your resume to be as aesthetically pleasing as possible. If you submit your resume on-line, be aware that any rules specified by companies for electronic resume submission (e.g., no boldface) need to be strictly followed.

Spell out numbers one through ten. Use numerals for 11 and above.

Be sure to add the page number and your name in the header or footer on each page of your resume. Do not bury your contact information in headers and footers, as these are routinely garbled or ignored by database import software.

## RECOMMENDED RESOURCES

***Universities Often Have Useful Career Services Web Sites, Such as:***

http://saawww.ucsf.edu/career/ (University of California at San Francisco)

http://www.vpul.upenn.edu/careerservices/gradstud/ (University of Pennsylvania)

***Resume Samples and Books Can Be Found at:***

http://saawww.ucsf.edu/career/studentpostdoc/lifejobkit.htm

http://www.quintcareers.com/resume_books.html (Quintessential Careers: Resume Books)

***Other Places to Find Good Advice Include:***

http://nextwave.sciencemag.org/ (Science's Next Wave)

http://www.medzilla.com/articles.html (Medzilla)

http://www.jobhuntersbible.com/ (the Web site associated with the book *What Color is Your Parachute?* by Richard Bolles)

The following is a composite example of a scientist's resume. Use it only as a template, and modify it as is most appropriate for your background.

**Gene Eric, Ph.D.**
1234 Oak Lane
Middletown, CA 91234
555-123-4567 (home) or 555-987-6543 (cell)
generic@emailaddress.com
555-135-2468 (FAX)
Date

**OBJECTIVE:** *The objective should include your background and the position you seek. State what you want to do and what kind of functional areas you are interested in; i.e., research, business development, project management, etc. Examples include the following:*

Biochemist seeking a senior scientist position in research and development.
Senior biochemist seeking a sales representative position.
Executive with five years of CEO experience seeking management opportunity.
Organic chemist with five years of process development experience seeking a position as a project manager.

**CORE EXPERTISE/SUMMARY**

*Consider these questions: What are you particularly good at? What is your core expertise? What are your most significant accomplishments?*

- Specialist in signal transduction pathways in the nematode *Caenorhabditis elegans*
- Extensive knowledge in molecular biology techniques, cloning, protein purification...
- Over five years of experience in stem cell therapy; designed the first...
- 16 peer-reviewed publications, reviews, and book chapters
- Ten years of clinical nephrology experience
- Awarded two patents on...
- Five years of project management experience in laboratory automation tools and chemical processing

**SUMMARY OF ACCOMPLISHMENTS**

*This summary is an alternative to the "core expertise" list. What are your most significant accomplishments? List them at the top of your resume.*

- Led, managed, and wrote sections for five INDs for X, Y, and Z drugs
- Developed new drug candidates that resulted in five INDs; two have progressed to Phase II trials
- Generated $3M in sales in one year, awarded Top Sales Representative
- The first person to isolate X gene involved in radiation therapy recovery

**PROFESSIONAL EXPERIENCE**

*List company names and dates, in order of most current position. Except for well-known companies, you should include a brief description of the company and the Web site address. Include the therapeutic area in which you specialized, e.g., oncology or inflammation.*

**Biomedical Company**

Biomedical is an emerging company developing diagnostic products to detect prostate cancer, www.prostatecancerdetectioncompany.com

-2-

Gene Eric

**Team Manager**
**November 2002–present**

- Increased sales by 20%
- Managed a project that progressed to Phase II clinical trials
- Built and developed entire...
- Wrote and awarded an SBIR grant for work on...
- Managed two direct reports and reported to the Vice President of Research
- **Net result:** brought in five new clients, resulting in $400M in revenues

**Postdoctoral Research Fellow**
**California University, Department of Biochemistry, Advisor Dr. John Smith**
**January 1999–October 2002**

*Include not only what you DID, but also what you ACHIEVED.*

- Wrote and was awarded an NIH research grant for work on...
- Established the first cell line to grow X, resulting in the ability to produce...
- Discovered and developed the first...
- Identified and characterized the pABC gene, which has been shown to be involved in breast cancer; this work was published in *Nature (even though this information is repeated in your publications list, it is good to mention it here as well)*
- Managed two direct reports
- Built and developed entire...organized and supervised...

**Doctoral Research Scientist**
**Massachusetts University, Department of Molecular Biology, Advisor Dr. Jane Doe**
**August 1993–December 1998**

- Managed projects, trained and supervised research staff, designed and analyzed molecular biological experiments and presented data at meetings
- Discovered several genes in signal transduction pathways, including one that encodes a novel protein kinase
- Awarded Fellowship from X Foundation (*okay to repeat this in the awards section*)
- Taught graduate students and served as guest lecturer; planned, organized, and conducted recitation sessions

**Intern, University of XYZ, June 1992–July 1993**

- Undergraduate research studying gene expression in *Aspergillus nidulans* in XYZ's lab

**EDUCATION**

*List in chronological order, with the most recent first. Include summer internships, if relevant.*

M.B.A., Northern California University, 2001. Focused on marketing.

Ph.D., Molecular Biology, Department of Molecular Biology, Massachusetts University, 1998. *Novel Protein Kinase Gene in Signal Transduction.* Graduate advisor: Dr. Jane Doe.

M.S., University of Boston, 1994, honors or awards.

B.S., Department of Chemistry, Illinois A&M University, 1993, Magna cum Laude.

**AWARDS**

*List as many as possible—you want to show a track record of success.*

Small Business Innovation Research Grant (SBIR), Principal Investigator, December 2003
NIH Postdoctoral Fellowship, October 2000

**-3-**

Gene Eric

**CONTINUING EDUCATION**
Project Management, Northern California University, 2004
Biotechnology Business Fundamentals, Southern California University, 2002

**TECHNICAL SKILLS SUMMARY**
**Biochemistry**
Six years in biochemistry, protein purification, SDS PAGE...
**Molecular Biology**
Five years of gene cloning, gene expression analysis, PCR, differential display...
**Cell Culture**
Five years of mammalian cell culture experience, including embryonic stem cell culture, DNA damage assays...
**Computer Skills**
*It is assumed that you are familiar with Microsoft Word, PowerPoint, Excel, and the Internet. It is not necessary to add this information unless you have specific computer science training, i.e., C++, Perl, Java, Oracle, etc.*

**PROFESSIONAL SOCIETIES**
*Include only those societies in which membership increases your likelihood of being hired.*
2005 President, Toastmasters International
2001–2007 American Society for the Advancement of Sciences
2000–2007 American Chemical Society

**VISA STATUS**
*If it is applicable, you might want to mention that you are a Permanent Resident.*

**LANGUAGES**
Fluent in Spanish, French, and Italian

**PUBLICATIONS**
*If this section is long (i.e., more than five pages), you might consider including it as a separate attachment. Believe it or not, people read this section. Some applicants boldface their name and each journal. Keep the publications separate from the list of meeting abstracts or invited presentations; otherwise it looks as if you are "padding" your publications list.*

**PATENTS**

**INVITED PRESENTATIONS AND ABSTRACTS**

# 5

# The Informational Interview

## Researching Your Options

IF YOU ARE EXPLORING NEW CAREER DIRECTIONS, an effective way to gather information about a prospective career choice is to conduct an "informational interview." An informational interview is simply an opportunity to meet professionals and ask them about their vocations and career paths. Short of actually working at a job, this is an effective way to determine whether a particular career choice is right for you. It can also save you time and money by preventing you from committing to an advanced degree for a career path that may not be your best fit.

Despite the value of this tool, most people have never conducted informational interviews as part of their career search. Although such an interview can be useful to anyone, it is most effective when considering a radical departure from your current line of work. It provides a risk-free opportunity to explore potential career paths.

When conducting an informational interview, it is important to remember that good business involves developing professional relationships. An informational interview is an excellent way to develop these relationships early in your career. If there is good rapport, the interviewee might become a mentor and serve as a future source of career, professional, and personal advice.

A word of caution: *Finding employment should not be the primary goal of an informational interview.* It is a purposeful, information-gathering conversation and an opportunity to learn about potential career directions. After you have identified a particular career option, you can then submit your resume to the company.

### THE PATH TO A SUCCESSFUL INFORMATIONAL INTERVIEW

#### Getting the Interview

- The best way to find interviewees is through your own personal network. Inform your friends and professional contacts that you are considering a career in a particular field, and ask if they know of individuals whom you can contact for an informational interview. The next best method for identifying potential interview subjects is to network at local scientific organizations and attend professional meetings.

- The best way to contact interviewees is by E-mail. A typical message might read as follows:

  Dear Dr. Jones,

  I have a Ph.D. in Molecular Biology and am currently a postdoctoral fellow at California University. Sids Page recommended that I contact you for an informational interview. I am exploring the transition from academia to industry. I was wondering if you have time to provide some advice about making that transition and discuss your career path in project management. It should take no longer than 20–30 minutes.

  Sincerely,
  Gene Eric
  generic@emailaddress.com
  555-123-4567

- If you do not get a response to your E-mail, try a phone call. Business etiquette dictates a limit of usually one or two E-mails or phone messages. If you do not hear back from your contacts, do not be too persistent—they might be in the middle of an important transaction, on vacation, or on maternity leave—you just never know!
- Be agreeable. If someone refuses to take part in an interview, thank them for their time and move on to the next person. In any field, expect professionals to be extremely busy. Sometimes their schedules do not allow extra time for an interview. Unless you have previous connections to your potential interviewees or have been referred by someone they know, expect only one in ten people to agree to an interview.
- Let your interviewees decide whether they would like to meet you in person. Keep in mind that you are likely to develop more rapport in person than over the phone. Try to pinpoint a date: "How does Friday the 26th sound? How about 2 p.m. at your office?" Provide your cell phone number and E-mail address in case his or her plans change at the last minute. If the interview is more than a week away, send a reminder in case their schedule has changed or they have forgotten about the appointment.
- The best times to call are Tuesday, Wednesday, and Thursday, from 9:00 a.m. to 11:00 a.m. and from 1:00 p.m. to 3:30 p.m. You can sometimes reach people during lunch or after business hours.

### Preparing for the Interview

- Follow the Boy Scout motto: "Be Prepared!" Formulate questions in advance. Read about the company and the interviewee's background. Bring pen and paper to take notes if needed.
- Identify your main objectives before the interview. Focus on what is most relevant to the interviewee's background. For example, if you are interviewing someone in Human

Resources, ask about the corporate culture and hiring practices. If you are interviewing a research scientist, you might ask about recent company publications, the global research focus, or product development.

- Arrive on time. The best way to make a bad first impression is to be late for an interview. Don't be too early, either! Be punctual.
- Wear professional attire and examine yourself in a mirror before the meeting to be sure that you are "pulled together."
- *Avoid the temptation to offer your resume unsolicited!* Remember, the goal of the informational interview is to gather information about a particular career option. If you offer your resume, you are undermining the premise on which the interview was based. It is appropriate to use your resume as a guideline to describe your career goals or to field constructive suggestions. You can also ask the interviewee for advice about improving your resume. If you are only interested in applying for a job, then do just that, and don't waste people's time requesting an informational interview.
- Bring a modest present as a token of your appreciation or, for example, offer to pay for coffee or lunch.

## During the Interview

- When you meet, extend a firm handshake, make eye contact, and smile. Be sure to exchange business cards. Try to be relaxed and think of it as meeting a friend. Try to make it as much fun as possible and develop an enjoyable connection with the interviewee.
- It is appropriate to make "small talk" at the beginning to break the ice, but only for a limited time. You want to be personable and friendly, yet genuine. Try to keep the conversation focused on your main objectives.
- Ask questions pertaining to the career path (see below, Suggested Questions).
- When you speak, be enthusiastic and maintain good eye contact. Avoid saying anything negative, especially about your current or former bosses! If there are potentially awkward situations that may arise during the conversation, be prepared with honest, positive, and brief responses.
- Avoid the temptation of talking too much. Make sure that you allow the interviewee to do most (80%) of the talking. Be an active listener.
- Be conscientious of the time. If the interviewee begins to look at his or her watch, it is time to end the interview or ask if he or she has to leave. If you go beyond 30 minutes, ask if the interviewee has more time to talk.
- Near the end of the interview, ask who else you might contact and if you can use the interviewee's name. Ask what other resources are available for your career search. Remember: *Do not ask whether they are hiring!*

### After the Interview

- Immediately send a thank you note. Reminisce about something special exchanged during the meeting and mention how useful the informational interview was in your career search.
- Continue with interviews until you are convinced you have found the most promising career choice. With this process, you will be better informed and more prepared for applying and interviewing for jobs.
- Most importantly, keep a file of the people with whom you have had informational interviews so that *when you do find a job, you can inform them of your new contact information.* Your interviewee is now a potential mentor. It is your responsibility to stay in contact with your interviewees and use them as resources in guiding your career path.
- As a final gesture of courtesy, inform the person who referred you to the interviewee that you have conducted the interview and thank them again for the helpful referral.

## SUGGESTED QUESTIONS

*About the Interviewee*

- What was your career path?
- Can you tell me about the transition from academia to industry and the differences between working in both?
- Reflecting back on your career, what would you have done differently if you could repeat it?

*About the Job*

- Describe a typical day of work.
- What do you enjoy most about your job? What do you enjoy least?
- How much time do you spend managing people?
- How much time do you spend "putting out fires"?
- What are the hours? Do you travel frequently?
- What advice would you offer about how to be successful in this position? What personality traits are needed to succeed at this job?
- How is success measured in this occupation?
- What career opportunities does this field offer? Where do people in your position typically go next?

- What are the important long-term trends affecting your industry?
- What are the typical pay ranges?
- What is the entry-level pay?

*About Getting a Job*

- What entry-level skills are needed?
- Is advanced education recommended?
- What steps would you recommend that I take to improve my chances of getting a job in your field?
- What is the job market like right now for this occupation? What about the future job market?
- What is the best way to learn about job opportunities in this field?
- What are the best places/ways to find employment in this field?

# 6

# The Biotechnology and Drug Development Industry

## An Overview

THIS BOOK PROVIDES A COMPREHENSIVE DESCRIPTION of the many different types of occupations in biotechnology and drug development. To help you navigate the chapters and to expand your career options, the general layout of industry and the stages of product development are described here.

## INDUSTRY OVERVIEW

### Biotechnology Versus Pharmaceutical Products

The narrow definition of the term "biotechnology" refers to companies that develop "large-molecule" drugs; i.e., biologics such as therapeutic proteins, antibodies, RNA, and DNA. Conversely, pharmaceutical companies develop chemistry-based "small-molecule" products. Over the years, however, these delineations have become blurred. For example, small drug development companies are now categorized under the "biotechnology and biopharma" umbrella, even if they are exclusively developing small-molecule products. Additionally, pharmaceutical companies are undertaking large-molecule biologic programs, and most large biotechnology companies have small-molecule programs. "Biopharma" is the collective term that currently refers to both biotechnology and pharmaceutical drug development companies.

### Career Alternatives to Biopharma

Most biopharma companies prefer to hire candidates with prior industry experience, making it a challenge for aspiring academic entrants. Fortunately, other options exist...

**Table 6-1.** Biotechnology and drug development overview... Where the jobs are

| Drug discovery and development | Biotechnology "tools" | Biotechnology services | Other pharma & biotechnology areas | Government institutions | Medical devices | Academia |
|---|---|---|---|---|---|---|
| Pharmaceutical companies | Reagents and chemical suppliers | Management consulting and accounting firms | Agricultural | Food and Drug Administration (FDA) and CBER | Medical devices | Technology transfer |
| Biotechnology therapeutic companies | Instruments (e.g., microscopes) | Law firms | Industrial biotechnology | Centers for Disease Control (CDC) | Diagnostic companies | Industry-supported labs and institutes |
| Vaccines | Platform companies (e.g., genomics, proteomics, nanotechnology) | Venture capital and investment banking | Molecular diagnostics | National Institutes of Health (NIH) | eHealth | |
| Drug delivery | Bio-IT | Executive search firms | Veterinary companies | U.S. Patent and Trademark Office (USPTO) | | |
| Molecular diagnostics | Software and hardware | Contract research organizations (CROs) | Foundations, social philanthropy | Research institutes and government labs | | |
| | Molecular diagnostics | Contract manufacturers (CMOs) | | Homeland Security & Defense | | |
| | | Research and clinical testing: clinical labs, customized antibodies, etc. | | CIA, FBI, and NASA | | |
| | | Bio-IT | | Trade commissions | | |
| | | Other agencies and niche providers: PR, advertising, market research, medical communications, etc. | | | | |
| | | Consultants | | | | |

### Biotechnology Tools and Services Companies

There are many biotechnology companies that do not produce drugs but instead provide drug development companies with tools such as reagents, instruments, platform technologies, software, and other products (see Table 6-1). There are also service companies, which include law firms, recruiting firms, management consultancies, contract research or manufacturing organizations, market research firms, and others. These companies often hire entry-level applicants. A job in a service firm will likely involve exposure to a broader array of technologies and therapeutics than a position in a small biotechnology company focusing on one product. Once you make it into a service firm, you can learn more about the industry and develop biotechnology contacts, both of which will make it easier to move to a biopharma company.

### Governmental and Research Institutes

Governmental and research institutes interact extensively with biotechnology extensively with biotechnology and drug development companies. Career transitions between the two are relatively common.

### Medical Device Companies

Medical device companies (e.g., companies that produce stents, defibrillators, point-of-care diagnostics, etc.) are not officially considered part of the biotechnology industry. This is changing, however, as more devices are being combined with drugs. Many of the positions, roles, and responsibilities that are part of the medical device sector can also be found in the biotechnology industry.

### Academia

Most people do not consider academia to be part of "industry" per se; however, there are some areas of study that lend themselves to positions in industry. In addition, there is a growing trend for more collaborations between biotechnology and academic institutions. Those who work in technology transfer departments can relatively easily move into business development or patent law industry positions, and people who work in laboratories sponsored by companies, particularly for clinical trials, are positioned to develop relationships that will facilitate transitions.

## PRODUCT DEVELOPMENT OVERVIEW

There are several types of biotechnology and drug development companies. Therapeutic, nontherapeutic, and medical device companies have different product development steps and, therefore, slightly different job responsibilities. A generalized overview of the key steps of product development in each of these company types is described below. For more details, please see the chapters that are listed in parentheses beside each step.

## Therapeutic Companies

### *Discovery and Development (Discovery Research, Preclinical Research, and Bio/Pharmaceutical Product Development)*

Discovery research scientists search for and develop biological or synthesized products (see Fig. 6-1). Chemists optimize products for efficacy, safety, and scale-up.

### *Preclinical Studies (Preclinical Research)*

Studies are conducted to determine the pharmacodynamics of the product in human cells, animal toxicity, formulation, and dosage before entering human clinical trials.

### *Chemical Process Development (Bio/Pharmaceutical Product Development)*

From the time that products are in preclinical studies up until manufacturing, chemists develop methods to optimize and scale-up production.

### *IND Filing (Regulatory Affairs and Clinical Development)*

Investigational New Drug Applications (INDs) are submitted to the U.S. Food and Drug Administration (FDA) to enter initial human clinical Phase I trials. Other countries require similar filings.

### *Clinical Trials I–III (Clinical Development)*

Human clinical trials are run to determine safety and efficacy.

### *Scale-up and Manufacturing (Bio/Pharmaceutical Product Development, Operations, and Quality)*

Before a product is approved, methods are developed for the synthesis and large-scale production of products.

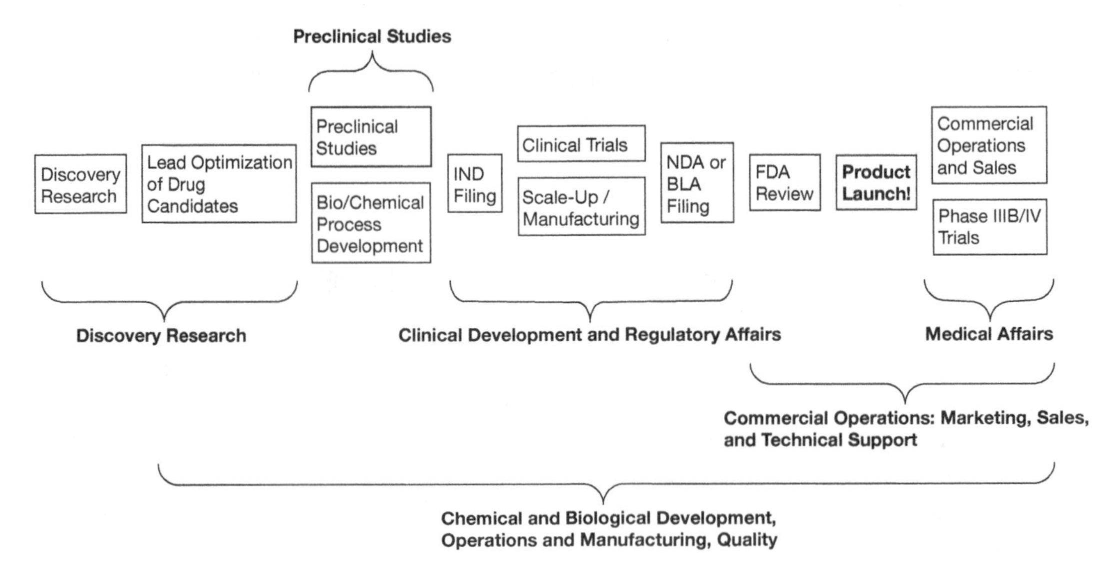


Figure 6-1. Key steps in product development: Therapeutic companies.

#### *NDA or BLA Filings (Regulatory Affairs)*

New Drug Applications (NDAs) or Biologic License Applications (BLAs) are the large regulatory filings submitted to the U.S. health authorities for marketing approval.

#### *FDA (Regulatory Affairs and Clinical Development)*

Regulatory filings are submitted to the U.S. FDA Center for Drug Evaluation and Research (CDER, for small molecules or pharmaceuticals) or Center for Biologics Evaluation and Research (CBER, for biologics).

#### *Launch (Marketing, Sales, and Medical Affairs)*

The product is launched and marketed, and sales professionals work to generate revenue.

#### *Commercial Operations (Marketing, Sales, Medical Affairs, and Operations)*

After the product is launched, brand management promotes the product line for the sales force. Medical science liaisons provide more information to prescribers, personnel in drug safety continue pharmacoviligance monitoring, technical support provides customer assistance, and operations helps define and implement cost-cutting measures to increase profitability.

#### *Phase IIIb/IV Studies (Medical Affairs and Clinical Development)*

Phase IIIb clinical trials are conducted after Phase III data have been submitted to regulatory authorities and before market approval. They are often conducted as continuations of Phase III trials and might further study safety and efficacy or focus on broader population samplings. Phase IV studies are conducted after the drug has been approved by regulatory authorities and typically involve continued drug safety monitoring and further testing on product applications for other disease indications.

### Nontherapeutic Biotechnology "Tools" Companies

Product development takes considerably less time and expense in nontherapeutic tools biotechnology companies, which produce instruments, reagents, diagnostics, and platform technologies, because human clinical trials are not required (see Fig. 6-2).

#### *Idea Generation, Product Concept Definition (Discovery Research)*

Product concept ideas are often developed in response to specific technical gaps. Concepts are frequently based on market research or the expansion of an already existing product, or for enabling drug discovery and development efforts. Eventually, companies specify parameters of the product for either a broad technology area or a defined market niche. These parameters can be determined by the sales and marketing groups, and generally have involved extensive market research and assessment.

#### *Feasibility Studies and Product Definition (Discovery Research and Operations)*

Starting with a general idea, feasibility studies are run to fine-tune the product design so that it includes the features needed by its user base. This step involves determining design

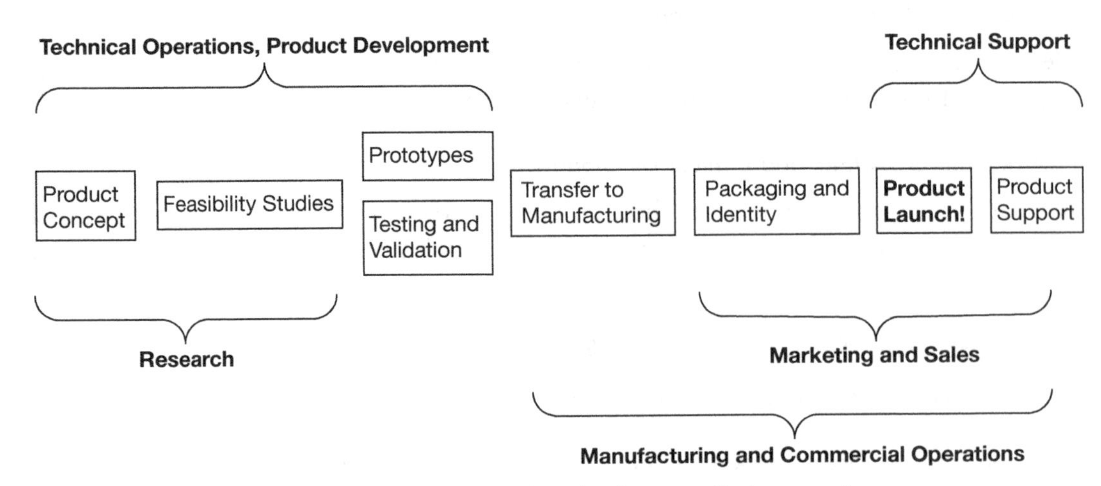


**Figure 6-2.** Key steps in product development: Tool companies.

specifications, experimentation, pursuing new ideas, troubleshooting, and more. Analyses that gauge scale-up potential and the cost of goods are performed to ensure profitability.

*Prototypes and Product Refinement (Discovery Research and Operations)*

Prototypes may be built as the product is optimized.

*Testing and Validation (Quality)*

During alpha testing, the product is tested by in-house scientists and/or external testers. Their feedback is used for further product or specification refinement.

*Transfer to Manufacturing (Operations and Quality)*

After final refinement, the product design is completed and is transferred from research to development, and then to manufacturing. This step involves scaling up manufacturing and ensuring that reproducibility and quality are built into the product. All of the documents that describe how to make the product, including which parts to buy and how to assemble them, along with the quality control testing protocols and specifications, are clearly defined before they are transferred to manufacturing.

*Packaging and Identity (Marketing)*

The packaging, logo, and identity of the product are developed by the marketing department.

*Product Launch and Brand Management (Marketing and Sales)*

Product launch and brand management include training the sales force and technical support teams, developing the marketing collateral, preparing media advertising and peer-reviewed publications, and speaking at conferences. Data that demonstrate performance of the system are collected, and experts are recruited to use and endorse products.

### *Post-launch Product Support (Marketing, Sales, Technical Applications and Support, and Operations)*

After the product is launched, continued brand and life cycle management, as well as technical support, is provided. The potentials for follow-on products and cost-reduction measures are explored.

## Medical Device Companies

### *Product Design (Discovery Research and Operations)*

A medical device idea usually originates with practicing doctors who are using currently available products (see Fig. 6-3).

### *Prototypes and Product Refinement*

New models and prototypes are developed. Some are modifications of existing devices, and some are brand new. Prototypes are studied for performance, conformance to standards, biocompatibility, and more.

### *IDE Filing (Regulatory Affairs)*

Companies submit an Investigational Device Exemption (IDE) application to the FDA before beginning clinical evaluation.

### *Human Clinical Evaluation (Preclinical Research and Clinical Development)*

Sometimes animal models are used before human clinical trials are begun. Clinical evaluation of a medical device's efficacy and safety tends to be much faster and cheaper than testing of biopharma products.

### *Transfer to Manufacturing (Operations and Quality)*

After the prototype has been tested and developed, it is transferred to the manufacturing department for production.

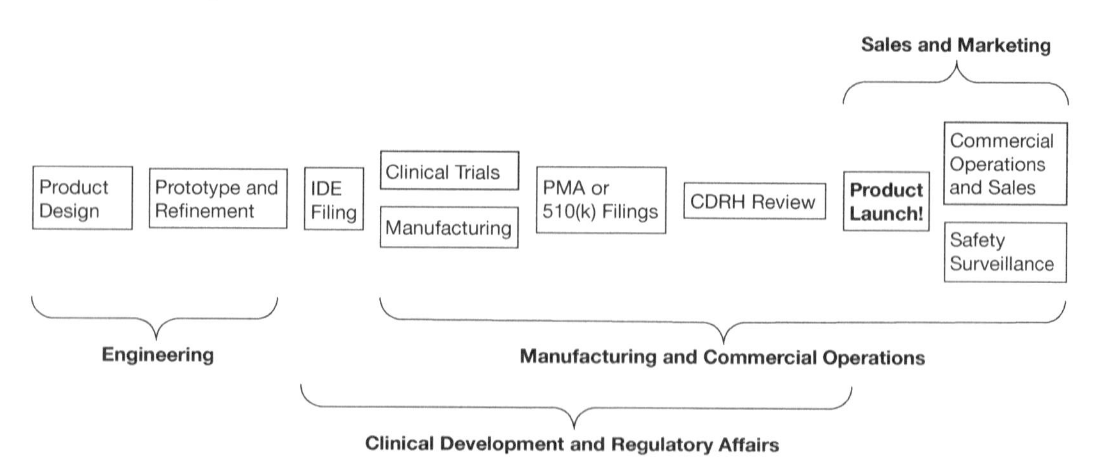


Figure 6-3. Key steps in product development: Medical device companies.

### *PMA or 510(k) Filings (Regulatory Affairs)*

There are two types of applications requesting market approval for medical devices. Premarket approval (PMA) applications are submitted to market new devices; additional clinical evaluation must demonstrate sufficient proof of efficacy and safety in humans. 510(k) applications are for products that are incremental improvements over substantially equivalent products previously approved and already on the market. 510(k)s must prove the product to be effective and safe.

### *CDRH Review*

Regulatory filings are submitted to the Center for Devices and Radiological Health (CDRH) at the FDA for review.

### *Product Launch and Brand Management (Marketing and Sales)*

As in tools and biopharma companies, marketing and sales groups promote the product and manage the brand.

### *Commercial Operations and Sales (Marketing, Sales, Technical Applications and Support, Operations, and Medical Affairs)*

After the product is launched, efforts are made to continue brand management and sales, provide technical support, reduce manufacturing costs, and produce follow-on products.

### *Safety Surveillance (Clinical Development and Medical Affairs)*

Companies continue safety surveillance after the product is approved.

# 7

# Discovery Research

## The Idea Makers

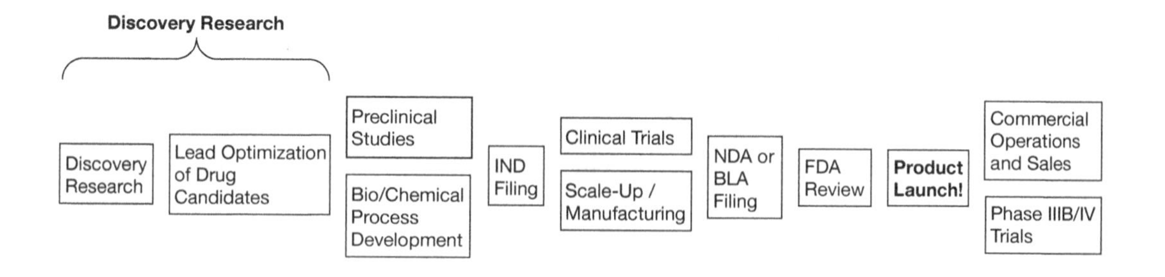


IF YOU THRIVE ON INTENSE SCIENTIFIC EXPLORATION and want to work with a sense of real and immediate purpose, consider applying your training to a career in discovery research. This field attracts bright and talented people who want to benefit humanity.

One of the major advantages of working as a researcher in industry is the team environment. The synergistic power of a team of dynamic and creative people is simply extraordinary: With the right mix of complementary skills and talent, problems that are seemingly impossible can be solved.

*What could be more satisfying than telling your relatives that you are working on a cure for cancer?*

For graduate students and postdoctoral fellows, discovery research is perhaps the easiest route to take from academia to industry. If your interests then evolve, it provides a convenient launching pad to new vocational areas and fields you may not even have considered before.

## THE IMPORTANCE OF DISCOVERY RESEARCH IN BIOTECHNOLOGY AND DRUG DEVELOPMENT

Discovery research is the first step in developing new products. Although this chapter focuses mainly on drug discovery research, there are also many research positions in life science companies that provide platform technologies, instruments, reagents, services, medical devices, and more (see Chapter 6). These products often have the advantage of a much shorter and less costly development path.

*Discovery research is about taking fundamental observations and discoveries and demonstrating their potential future value as drugs for the treatment of human disease.*

## CAREER TRACKS IN DISCOVERY RESEARCH

There are two main career paths in discovery research: research fellow and management tracks. Most people start out at the bench, and as their careers advance, they decide whether to move up the management track or to continue doing research as a specialist and innovator.

### *Management Track*

Line managers typically organize, manage, and lead scientists in a particular discipline. They are often responsible for the budget, hiring, staff performance and development, and performance evaluation. They usually advance from group leader (overseeing 4–6 scientists at various levels) to manager (overseeing more scientists and perhaps more than one group), director (leading one or more departments), and eventually, vice president levels.

### *Scientific or Fellow Track*

Companies recognize that outstanding scientists may not want to step away from experimentation to manage people. As a consequence, they have designed the equally prestigious "staff or principal scientist" or "fellow" titles. This dual career ladder was created to retain exceptional scientists and create a pathway where top-notch performers can continue to advance technically. A researcher in this track generally begins as a scientist and progresses through positions such as senior scientist and staff scientist to fellow, senior principal, or distinguished fellow. Note that such fellow positions are not easy to obtain.

### Discovery Research Interdepartmental Functional Areas

The organization of discovery research departments varies depending on their size and product focus. Some are organized according to therapeutic areas and others according to project objectives or scientific disciplines. The most common configuration is a mixture of the two ("matrixed"), as diagrammed in Figure 7-1. The various roles in a matrixed environment are described as follows:

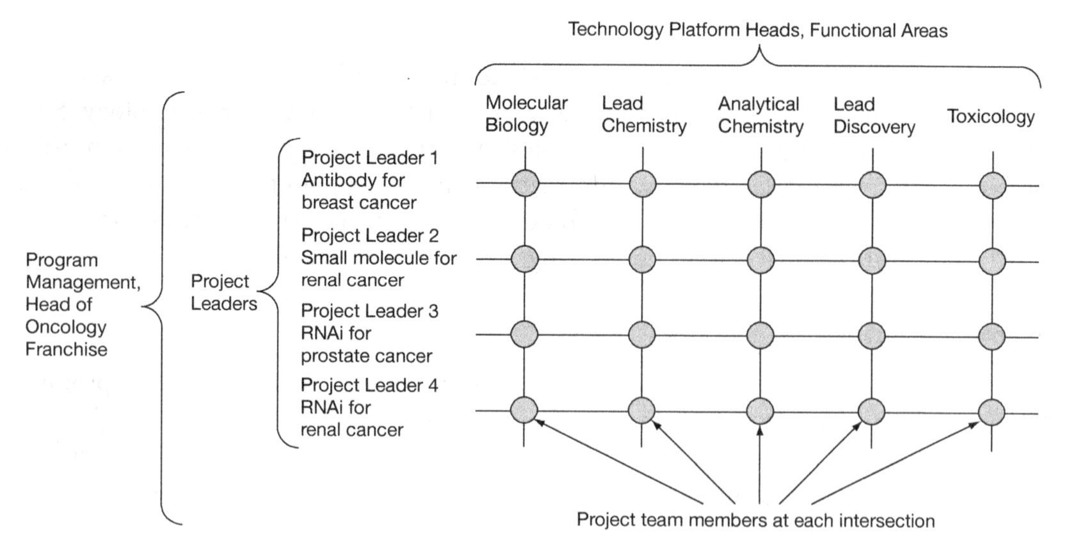


**Figure 7-1.** An example of a matrixed organization.

### *Basic Research*

Basic researchers, including those on the fellow career track, conduct laboratory experiments and remain up to date with the literature. They may or may not serve as members of project teams. This department most closely resembles academic science.

### *Program Leadership*

Program leaders or heads of therapeutic areas are responsible for entire therapeutic areas or product lines, and they bring specific biological disease knowledge to the project (see Fig. 7-1). Each program might have numerous projects. Program leaders move projects forward and communicate with individuals in multiple technical disciplines.

### *Project Management*

Project management is used sparingly and inconsistently in discovery research. Project managers most commonly oversee projects in clinical development or serve as alliance managers with important pharmaceutical partners. Project team managers oversee individual projects in matrixed companies (see Fig. 7-1). They focus on timelines and budgets and help multidisciplinary teams move projects forward. For more details, see Chapter 9.

**_Program leadership versus project team leadership_**

A program might include all the drug discovery efforts for an entire therapeutic area, such as oncology, whereas a project team might be working on a promising drug candidate for breast cancer.

#### *Functional Line Management*

As shown on the horizontal axis of Figure 7-1, technology platform heads are responsible for the line management of a given specialty in a particular science or technology. Some typical functional categories include genomics, structural biology, gene expression profiling/biomarkers, molecular biology, informatics, high-throughput screening, medicinal, computational or hit-to-lead chemistry, biochemistry, cell biology/assays, pharmacology, analytical chemistry, pharmaceutics, imaging, and formulations.

#### *Portfolio Management*

Some large biopharma companies designate a team to manage their portfolio of products. They track and compare the performance of projects against goals, perform risk assessment and management, conduct strategic marketing, and benchmark product performance compared to peer groups and competition. For more details, see Chapter 17.

### Summary of Steps in Small-Molecule Drug Discovery Research

*Drug discovery is a partnership between chemistry, biology, pharmacology, and clinical development.*

The main goals of drug discovery research are to identify appropriate disease-related molecular targets and to discover new drug candidates that alter the molecular activity of the targets. Drug candidates are then improved upon with medicinal chemistry and tested in preclinical studies on animals before clinical evaluation in humans. The ultimate goal is to provide sufficient evidence of the potential for efficacy, safety, and market approval.

#### *Target Identification*

Starting at the most fundamental level, scientists initiate studies to identify potential targets. A target could be a receptor, an ion channel, or an enzyme that is thought to be a causative agent in the disease state being studied.

#### *Target Validation*

Once the company has a collection of targets, discovery researchers conduct experiments to provide evidence that the targets are associated with a particular disease state and that modulation of target activity could potentially provide therapeutic benefit. Target validation may involve using knockout mice, RNAi technology, or gene expression profiles. Note that a target is not truly "validated" until the drug candidate is shown to actually cure the disease in the affected population.

#### *Assay Development*

Cell-based, binding, or biochemical assays are created to test whether the new chemical entities (NCEs) synthesized as part of a research project have the desired effect on the specified target. Cell-based assays are also developed to determine how much of an NCE is needed to modulate the activity of the target in order to generate the desired outcome. An additional goal is to develop assays that are so robust and consistent that they can function in high-throughput screening mode.

### Lead or New Chemical Entity Discovery

After targets have been identified, scientists use the assays described above to find and evaluate NCEs as leads for further steps in development. If a company is developing small molecules, they typically use high-throughput screening (HTS), in which millions of compounds can be tested against the target to identify "hits" (NCEs that affect the selected target). HTS is a somewhat random process, basically involving a brute force approach using libraries containing millions of NCEs. It is, however, unbiased and has revolutionized drug discovery. For biologics, companies generate and screen monoclonal antibodies or other potential therapeutic biologics against molecular targets. "Virtual screening" is sometimes used as a supplement to experimental HTS.

### Lead Optimization

During lead optimization, functional characterizations are run to ensure that each NCE has the right pharmacokinetic and Absorption, Distribution, Metabolism, Excretion, and Toxicity (ADMET) profiles, plus the right biochemical properties for further development. Medicinal chemists work to improve pharmacokinetic profiles by making products more potent, less toxic, less reactive, and more selective. As soon as NCEs are deemed potent, they are analyzed to determine whether they have the desired effect on the disease in animal models in vitro and in vivo.

### Preclinical Research

After an NCE has been shown to be biologically active and to have promising drug pharmacokinetic characteristics, it is passed along for even more stringent scrutiny in preclinical studies (see Chapter 8). Potential products are analyzed in complex animal disease models, and more detailed pharmacokinetic, toxicology, and dosage optimization studies are run. If this information is promising, it will eventually be included in the investigational new drug application (IND) required for entrance into Phase I clinical trials.

### And Then Back Again...

These are iterative processes. Potent NCEs may go back and forth repeatedly between the medicinal chemistry group and the preclinical research group until a set of compounds with improved pharmacokinetic profiles emerges and is ready for further scrutiny.

*Drug discovery is an iterative process of improvement.*

*Small-molecule timelines*

Starting with hundreds of thousands, sometimes millions, of compounds in screening libraries, fewer than 50 are usually identified during the evaluation stages before progressing forward into the lead optimization stage. Several hundred structurally related compounds are typically synthesized during lead optimization. A few (3–5) preclinical development candidates are selected and then proceed into clinical development and finally make it to market approval. On average, the timeline for small-molecule drug discovery and development is 10–15 years.

### Biological Versus Small-Molecule Discovery Research

Small-molecule discovery requires serendipity. Before screening, there is no way of knowing whether your chemical library includes a molecule that will have the desired effect on your target. Once such a compound is discovered and improved upon, large-scale production is usually easy and inexpensive. In biological discovery, however, the approach is more targeted. Developing monoclonal antibodies that inactivate a specific receptor, or isolating protein hormones, are typical goals.

*Research in nontherapeutic biotechnology companies*

Research in biotechnology tools, medical devices, and service companies can be similar to the initial steps of discovery research (see Chapter 6). In tools companies, the first step is product concept definition, whereby the product and its parameters are defined. Ideas for products are often based on market research and assessments of drug discovery needs. Feasibility studies are then run to fine-tune the product so that it meets the needs of its user base. The product is modified accordingly, and analyses that determine cost and other characteristics are performed. After a prototype is developed, it is optimized and refined, and after testing and validation, documentation about how to build the product is transferred to manufacturing for scale-up production and shipping.

## DISCOVERY RESEARCH ROLES AND RESPONSIBILITIES

*The roles and responsibilities of discovery researchers vary with the level of the position. Basic research is similar to academic science, so this section describes mostly senior management responsibilities to provide a more insightful career development perspective.*

The scientific investigation that takes place in industry is every bit as good, cutting-edge, and challenging as that in the best academic institutions.

### *Scientists, Fellows, and Team Members*

#### *Conducting Research*

Bench scientists invent and test hypotheses, conduct laboratory work, analyze and write up data, record information in notebooks, and report the results to the team. Ultimately, they may write up data for publication, patents, and regulatory filings.

### *Team Participation and Management*

*You cannot individually process all of the information that exists in a drug discovery program—a team of individuals is required.*

In discovery research, every piece of information that is potentially relevant to the mission of the organization should be considered. This is best accomplished when researchers work together as teams, leveraging their members' cumulative breadth of experience and knowledge when approaching scientific questions. Project team leaders or functional or program heads typically orchestrate this process. They encourage team members to enthusiastically contribute their ideas and continue pressing forward.

## *Managers, Directors, and Vice Presidents*

### *Providing Scientific Oversight: Picking the Most Promising Projects and Product Candidates*

Senior members in discovery research are responsible for the ultimate success of a program. They organize it and set aggressive but realistic goals and objectives. They ensure that the program makes scientific sense, that the work supports company goals, and that everyone is working toward a marketable product in accordance with timelines and budgets.

### *Picking the Best-Suited People*

*Managers hire the most promising scientists and let them do their best—and remain ready to help, mentor, commiserate, and celebrate!*

Senior members are responsible for recruiting and building appropriate and qualified teams so that there is suitable technical talent for solving scientific problems and accomplishing program goals. Good personality matches are imperative so that teams function smoothly.

### *Providing an Environment in Which Creative Scientists Can Flourish*

Drug discovery is not an assembly-line operation. There are no prescribed recipes to follow in drug discovery, only general principles, and each effort follows a different path. If you want people to do things in new ways, you need to establish a creative environment and give scientists the maximum freedom in which to do their work.

### *Managing People*

Managers are responsible for communicating expectations clearly and providing scientists with the tools and resources they need to accomplish their goals. Line managers are also involved in employee development, annual evaluations, and making sure that morale is high and that personnel are working well together. They have the important role of serving as internal "barometers" of the difficulty of projects or scientific tasks, so that they can estimate project timelines, decide how to assign projects, and recognize employees who are performing well.

### *Establishing a Culture that Is Conducive to Sharing Ideas*

*The whole of what people can contribute as a team is much greater than the sum of what each person can contribute.*

One key responsibility of managers is to set and manage the corporate culture. Leaders work to create an environment that promotes the free and open exchange of ideas so that there is a wider range of ideas and solutions to problems. Ideally, team members should feel free to disagree with their superiors without fear of being punished or humiliated. For more details, see Chapter 22.

#### *Spokesperson for R&D*

Senior members communicate the goals and progress of programs to upper management and employees. Broader responsibilities include integrating with other parts of the company, such as business development personnel, intellectual property lawyers, contractors, and others.

Externally, senior members routinely speak at scientific meetings and conferences. They give presentations and update investors and analysts on the most current company information.

#### *Resource Allocation and Portfolio Management*

Senior-level managers pick the most promising product candidates, prioritize projects, and allocate resources based on the budget. They may also have the principal responsibility for developing the intellectual property portfolio and coordinating grant writing.

#### *Business Development and Alliance Management*

Senior members often establish and manage business development relations, collaborations, and research alliances with other companies and academic labs.

#### *Public Service*

Some senior researchers spend extracurricular time as members of learned societies, as review panelists, editors for journals, and board members for other organizations.

## A TYPICAL DAY IN DISCOVERY RESEARCH

*A typical day might include some of the following, depending on one's level and position:*

#### *Bench Researcher, Scientist, and Fellow Positions*

- Working irregular hours at the bench doing hands-on experimentation.
- Reviewing and interpreting data and discussing its implications.

*For many people, discovery research is fun—it's like being paid to play in the sandbox all day!*

#### *Management Positions*

- Attending lots of meetings. These could include status updates, one-on-ones with direct reports, meetings to discuss specific issues dealing with programmatic aspects, corporate objectives, budgets, and high-level strategy, meetings with other companies, etc.

- Managing scientists and teams, regularly checking with them and attending to administrative details.
- Dealing with personnel issues, including career development, mentoring, performance evaluations, or conflict resolution.
- Attending scientific seminars presented by internal and external speakers.
- Traveling; speaking with potential customers or collaborators or speaking at national scientific meetings.
- Reading scientific journals.

## SALARY AND COMPENSATION

Discovery research is, in general, on the lower end of the salary range compared to other careers in biotechnology and drug development. The general rule of thumb is that the closer you are to revenue, the better you are compensated. Thus, employees who are further along in the product development steps, such as those in clinical and regulatory affairs, are paid more. Personnel in drug safety, process chemistry, sales, business development, legal affairs, and marketing are generally also paid more. Salaries in discovery research, however, may be slightly higher than in manufacturing, technical support, and other areas.

*Scientists in general are motivated by science. If they are given uninteresting, inane projects, they will leave...regardless of the money.*

Part of the reason for the lower compensation is simply supply and demand: Discovery research is not a rare skill. Regulatory affairs personnel or pharmacology experts, for example, learn their trades on the job, whereas molecular biologists are educated in academia, and there are many more of them. Researchers in industry, however, earn much more than those in academia. People in industrial positions can earn twice as much as their academic counterparts.

***How is success measured?***

Success as a researcher in industry is measured by how well you contribute to the organization. This may include whether you meet goals on time; are collegial, innovative, and skillful; whether you take the initiative, and more.

*Success is measured by three things: compounds, compounds, and compounds.*

For discovery research departments, the number and quality of candidate products that graduate to development is one measure of success. Another is the making of decisions to advance or terminate projects.

## PROS AND CONS OF THE JOB

### Positive Aspects to Discovery Research as a Career

- It is immensely gratifying to work in a team-oriented environment with intelligent, motivated people from a variety of disciplines. A strong sense of camaraderie develops as you tackle problems together. As a team, you will be able to accomplish much more than an individual researcher.

*Working with extraordinarily creative, intelligent, and enthusiastic people in a team environment is very rewarding.*

- It is fundamentally satisfying to work on a project that may benefit mankind. There is a daily sense that you are working on something important and really worthwhile.
- There is a great deal of intellectual stimulation and broad exposure to high-quality science in discovery research. It is an ideal setting for a career of lifelong learning. Drug discovery is a complex, challenging process: an art as well as a science. There is no single approach; innovation is encouraged. Solving difficult problems each day and applying innovative approaches is extremely gratifying.

*The more you learn about discovery research and drug development, the more you realize how much more there is to learn.*

- Academic research often restricts you to a fairly narrow and defined area of science; whereas in industry, you can work on dozens of molecular targets and expand your interests into other therapeutic areas.
- Industry has access to the latest cutting-edge technological tools, large resources, and an infrastructure that make possible huge, multidisciplinary, expensive projects, like the Human Genome Project. These "big science" programs are much more difficult to establish in academia.
- There is satisfaction in building complex projects and seeing them progress. In industry, projects don't languish; those that fail are quickly terminated. Whereas academic investigators try to attain complete understanding of a particular issue and aim at more elusive basic concepts, industry researchers are more focused on immediate goals. Your efforts in discovery research will yield tangible results.
- There is much job variety, and things change quickly. The latest set of data can have profound impacts on a program. New surprises and challenges come with each step a product takes along the development path.

*Discovery research is an environment in which change is an absolute constant—it is impossible to be bored.*

- Those with difficult personalities have limited options in industry and often are not tolerated in managerial positions, whereas the academic environment accommodates a much broader variety of behavioral phenotypes.

- Discovery researchers don't have to chase grants the way their counterparts in academia do. Some small companies do rely on small business innovative research program (SBIR) grants, but even then the grants are often written by consultants.

### The Potentially Unpleasant Side of Discovery Research

- When business decisions are made, you need to support and embrace subsequent changes in your work objectives, even if you don't agree with them. Common examples include having to terminate projects if they are not commercially viable or downsizing a research department in order to focus resources on clinical development programs.

*Science is not the only thing that drives decisions.*

- The failure rate in discovery research is high. Even projects with promising science behind them will be terminated if their products are unlikely to be commercially viable. Failure and its consequences can be disappointing after you have invested so much time and effort.

*The high failure rate of projects is an unfortunate fact of life in discovery research.*

- Timelines in discovery research are long, and it can be challenging to stay motivated.
- Expect to work long hours, as many as 50 hours per week in an established company, and more than 60 hours per week in a start-up. Depending on the corporate culture, you may be working weekends, too.
- There are many meetings to attend. The higher your position, the more meetings and the more paperwork.
- Discovery research jobs are inherently unstable. Departments can close unexpectedly, and layoffs are common, irrespective of one's performance (although abrupt, cataclysmic layoffs are mercifully rare). For small companies in particular, as soon as projects move into development, their resources follow. Likewise, a merger may result in the elimination of entire research groups. You can be laid off for simply working in a successful, yet ill-fated, department.

*In discovery research, "Layoffs happen," as do mergers, acquisitions, and reorganizations.*

- You are less likely to be able to work in a particular scientific area for the rest of your life, as you can in academia.
- Autonomy and independence are not easily acquired in industry. Drug development is for profit, so projects are run from a commercial perspective and not because they are interesting. Management can often make capricious decisions, and you may lack some control over what you are working on.
- Managing people is challenging, and some team members can be difficult to work with. It takes skill to tell people, without offending them, what they don't want to know or to inform them that they are not performing well.

- You might not be free to publish if your scientific results are being used for a patent or if competition is fierce.

## THE GREATEST CHALLENGES ON THE JOB

### *Successfully Developing Product Candidates on Time and on Budget*

Developing promising drug candidates and achieving market approval is arduous. The process is incredibly complex and costly, and many things can go wrong along the way. It takes great insight, talent, and creativity to develop innovative breakthrough products, but when everything works well and you make it to market, it's a fantastic feeling and well worth the effort.

*Projects don't grow on trees. Their development requires steadfast diligence over a long period of time.*

### *Managing and Interacting with People*

*An environment that fosters creativity is of the utmost importance...and is also one of the most challenging to manage.*

Many people initially expect that the biggest challenge in drug discovery is determining the program's scientific direction. However, managing the personnel involved can be an even bigger challenge. It is often difficult to ensure that a group of people with disparate skills can work in a smoothly running team. This problem is exacerbated in global companies with employees who have different cultural values and work ethics.

### *Keeping up with the Science*

The quantity of scientific information is vast and growing exponentially. Just staying current and figuring out what is important requires time. You must be committed to lifelong learning, or else you will soon be out of the game.

### *Ethical Responsibility: The Financial Reality Versus Obligation to Patients*

The ethics of drug discovery and development are complicated. Every time you choose a target and allocate resources for a project, you have to balance the financial realities of product development against the world's medical needs. If you choose to develop a product that would benefit a third-world country but lacks a significant market, the company might go under, in which case neither patients nor stakeholders benefit. On the other hand, producing a "me too" pharmaceutical drug lowers drug costs for the consumer by increasing competition, provides a return on stakeholders' investment, and potentially funds development of less profitable drugs.

## TO EXCEL IN DISCOVERY RESEARCH...

Drug discovery is a tough business. Even though project failure rates are high, some people lead drug discovery programs more successfully than others. Below is a list of qualities shown by those who have demonstrated excellence.

#### *Scientific Leadership that Is Visionary, yet Realistic*

Those who excel understand the many nuances of product development. They combine technical expertise and scientific brilliance with extensive experience to identify drug candidates that are promising not only from a scientific point of view, but also from a business perspective. They can apply sound, fundamental scientific knowledge and a realistic sense of the totality of what is needed to identify promising drug targets and ultimately produce an approved drug.

#### *Passion and Persistence*

Drug discovery takes time, and there are constant challenges to progress, including technical problems and organizational pressures. Those who excel are persistent. They are committed to bringing projects forward and can be fantastically productive.

#### *Exceptional Management*

*Exceptional leadership gives scientists the freedom to innovate.*

A key part of managing a scientific organization is providing a creative environment where talented and bright scientists can do their best work and feel empowered and respected. Successful leaders know how to encourage progress while maintaining high scientific standards and a motivated, first-rate staff.

#### *Those Who Embrace the Concept of "Lifelong Learning"*

*Historically, serendipity has played a big role in drug discovery and development.*

Exceptional scientists treat every new piece of information that crosses their path as an opportunity to learn and grow. They internalize the information and make connections to it in other contexts. The drug development path is rarely clear or simple: A lipid-lowering drug, for example, was found to have anticancer activity (Raloxifene), and a product developed for cardiovascular disease became the leading erectile dysfunction product (Viagra). Those who embrace lifelong learning are more fundamentally prepared for serendipitous findings.

## Are You a Good Candidate for Discovery Research?

*People who flourish in discovery research careers tend to have...*

*A good researcher will tell you what the problem is. A great researcher will get seemingly impossible tasks done effortlessly.*

***A strong research background.*** You should have a solid foundation in science and a love for research. Good scientists are observant, objective, fully informed in their field, driven by data, and able to ask the right questions. Early on in your career, you should demonstrate adeptness with laboratory research. If you want to move into management, you first need to gain credibility by being a good scientist and demonstrating that you can supervise people and manage a scientific program.

***A collegial attitude and ability to work in teams.*** Discovery research is a team endeavor (see Chapter 2, "Team Player"). To be successful, you should be able to work well in teams.

This includes being sincere in your interactions, considerate and respectful of the other members, willing to share your knowledge, able to negotiate and compromise, and willing to tackle difficult problems. There may not be as many individual accolades in industry as there are in academia, so you will need to share rewards as a group member. You may need to override your own ego for the overall benefit of the team.

*Tenacity and perseverance.* Timelines in discovery research are long. Your efforts from idea to final development can take ten or more years, so you must be able to stick with projects for substantial periods of time.

*A high level of energy, courage, persistence, and tenacity separate the great from the good in discovery research.*

*The ability to tolerate frustration and disappointment.* A common scenario is for scientists to toil on a project for several years and then receive that dreaded call from someone in the toxicology department saying that the project will be terminated. This happens more often than not and can be difficult to accept. Although it is natural to become intellectually and emotionally invested in a project you've spent several years on, you need to be able to handle disappointment and move on to the next project without rancor or resentment.

*Broad knowledge.* Discovery researchers generally work on multiple discovery areas and have projects in various stages of development.

*The ability to work in a goal-oriented, time-constrained, research environment.* Discovery research is product-oriented. You need to be comfortable working within timelines and sustaining a continuous sense of urgency.

*A receptive attitude toward feedback.* It is important to learn how to manage your relationship with your supervisor. This includes being comfortable with receiving positive and negative feedback and with the fact that your salary and promotions are directly tied to not only whether you met your goals, but also *how* you met your goals; i.e., how you interacted with colleagues.

*Excellent problem-solving and analytical skills.* Critical thinking in combination with intelligence, energy, and optimism are needed to find innovative solutions.

*Leadership skills (for management track).* Managers need to respect other people's abilities and skills, delegate well, and have clear insights into what motivates people. They need to understand how to motivate and retain employees by challenging them and fostering a stimulating environment.

*A flexible attitude.* Often the team approach is used for selecting the best scientific direction, and the way forward is usually not obvious. You must be flexible enough to listen to your colleagues with an open mind. In addition, it is important to realize that after decisions have been made, you need to accept and embrace the changes and move forward, even if you disagree.

*Excellent communication skills.* It is important to be able to write and speak well in discovery research. To inform others about the science is as important as conducting the experiments.

***A drive to succeed.*** Some people are content to operate as individual contributors, while others aspire to become vice presidents or CEOs. It is important to decide early in your career how you define success and to determine what you need to do to achieve your goals.

***The ability to gauge how long experiments will take.*** With more experience, it becomes easier to estimate over longer time periods and for more complex projects. Whereas entry-level B.S./M.S. scientists gauge and plan their day's activities, a Ph.D.-level researcher with experience can plan months ahead of time. A project leader makes plans for entire departments in terms of months and years.

***Intellectual honesty.*** It is important to interpret data objectively, even when the conclusions are not what you expect.

***You should probably consider a career outside of discovery research if you are...***

- A micromanager, unable to delegate (for managerial positions).
- An academic prima donna, unwilling to share your intellectual knowledge or give credit to others.
- Only interested in publishing high-quality, cutting-edge papers.
- Unable to work well in teams.
- One who takes criticism personally.
- Only superficially interested in discovery research; a scientist who lacks passion or common sense.
- Unable to adjust to the corporate culture and drive of industry.
- Motivated mostly by money.
- One who follows trends and does not think critically and independently.
- Hoping to avoid decision making and responsibility.
- Disorganized or not observant.
- Unwilling to take the initiative or work hard.
- Someone who needs immediate gratification.
- Someone who wants to explore fundamental scientific questions without regard to their commercial worth.

## DISCOVERY RESEARCH CAREER POTENTIAL

Discovery researchers can move up the ranks to chief scientific officer or vice president of R&D, and sometimes even to CEO positions (see Fig. 7-2). There are numerous alternative pathways for which training in discovery research prepares you—in fact, it is a launching pad for just about any career direction in biotechnology companies. Scientists can move into other areas of drug discovery and development, including preclinical

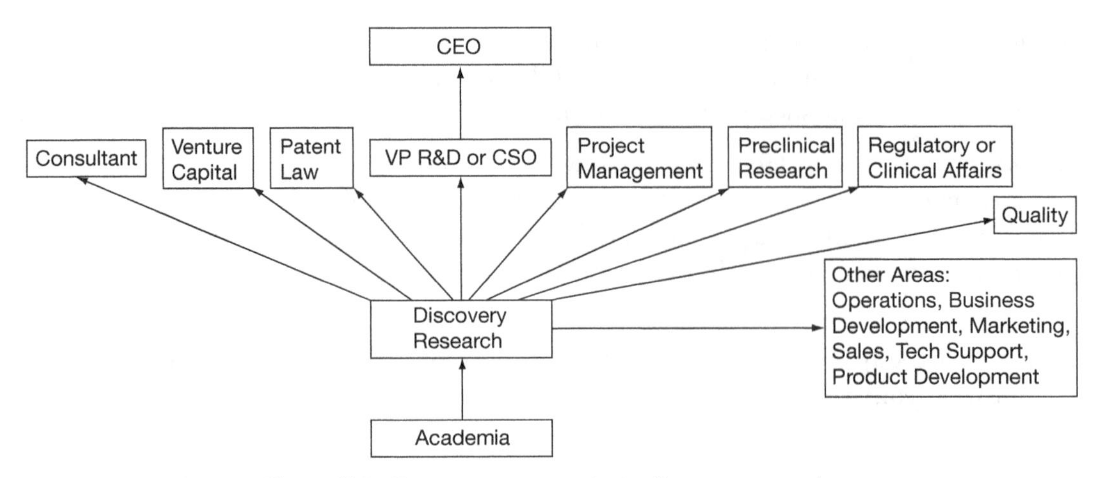


**Figure 7-2.** Common career paths in discovery research.

research, project management, clinical development, regulatory affairs, and quality. On the business side, field application scientists, business development, patent law, venture capital, marketing, and technical support are relatively common transitions.

Project management is a particularly attractive career for scientists who want to move out of the lab, because it requires the ability to understand fundamental science and to transition research projects into development programs. Research-trained scientists also make good human resources personnel, serving as recruiters or managing benefit programs, for example.

## Job Security and Future Trends

*Employability is not a company responsibility anymore—it is a personal responsibility.*

The larger, more established companies generally provide better job security, because they have deeper resources for tolerating failed products and setbacks. In start-ups, job security is tenuous. Continued learning is one of the keys to ensuring your employability: As things change, you have to change with them.

Although there currently seems to be a glut of Ph.D.s working in the biological sciences, the strong demand for specific talent and experience remains. The biggest demand is currently in pharmacology (particularly in vivo and integrative pharmacology), toxicology, enzymology, medical chemistry, and organic chemistry. There continues to be a constant demand for talented research associates and for senior-level executives with outstanding managerial and leadership skills.

The number of available research positions in the United States seems to be declining. There are numerous explanations for this phenomenon: Venture capitalists are fueling fewer early-stage start-ups; companies reduce or eliminate research programs to reserve their cash when projects move into clinical development; more positions are moving overseas; and with more mergers, there are fewer companies and thus fewer jobs across the board.

## LANDING A JOB IN DISCOVERY RESEARCH

### Experience and Educational Requirements

*Being "house-broken" is a big plus in industry—the learning curve is not as steep for those with prior industry experience.*

Most people entering careers in discovery research arrive from academia with a bachelor's, masters, Ph.D., or M.D./Ph.D. degree (see Fig. 7-2). Those with bachelor's or master's degrees start as research associates and advance to senior research associate positions. There is a "glass ceiling," however, and after six to seven years, research associates without advanced degrees will likely experience limited career potential. Many people start in industry as research associates and return to school for a Ph.D. They then come back to industry, this time with an excellent combination of qualifications: the advanced degree *and* industry experience.

Three or more years of industrial experience is the ideal background for applying for scientist positions. Entry-level graduates usually enter at more junior levels compared to others with more experience, regardless of their educational level. To obtain a research associate position with only a B.S. degree, it is advisable to gain laboratory experience as an undergraduate.

Ph.D. chemists coming directly from academia usually can qualify for jobs immediately, whereas biologists are expected to obtain at least two years' postdoctoral research experience before beginning an industrial job.

### Paths to Discovery Research

- When choosing a lab in which to work in academia, make sure it is in a hot area of biotechnology and drug development that is relevant to human disease.
- Do something in your postdoctoral fellowship that is different from what you did in graduate school. This shows flexibility and a willingness to tackle different problems and build new skill sets.
- Learn how to give concise and professional presentations. Learn how to create coherent, audience-friendly slides and how to speak within an allotted time.
- Publish and patent as much as possible. Hiring managers look for solid, substantial papers in peer-reviewed journals. The ability to publish is tangible evidence of your productivity, goal-orientation, and proficiency in your field.
- Make yourself visible in the scientific community. Actively participate in meetings, symposia, and professional organizations.
- Network at meetings attended by people in industry. Get to know them; ask who is hiring and who is firing.
- Talk to the local school faculty about job opportunities. Often academics serve as company consultants; they may know of job leads and provide industry contacts.

- Find a mentor in industry who is willing to coach you through the job hunting and interviewing process, and perhaps beyond.
- Contact biotechnology and pharmaceutical liaison counsels or outreach programs in your university.
- Apply for summer internships and postdoctoral fellowships at companies. They build your network and provide a first-hand view of working in industry. Apply around March or April for summer internships on company Web sites.
- Find temporary employment through staffing firms to obtain initial entry-level experience. If there is a personality match and your performance is good, temporary jobs often turn into permanent positions.
- Sharpen your computer skills. Become adept with programs such as Word, PowerPoint, and Excel. These essential skills can be learned on the job, but it's expected that you will already be familiar with them.
- Learn nontechnical skills, such as team and project management, effective writing, and decision making. You may be assigned managerial responsibilities early in your career, and classes in effective management and leadership will prove invaluable.
- Consider starting out at large companies before going to smaller ones. Working in a start-up can be highly rewarding, but large companies have many more programs where you can gain experience in all phases of drug development and in multiple therapeutic areas.

*Career tip: Go from big to small.*

- Interviewing in industry is different than in academia. During an interview, take an active interest in the opportunity—do not be passive or aloof. Be energetic, and tell them how much you want the job and what a great company it would be to work for. It really does make a difference. If you don't express your interest, the hiring manager may be less inclined to hire you.

*Hiring managers look for people who are enthusiastic and eager to get their work accomplished—not the "9-to-5er" types.*

- Join companies that have both scientific and management career tracks. Try to join companies with a healthy, scientific career ladder.
- When interviewing, ask about publication policies. The best companies have a vigorous publication record, but some do not allow scientists to publish.
- When interviewing, inquire about the company's corporate culture; it can vary profoundly from one company to another. Top-down management styles, for example, in which scientists are treated like cogs in a huge wheel, will not attract the best scientists. Ask to meet with management and scientists, and get a sense of how happy they are, how empowered and respected they feel; their energy level, commitment, drive, and passion. To learn more about assessing corporate cultures, see Chapter 3.

## RECOMMENDED TRAINING, PROFESSIONAL SOCIETIES, AND RESOURCES

*Courses and Certificate Programs*

Courses in:

- Drug development and clinical trials
- Aspects of discovery research
- Finance and business fundamentals
- Management and leadership training
- PowerPoint and advanced Excel
- Project management

*Societies and Resources*

Most professional societies provide meeting announcements and job postings on their Web sites.

American Chemical Society (www.acs.org). The ACS offers job postings, short courses, and national meetings.

American Society for Cell and Biology (www.ascb.org)

American Society for Pharmacology and Experimental Therapeutics (www.aspet.org)

American Association for the Advancement of Science (www.aaas.org)

The Society for Biomolecular Sciences (www.sbsonline.org)

BioSpace (www.biospace.com) and Biotechnology Industry Organization (BIO, www.bio.org) lists scientific meetings and job postings.

Job opportunities are posted in the journals *Nature* and *Science* and specialized journals.

Tufts Center for the Study of Drug Development (http://csdd.tufts.edu)

*Journals*

*Nature Reviews Drug Discovery.* This journal is frequently recommended as the leading authority in drug discovery.

*Nature*

*Nature Biotechnology*

*Science*

*Journal of Medicinal Chemistry*

*Current Opinion in Investigational Drugs*

*Drug Discovery Today*

*Journal of Pharmacology and Experimental Therapeutics*

*The Scientist*

*Books*

Cohen C. and Cohen S. 2005. *Lab dynamics: Management skills for scientists,* Cold Spring Harbor Laboratory Press, Cold Spring Harbor, New York.

Medawar P.D. 1979. *Advice to a young scientist,* Alfred P. Sloan Foundation Series, Harper and Row, New York.

Sapienza A.M. 2004. *Managing scientists: Leadership strategies in scientific research.* Wiley-Liss, New York.

***Free On-line Subscriptions***

Fierce BioResearcher (www.fiercebioresearcher.com)

Fierce Biotech (www.fiercebiotech.com)

BioSpace (www.biospace.com): sign up for their daily "GenePool" E-mails

# 8

# Preclinical Research

## The Bridge between Discovery Research and Clinical Development

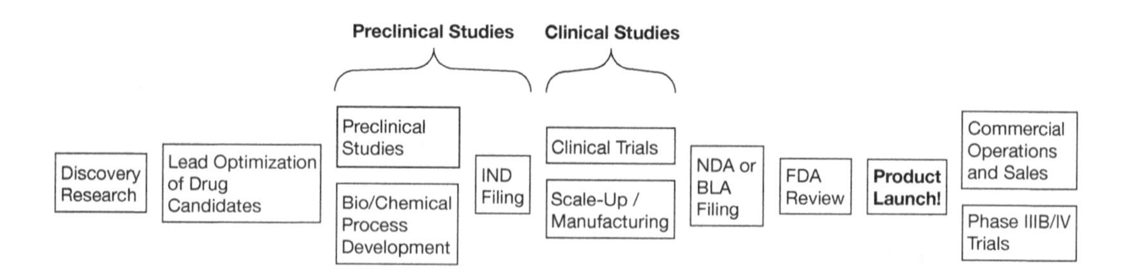


*Preclinical research is about developing drug discovery products into viable clinical therapeutics.*

IF YOU ENJOY DISCOVERY RESEARCH BUT SEEK A POSITION where the results of your work are closer to becoming a commercial product, consider preclinical studies. Preclinical research combines the appeal of working in a discovery research environment with the satisfaction that comes from progressing beyond discovery research results. Biological scientists, chemists, toxicologists, and pharmacologists enjoy this career because they are more intimately involved in advancing the products of basic research to clinical development.

### THE IMPORTANCE OF PRECLINICAL RESEARCH IN BIOTECHNOLOGY AND DRUG DEVELOPMENT

*It's better to be aware of a drug candidate's negative attributes BEFORE entering clinical trials.*

In discovery research, the goal is to identify, synthesize, and characterize new drug candidates ("new chemical entities" and biological therapeutics). In clinical development, drug candidates are tested for their safety and efficacy in humans. Because clinical trials are exorbitantly expensive, it is important to choose only the candidates that are most likely to succeed. The role of preclinical research is to mitigate clinical risk by conducting as much investigation as possible in animals and human tissue cells before the company commits to clinical trials.

In the earliest stages of drug discovery, many molecules will undergo testing with a few, relatively easy-to-implement (usually in vitro) assays. During the preclinical stage, the most promising drug candidates undergo much more extensive testing with a large battery of in vitro and in vivo assays, many of which are performed using good laboratory practice (GLP) standards. Preliminary safety and efficacy studies are conducted to determine whether a reasonable therapeutic index in animals and animal models justifies a path into the clinic. The information gathered during preclinical studies will ultimately be included in the investigational new drug application (IND) that is filed with the U.S. Food and Drug Administration (FDA). The IND is a request for permission to do first-in-human Phase I clinical trials.

*It's cheaper to ask the hard questions in the preclinical stage of development than in the clinical stage.*

## CAREER TRACKS IN PRECLINICAL RESEARCH

Although the organizational structures of companies vary greatly, preclinical research typically includes four main disciplines: pharmacokinetics, toxicology, pharmacology, and chemistry. Groups such as translational medicine and laboratory animal sciences are also sometimes part of this division.

### *Metabolism and Pharmacokinetics, ADMET and DMPK*

To continue as a candidate in clinical development, a potential drug needs to have a good "ADMET" profile. ADMET stands for *a*bsorption, *d*istribution, *m*etabolism, *e*xcretion, and *t*oxicity. Preclinical researchers who study ADMET or DMPK (*d*rug *m*etabolism and *p*harmaco*k*inetics) examine what the body does to drugs. For instance, if a drug is eliminated from the body too rapidly, it won't have enough time to produce a therapeutic effect. On the other hand, if it lasts too long, it may accumulate and cause toxic side effects. Scientists also determine whether a drug's breakdown products (metabolites) are toxic or have additional effects. They study the drug's distribution in the body to ensure that it reaches its intended target. Animal models and tissue culture cells are used to help define all of these properties and to determine the optimum therapeutic dose in animals and, ultimately, humans. Some scientists in this group work with chemists to modify the drug candidates and optimize their ADMET profiles.

*Preclinical researchers explore what the drug does to the body (pharmacodynamics; safety and efficacy) and what the body may do to the drug (pharmacokinetics; ADMET and DMPK).*

### *Toxicology, Drug Safety Evaluation, and Pathology*

Tissue damage and relative safety margins are studied in animals to help estimate the dose for humans and to determine possible side effects and danger signals that clinicians should be aware of when running human trials. This information is also used in part to determine the "therapeutic window" of the drug. The therapeutic window is basically a benefit-to-risk ratio: How much drug is needed for the desired therapeutic effect, and what dose produces toxic effects?

#### *Pharmacology*

In pharmacological studies, the drug candidates are tested in the most appropriate animal and cellular models of human diseases to further develop and characterize their activity. On the basis of the results, preclinical scientists can anticipate how the drug will interact with molecular targets in humans.

*Pharmacodynamics is the study of how a drug affects the molecular targets in the body and the kinetics of that activity: Does the drug reach the target in the diseased tissues, does it inhibit its target, and for how long?*

In vitro or molecular (also called cellular) pharmacology involves studying the action of drug substances on cells, using a variety of techniques including receptor-binding assays, cell-based functional assays, calcium efflux assays, and many more. Hundreds of different types of assays can be applied, depending on the biology of the drug target. In vivo pharmacology refers to experiments conducted on live animals.

#### *Translational Research and Medicine*

Translational medicine involves the steps of preclinical research up to and including the first clinical trial and the submission of an investigational new drug application to the FDA. Translational medicine may encompass formulation, development of analytical methods, ADMET, toxicology, dose range-finding, pharmacological screening, and preclinical toxicity testing.

Translational medicine may involve identifying and developing tests for biological markers (i.e., protein and RNA levels, psychometric tests, etc). Biomarkers are used to select patients who are more likely to respond to therapy, to measure whether the treatment is working in patients, for diagnostic purposes, and more. Translational medicine can also be found within the clinical development department (see Chapter 10).

#### *Laboratory Animal Sciences and Welfare*

In some organizations, there may be a department specifically devoted to the procurement and care of animals for preclinical research purposes.

#### *Chemistry Manufacturing and Controls (CMC) and Biologics*

Most companies have a CMC group, which is usually a separate department or, in some companies, in the preclinical division. In this group, chemists optimize a drug substance's ADMET profile by modifying its structure and formulation. CMC includes several activities required for IND filings: chemical development, pharmaceutical sciences, and analytical sciences. For companies developing biologics, a bioprocessing department develops and optimizes biological techniques, such as cell line construction, cell culture scale-up, purification, and analytical assays. These departments are described in more detail in the bio/pharmaceutical product development chapter (see Chapter 15).

*A new chemical entity needs to be optimized so that it is neither too short-lived nor too long-lasting in the body.*

## PRECLINICAL ROLES AND RESPONSIBILITIES

### *Characterizing Drug Candidates*

*The primary function of preclinical studies is to put drug development candidates through the wringer.*

Preclinical research primarily involves putting new drug candidates through a first-degree examination to determine whether they will have promise in the clinic. New drug candidates and their metabolites are evaluated and characterized in animals and tissue culture, and the resulting information is used to determine maximum tolerable doses in humans and to predict the drug's behavior in clinical trials. Preclinical researchers also work to optimize yields of drug synthesis reactions, minimize the cost of goods, and define the best formulation.

### *Evaluating and Selecting the Most Promising Drug Candidates*

Approximately 2–10% of all drug candidates move forward from discovery research into clinical development. Only a small fraction of those make it from the clinic to the market, so decisions about which candidates to promote must be made carefully. During the preclinical phase of drug development, senior management and product champions tout each particular product's good points and assess potential pitfalls. Criteria include whether the drug candidate will fill a particular niche for an unmet medical need, the ease and cost of producing the product, the formulation (e.g., oral or injectable), and whether the compound has the potential to provide benefits that are superior to the competitors' products. In a team-based approach, budgetary restrictions and corporate goals are applied to determine whether more research is needed, or whether the candidate should be sold to another company with better capacity to develop it.

### *Preclinical Studies during Clinical Development*

Clinical trials often lead to more questions than answers. Preclinical studies are sometimes continued to address new questions that cannot be answered by testing on humans, even when products are already in clinical development. Frequently, expensive preclinical studies such as long-term animal carcinogenicity tests are delayed until Phase III clinical trials are under way. Preclinical representatives remain on project teams even after the product enters clinical studies so they can determine whether animal models can be applied to expedite progress or answer newly generated questions or concerns.

## A TYPICAL DAY IN PRECLINICAL RESEARCH

*Depending on his or her level and position, a preclinical researcher might expect some of the following activities on a typical day:*

- Attending project team and group meetings.
- Discussing and interpreting data with colleagues.
- Conducting experiments.

- Performing autopsies on animals; reading tissue slides.
- Evaluating and reviewing laboratory results and designing new experiments.
- Investigating new technologies and methods to improve models and expedite results.
- Creating presentations for team meetings.
- Reading, keeping current with scientific literature, and attending conferences.
- Writing reports or manuscripts for publication.
- Writing sections for IND filings.

## SALARY AND COMPENSATION

Compensation in preclinical research is about the same as in discovery research, and lower than in clinical development.

***How is success measured?***

Success can be measured in both time and dollars by how efficiently a product moves through the preclinical stage. Other, more indirect measurements include the number of accepted publications and funded grants, and how quickly problems are addressed.

## PROS AND CONS OF THE JOB

### Positive Aspects of a Career in Preclinical Research

- Your work can encompass the scientific and product development issues of discovery, preclinical and clinical research, and chemistry. Preclinical research offers the opportunity to have a large, more direct effect on the drug candidates that will move forward into the clinic.
- Decisions in preclinical research are typically driven by data, not politics. Problems tend to be of scientific, technical, and logistical nature.
- You will have plenty of chances to "talk science" with your colleagues. Hallway conversations are common, and discussions about experimental results and future plans are interactive and dynamic. Rank is generally not as important as it is in other settings: If you have a new insight, others will listen and consider your opinions.
- The environment in preclinical research is often stimulating. Colleagues tend to be bright, motivated, and fun to work

*Preclinical studies departments provide a nurturing environment for people who are truly interested in science.*

*It's fun to work in an environment where coworkers are passionately arguing the pros or cons of a drug candidate in one minute and drinking beers together in the next.*

with. The discovery researchers who originally developed the products frequently show a great deal of enthusiasm for their drug candidates.

- You can advance within your particular technical discipline. For example, if you enter preclinical research as a chemist, you may spend your entire career building an expertise in chemistry. In clinical development, however, you may spend many years building clinical expertise in several therapeutic areas.
- Preclinical work provides a broader exposure to drug discovery and development than does discovery research. Preclinical activities cover everything from small- to large-molecule products and include multiple therapeutic areas.
- Preclinical work provides good training for those interested in going into discovery or clinical development careers.
- Preclinical research is subject to far less FDA regulation than clinical development, and there is more opportunity to explore basic science.

*There is less paperwork and there are fewer strict guidelines than in clinical development. You can't easily explore mechanisms of action in clinical development.*

### The Potentially Unpleasant Side of Preclinical Research

- Pharmacology involves animal work, and research animals are often euthanized. It's not for people who are squeamish or members of People for the Ethical Treatment of Animals (PETA).
- Sometimes it can be difficult to inform upper management about negative data—a delay or negative result may not be good news for the company's ability to reach corporate milestones. It can be even more difficult to relay negative information to the discovery research champion, who is eager to see his or her product enter into clinical development and may prematurely envision an IND filing and product success.

## THE GREATEST CHALLENGES OF THE JOB

*Answering Unmet Medical Needs*

Introducing new drugs for unmet medical needs is one of the greatest challenges in preclinical research. There is much risk and scientific fortitude involved in being a pioneer and tackling new technical hurdles for "first in class" (yet to be proven) products.

*Generating Promising Drug Candidates for the Company's Pipeline*

Providing the company with viable candidates for clinical evaluation is a constant challenge. This is difficult work and includes putting together a well-prepared package with a convincing story for each promising new product.

*Developing a robust pipeline of new products to fulfill medical needs is what preclinical research is all about.*

*Working with a Limited Pool of Publicly Available Negative Data*

One of the challenges in preclinical research is learning how to effectively and efficiently predict whether a product will be a good candidate or not. This task is made more difficult by the limited availability of negative data. Companies and academia rarely publish negative data, so each company must generate and rely on its own unpublished proprietary information and accumulated knowledge. As a consequence, experiments that have failed in one company may be repeated in others, and the lack of publicly available information retards the advancement of new products.

*Bridging the Communication Gap between Discovery Research and Clinical Development*

Working with multidisciplinary teams that include both discovery research and clinical development can be difficult, because the terminology and goals of the two sides are sometimes very different. Getting people to communicate quickly and effectively and being able to address problems in a scientific and sound manner can be challenging.

## TO EXCEL IN PRECLINICAL RESEARCH...

*Insight and Speed*

Great preclinical executives apply their extensive knowledge of basic biology and chemistry to the analysis of new drug candidates. They quickly and efficiently evaluate data and sift through the slate of candidates to find products that have the highest probability of success. The best preclinical researchers consider "killing a project" a good thing, so that the company doesn't waste money on projects that are likely to fail.

### Are You a Good Candidate for Preclinical Research?

*People who flourish in preclinical research careers tend to have...*

*A solid knowledge of science.* Preclinical researchers tend to be highly trained, scientific, and focused. Most are comfortable with disciplines outside of their fields of expertise.

*The ability to be comfortable with standardized procedures while remaining capable of innovation and novelty.* Work in preclinical development can be more routine than in discovery research and marketing, where the freethinkers congregate, but less routine than in quality, manufacturing, and clinical development. Preclinical researchers need to be capable of creative, scientific, intellectual thought, yet comfortable with applying processes and standardization to their work.

*The ability to be open-minded and creative problem solvers.* Experiments on drug candidates can generate unexpected results, and there is often an element of surprise to preclinical research. New products sometimes act differently in vitro than

*A little chaotic creativity and the ability to make loose associations are valued skill sets in preclinical research.*

they do in vivo, for example, and some of the most successful products were initially developed for diseases and conditions entirely different from their original targets. Viagra, for example, was originally developed for cardiovascular purposes.

*The ability to work with people in a team environment.* People skills are a necessity in this career, and a sense of humor certainly helps. You must be able to communicate with both the discovery researchers and clinical development teams.

*The ability to champion other people's ideas.* Discovery researchers usually generate the ideas for drug candidates, but preclinical researchers should be able and willing to work with other people's drug concepts.

*The ability to be scientifically goal-oriented.* It's important to maintain a balance between digging for scientific details and ensuring that projects continue to move forward. You can and should hone your skills in your field of expertise, but stay mindful of time constraints.

> *You need to be able to identify and answer questions that are relevant to the evaluation of drug suitability, but avoid the quest for basic knowledge in and of itself.*

*A critical and inquisitive mind.* It is important to question data and to keep digging until you are satisfied.

*Flexibility.* Things change quickly in preclinical studies. You need to be able to handle sudden adjustments and changes in priorities.

*The ability to derive valid conclusions from complicated information.* You need the vision to distill complex data sets and derive meaningful conclusions. This includes the ability to keep the "bigger picture" in mind and envision how a drug might perform in clinical trials and, ultimately, in patients.

*Good communication skills.* Preclinical positions require the ability to clearly articulate research results. In some companies, even research assistants present their data at group meetings.

*Hands-on technical skills.* Most preclinical researchers spend time in the laboratory, except for VPs and directors, who tend to work more with strategic issues or management of the science.

*Project management skills.* Those in managerial positions need to coordinate large groups to accomplish the needed work in the shortest amount of time.

*An optimistic attitude.* These jobs will be easier if you can tolerate the high failure rate of projects.

*Attention to detail.* Preclinical studies require thoroughness. Cutting corners is not well tolerated; one needs to be scientifically vigilant.

> *It is important to be observant, to be open to new ideas or applications, to enjoy a little contradiction in results, and to be able to persevere with challenging programs.*

***You should probably consider a career outside of preclinical research if you are...***

- Someone with a big ego who wants to be treated like a VIP.
- Unable to communicate effectively.
- Not interested in developing scientifically in your area of discipline.
- The inventive type who is more interested in watching your own ideas develop into drugs (consider discovery research instead).
- One who gets too involved in the hunt for scientific details and needs to pursue answers to unlimited depths (consider academia or discovery research instead).

## PRECLINICAL RESEARCH CAREER POTENTIAL

Once in preclinical studies, you can stay at the bench or be promoted up the management track, starting with scientist and continuing to manager, to director, and then to VP of preclinical research. Successful individuals can eventually be promoted to the head of drug discovery or VP of Research and Development.

Preclinical research offers a launching pad into just about any area of drug development (see Fig. 8.1). The most common transitions are into regulatory affairs, quality, manufacturing, business development, project management, and patent law.

### Job Security and Future Trends

A preclinical research background provides more job security than discovery research, because preclinical research is considered more essential. Some areas of preclinical research

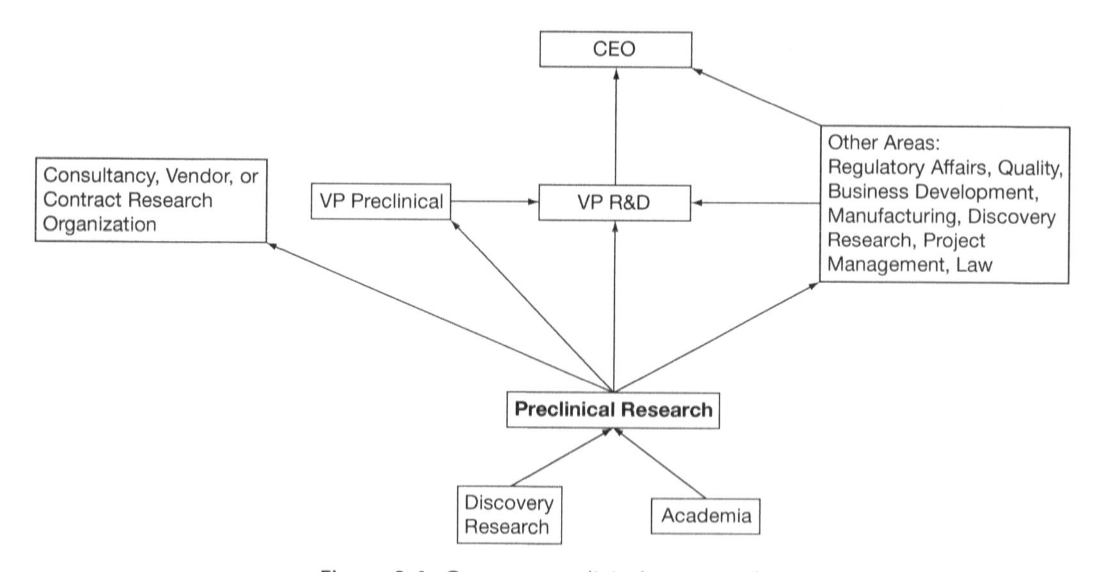


Figure 8-1. Common preclinical career paths.

are in more demand than others. Currently, there is an urgent need for those with expertise in in vivo pharmacology, toxicology, pharmacokinetics, and bioanalytical chemistry.

## LANDING A JOB IN PRECLINICAL RESEARCH

### Experience and Educational Requirements

Most people enter preclinical research from discovery research or academia. Most possess a basic science background in the biological sciences or chemistry, and many have specialized expertise in areas such as pharmacology or toxicology. Many employees hold veterinarian, M.D., Ph.D., or master's degrees. Many have backgrounds in pharmaceutical toxicology, pharmacology, pathology, chemistry, biology, biochemistry, or analytical chemistry. The DMPK or ADMET group is usually headed by bioanalytical chemists.

Those with bachelor's degrees start at technician or research associate levels. They normally max out as senior research associates or senior technicians. Occasionally, exceptional performers can make the transition to scientist titles, but an advanced degree is typically required for managerial positions.

### Paths to Preclinical Research

- The most common path into preclinical research is from discovery research. You may want to initially join a discovery research department and then transfer into preclinical research.
- Develop hands-on skills working with laboratory animals. Work with animal disease models when appropriate.
- If you are still in college, consider a summer internship at a drug discovery and development company.
- Gain more understanding of drug discovery and development and the various therapeutic areas. Specialize in one therapeutic area and become an expert.
- Network at professional societies specializing in therapeutic areas that interest you.
- If you are interested in toxicology, consider becoming board certified. There are undergraduate degrees and certificate programs offered in toxicology.
- Consider joining a contract research organization (CRO) that has a preclinical studies arm, and then transfer into a biotechnology company after you have gained experience. Service organizations are frequently an easier entry point into industry and also offer a wider array of projects.

## RECOMMENDED TRAINING, PROFESSIONAL SOCIETIES, AND RESOURCES

*Courses and Certificate Programs*

Board certification is required for toxicologists.

*Societies and Resources*

Society of Toxicology (www.toxicology.org)

Professional societies relevant to specific therapeutic areas, such as American Association for Cancer Research (www.aacr.org)

The American Physiological Society (www.the-aps.org)

The American Chemical Society (www.chemistry.org)

American Association of Pharmaceutical Scientists (www.aapspharmaceutica.com)

The Pharmaceutical Quality Group (www.pqg.org)

Drug Information Association (www.diahome.org) has a center for career and professional development for biotechnology and drug development.

*Books and Magazines*

Benet L.Z. 1984. *Pharmacokinetic basis for drug treatment.* Raven Press, New York.

Ferraiolo B.L., Benet L., and Levy G. 1984. *Pharmacokinetics.* Springer, New York.

Any books written by Malcolm Rowland, such as:

Rowland M. and Tozer T. 1995. *Clinical pharmacokinetics: Concepts and applications,* 3rd edition Lippincott Williams & Wilkins, Philadelphia.

*Nature Biotechnology* (www.nature.com)

*Science Magazine* (www.sciencemag.com)

*Genetic Engineering & Biotechnology News* (www.genengnews.com)

# 9

# Project Management

## The Product Development "Orchestra Conductors"

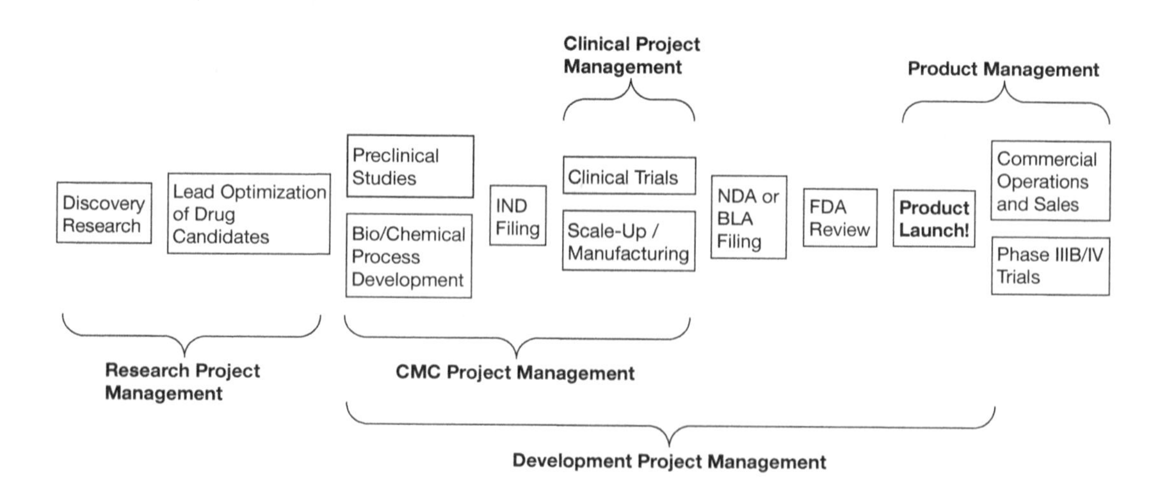


DO YOU PREFER TO THINK ABOUT THE BIG PICTURE rather than specialize in a particular scientific discipline? Do you enjoy facilitating the decision-making process and helping to define options by pulling together information from the many groups working on a project? Would you like to serve as the highly visible, primary representative of a project? If so, project management might be the career for you. It requires exceptionally good communication, interpersonal, and diplomatic skills, as well as the ability to manage people without having direct authority over them. You must be able to view a project as a whole while also possessing the technical know-how needed to remove obstructions that might impede project development.

*The three most important components of project management are communication, communication, and communication.*

## THE IMPORTANCE OF PROJECT MANAGEMENT IN BIOTECHNOLOGY AND DRUG DEVELOPMENT

The project management role in biotechnology was created because of the need for someone to coordinate the activities of the many scientific disciplines that are frequently involved in projects. Project managers ensure that projects are moving forward according to pre-established timelines, scope, and budgets. A project manager works as part of a team of people that also includes technical specialists. The manager does not have direct authority over these functional team members; instead, he or she helps to coordinate the tasks of the project so that the team works more effectively and efficiently.

*Project managers take part in the decision making but are not the decision makers.*

This close involvement with the team members requires the project manager to learn about different departments, their interrelationships, and how products are developed. A project manager needs to be familiar with many diverse technical areas and essentially becomes a Jack or Jill of all trades.

## CAREER TRACKS: PROJECT MANAGERS AND PROJECT TEAM LEADERS

*The titles and corresponding roles of people in project management vary depending on the size and type of company. For the sake of simplicity, discussion here is limited to the roles of project managers and project leaders.*

### *Project Managers*

In therapeutic drug discovery and development companies, the vast majority of project managers handle drug development programs that are in, or are approaching, clinical trials. There are also an increasing number of research project managers, who work in earlier stages of drug development, including early discovery research, late-stage research, and preclinical projects. In addition to the program-wide project managers, there can also be project managers who are dedicated specifically to functional areas that tend to be especially complex, such as clinical research and manufacturing. These positions can often serve as a "training ground" that can lead to an eventual role as a program project manager.

Not all project managers work in drug development companies. Project managers are also needed in life sciences companies that develop products such as instruments, reagents, tools, diagnostics, technology platforms, and medical devices (see Chapter 2).

The roles and responsibilities of project managers depend on the company, project, and product, and they range from recording meeting minutes to leading an entire project. In general, however, the project manager has a more tactical or operational role and serves as a team member with project management responsibilities.

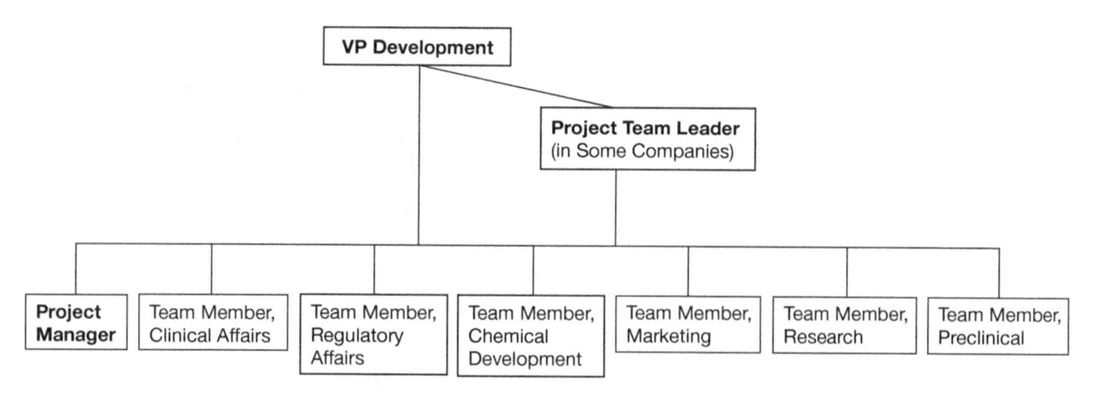


**Figure 9-1.** Typical team structure for PM.

### *Project Team Leaders*

Project leaders have a more global and strategic role than project managers (see Fig. 9-1). They may be responsible for a program that includes the entire development and testing process of a drug candidate. They communicate the vision and inspiration that drive the project goals, and they ensure that the company's operating values are guiding the project teams. They provide scientific and business leadership and are typically high-ranking, influential people who can speak with authority about the project. Project leaders are frequently M.D.s or Ph.D.s and may be in upper management (e.g., as high as Vice President of Clinical Research). The person designated as project leader can be changed with the stages of product development. For example, a senior research scientist can be the project leader until the drug is in the clinical stage, after which the project leader position shifts to a clinician or senior project manager. Project leaders tend to have project managers on their team, but some smaller companies do not have both project leaders and project managers.

## PROJECT MANAGEMENT ROLES AND RESPONSIBILITIES

*The roles and responsibilities of project managers or project leaders may include the following:*

*Project management is about leading with vision and facilitating teamwork. It takes many different disciplines to develop a drug; if they are not coordinated, time and resources can be squandered on unproductive activities.*

### *Leadership*

Project managers (and primarily project leaders) provide vision and inspiration to the project's members. By conveying enthusiasm for the project to the team, they try to create a more productive working environment. The project manager works as a member of the team to help define the strategy, goals, and metrics for the project.

In addition, the project manager helps to define the roles and responsibilities of team members so that they clearly understand what they are supposed to be doing. Project managers help resolve personal and cross-functional conflicts so that the team functions smoothly. They serve as psychological team boosters by listening to individual team members, reassuring them, and motivating them to best fulfill their team functions.

#### *Communication*

In most companies, one of the most important roles for project managers is to facilitate communication in many directions—upward to senior management and corporate partners, sideways to project team peers, and downward to the technicians and others who support the project. The team develops the goals, and the project manager, as a representative of the team, presents them to the company.

#### *Meeting Management*

Project managers spend a significant percentage of their time scheduling and running meetings, creating agendas, identifying action items, recording the minutes, and sending follow-up messages afterward.

#### *Resource Allocation*

Project managers manage project timelines and budgets. They work closely with finance department members to calculate how much money is required to run the project. They track expenditures and adjust the projected budget and staffing requirements as needed over time. One of a project manager's key responsibilities is to anticipate budget or staffing shortfalls that threaten the project, to notify upper management of the risk, and to request additional funding if needed.

#### *Strategy and Decision Management*

The project manager tries to ensure that the team addresses cross-functional issues and considers all pertinent information when it makes decisions. He or she facilitates discussions to develop consensus among team members. When a consensus cannot be reached, the project manager sometimes has the responsibility to make the final decision, but he or she also should know when it is appropriate to bring upper management into the discussion. Project managers are expected to represent the views of team members and to communicate relevant issues to upper management.

#### *Risk Mitigation and Contingency Planning*

Because of the complexity and difficulty of many projects, there are multiple chances for disaster to strike. Among other things, the project can fail to meet its clinical objective, supplies can be exhausted during a clinical trial, or senior management can decide to terminate a project. With the team's help, the project manager identifies potential risks, conducts analyses, and develops contingency plans to mitigate those risks so that the team and upper management can assess the probability of success.

#### *Problem Solving*

When something goes wrong with a project or when progress stops, the project manager is the person who needs to get things moving again. He or she should either possess enough technical know-how to help resolve the problems or know whom to contact for help. The project manager needs to make sure the functional areas take problems seriously, identify solutions, and act on them.

#### *Alliance Management*

In some companies, project managers coordinate projects with corporate partners. They are responsible for promoting good relationships and effective communication with corporate partners to make sure that the goals of the two companies are aligned and the teams are working together effectively.

#### *Documentation, Processes, and Procedures*

Project managers are responsible for creating, maintaining, and documenting development plans, as well as tracking the project's progress. Other management processes that may need to be established and tracked include cost, quality, risk, and procurement.

## A TYPICAL DAY IN PROJECT MANAGEMENT

Because of the unpredictability of product development and the broad range of roles and responsibilities in project management, there is no such thing as a "typical day." In general, project managers and project leaders spend most of their time talking to people in the form of one-on-one or group meetings. Time is spent discussing upcoming milestones with team members and addressing any problems with meeting those milestones. If there is an alliance partner or a significant number of team members at a different site, project managers and project leaders may spend time traveling.

*A project manager or project leader can expect some of the following activities on a typical day:*

- Preparing project reports, budgets, timelines, and analyses, and presenting these reports to project members and senior management.
- Arranging meetings, creating presentation materials, recording meeting minutes, and distributing minutes after meetings.

- Meeting one-on-one with project members and department heads to solve problems and ensure that priorities are uniformly established.
- Managing alliances with corporate partners, traveling, and networking.

## SALARY AND COMPENSATION

In general, the salary of a project manager is comparable to that of a discovery research scientist, which is concomitant with title and responsibility, years of experience, and expertise. Consultants, vice presidents, and project leaders can earn higher incomes and are often at the top of the pay scale. Project managers and project leaders who have overseen successful product developments should be able to demand higher compensation.

***How is success measured?***

Success is usually subjectively measured by how efficiently the project moves forward, the quality of execution, and how well the team functions together. Other metrics include project success, timeliness, and how well the project remained within budget. Project managers should gain satisfaction from the success of the team as opposed to recognition for their own individual contributions.

## PROS AND CONS OF THE JOB

### Positive Aspects of a Career in Project Management

- Project management is a very dynamic job; there is "never a dull moment."
- It provides an excellent opportunity to discover the ins and outs of product development, including operational, financial, clinical, scientific, regulatory, and legal issues. Successful navigation of these waters can lead to other career opportunities.
- A project manager interacts with people throughout the company... all the way from laboratory scientists to the CEO.
- Senior project managers and project leaders may be highly visible in a company and have decision-making responsibilities.
- Managing complex and technically difficult projects can be intellectually stimulating.
- Project management does not require bench work. There are usually no direct reports to manage, either.

- Seeing a project to its completion and watching a team perform well as a consequence of one's efforts can be highly rewarding.

### The Potentially Unpleasant Side of Project Management

- The challenge of influencing people without having direct authority over them can be frustrating. Much time is spent convincing and coaxing team members (see "Greatest Challenges"). Senior managers may make strategic decisions without the project manager's involvement.

*Project managers have much of the responsibility and little of the authority.*

- Frequent travel may be required.
- Scientific expertise might be sacrificed to the time-hungry demands of becoming familiar with many disciplines across the company.
- Long-term projects can try one's patience and sap enthusiasm.
- Daily progress can be hard to measure—project management is unconventional and exciting but also may lead to little immediate gratification.
- When things go well, the functional team members usually receive the credit. The reverse, however, is not true: When things go badly, project managers frequently suffer the blame. Many things can go wrong, including things beyond the project manager's control, and projects can easily fail and be terminated.
- Day-to-day activities such as arranging meetings, writing minutes, etc., can be mundane. There can be a lot of paperwork.
- It can be frustrating to manage alliances when corporate partners have different cultural values.
- If there is bad news, it is often the project manager's job to tell the team.

## THE GREATEST CHALLENGES ON THE JOB

#### *Responsibility without Authority*

The biggest challenge is the project manager's lack of direct authority over the team members he or she manages. When team members have aims and responsibilities that prevent them from contributing effectively to the project, the project manager cannot order them to change their agendas. He or she can try to persuade them by explaining the corporate culture and the priorities or overall goals of the project or company, and by convincing them that their work is needed for a particular function. The project manager can also resort to speaking with a team member's manager or with upper management. All of these options require very good interpersonal and diplomatic skills. Much time is spent coaching, reassuring, and motivating people.

#### *Perseverance*

Maintaining a consistent point of view and keeping team members motivated on lengthy projects can be challenging. Tenacity should not be underestimated as a key personal attribute needed for success.

#### *Diplomacy*

*There are great project managers with mediocre scientific skills, but there are no great project managers with mediocre interpersonal skills!*

Project managers need to be able to make or facilitate decisions based on varied points of view without causing conflict. Maintaining positive team relationships while working to advance the project can sometimes be a delicate balancing act.

#### *Objectivity*

Whereas project managers need to keep the team motivated and excited about the project, it is important that they also objectively evaluate the project's potential. As driver of the project's decision management, knowing when to end a project is just as valuable as deciding to push forward with it.

## TO EXCEL IN PROJECT MANAGEMENT...

#### *Years of Experience*

*You won't be a successful project manager if you think you know everything.*

Ultimately, the combination of exceptionally good interpersonal skills and years of experience is what separates the good from the great. With experience, project managers develop the ability to anticipate potential issues before they arise. They also cultivate a good understanding of, and appreciation for, the different functional areas and their cross-functional interdependencies.

## Are You a Good Candidate for Project Management?

***People who flourish in project management careers tend to have...***

*Excellent communication and interpersonal skills are essential.*

***Superb interpersonal skills.*** This is probably the most important factor for success in project management. Good interpersonal skills allow you to develop positive, collaborative, and productive relationships with team members and other coworkers (see Chapter 2).

***Excellent communication skills.*** Many of a project manager's duties center around the ability to communicate with multidisciplinary team members. You must be able to speak and write clearly and in such a way that you can accomplish your goals while avoiding being confrontational or alienating individuals.

***The ability to simultaneously see the big picture and pay attention to the details.*** Understanding and thinking strategically about a project as a whole is as important as taking care of the minutiae.

***An ability to foster a collaborative and positive work environment.*** Sometimes success is measured by how well a team worked together. The ability to understand and tolerate different perspectives and to be able to formulate and implement a plan that is agreeable to the team helps foster a collaborative environment... after all, happy coworkers are more productive!

***A "team player" attitude.*** This is a must in project management (see Chapter 2). Project managers tend to be gregarious, yet willing to face disagreement for the good of the team.

***Excellent organizational and time management skills.*** Often project managers work on multiple assignments simultaneously. Good organizational and prioritization skills must be applied to save time, manage the volumes of information, and keep track of technical details.

***Strong leadership skills.*** It helps to be assertive, action-oriented, and self-confident if you want to convince your fellow team members to move projects forward, but you also need to be diplomatic at the same time.

***Proactive and analytical thinking skills.*** You must be able to anticipate difficulties and develop contingency plans before problems become obstacles. It helps to be analytical, consistent, and level-headed.

***Creative problem-solving skills.*** Project managers are constantly faced with the need to solve problems. The ability to think objectively and flexibly and to quickly evaluate alternative solutions makes it easier to overcome technical obstacles and internal conflicts that might slow a project's progression.

***Good judgment when making difficult decisions.*** Often there is not enough information available to make the best-informed decisions, so you need wisdom and intuitive judgment to select the most promising choices based on limited data.

***You should probably consider a career outside project management if you are...***

- Frequently unable to move forward because you get stuck on details.
- Too aggressive.
- A micromanager or someone who needs to be micromanaged.
- Someone who tends to take disagreement too personally.
- A person who needs immediate gratification and personal recognition.
- Unable to function within an unstructured environment or with uncertain outcomes.
- A person who manages by using negative reinforcement.
- Too easygoing.
- Someone who likes to work alone.

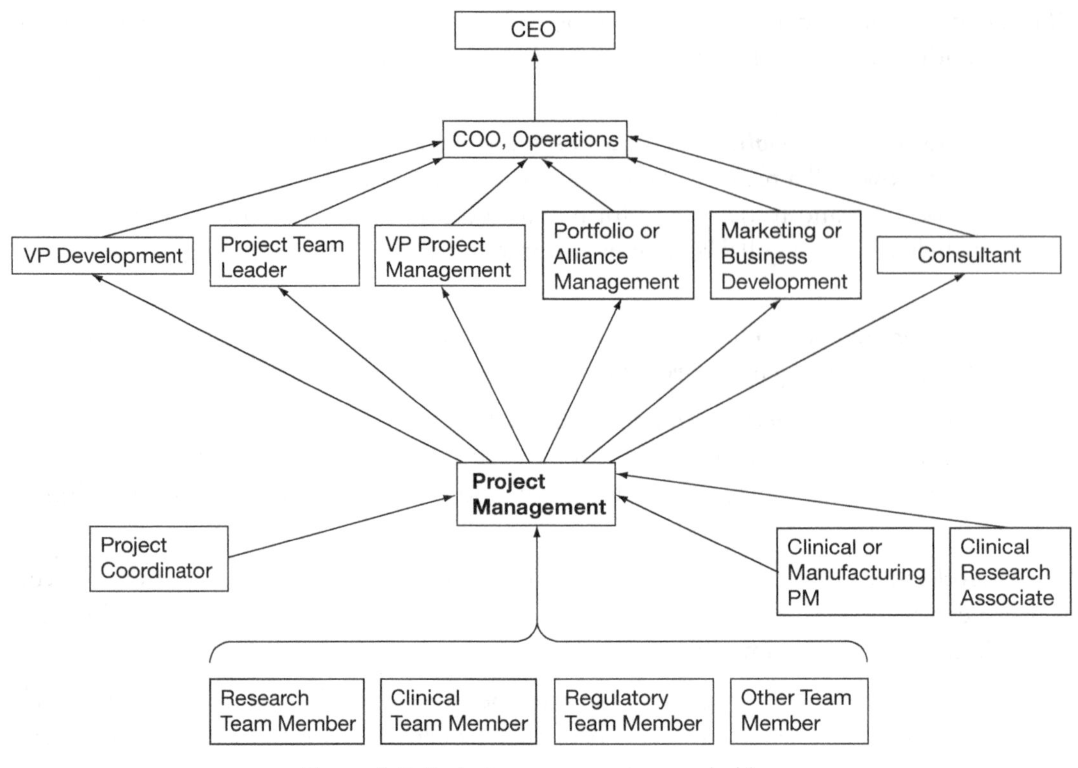


**Figure 9-2.** Project management career ladder.

## PROJECT MANAGEMENT CAREER POTENTIAL

Project managers become technically knowledgeable in multiple functional areas, so this occupation can lead to a diverse set of career options (Fig. 9-2). Within project management, career possibilities include Project Team Leader and Vice President of Project Management. Project managers with more general skills often go into consulting, business development, alliance management, organizational learning, or portfolio management. Those with more technical expertise can transfer to positions as functional heads of product or clinical development, operations, manufacturing, or regulatory affairs.

Project management experience is excellent preparation for eventual COO/CEO leadership roles. Project managers hone their organizational skills, become adept at motivating and managing people, and understand how to move a product through development and into the market.

### Job Security and Future Trends

The demand for talented project managers is strong and is predicted to increase. As the biotechnology industry has shifted its focus toward more development-related activities, project management has become a more visible and marketable discipline. Companies have recognized the valuable role of project management in expediting and coordinating product development. The combination of highly sophisticated technical knowledge and strong interpersonal skills gives project managers a unique skill set, so in general, they enjoy relatively long-term job security. It should be noted, however, that a project manager in a smaller company could be an early victim of layoffs during economic downturns.

## LANDING A JOB IN PROJECT MANAGEMENT

### Experience and Educational Requirements

Project management requires a broad technical and operational background and strong science acumen. A project manager must be technically credible, provide reasonable input, ask appropriate questions, and be able to resolve problems strategically and creatively. For these reasons, a project management position typically requires at least three years of industry experience to develop insight into how teams work. Most project managers have previously been team members from one of the various functional areas such as discovery research, manufacturing, process development, clinical research, or regulatory affairs. Project leader positions require extensive involvement in product development as both a team member and a manager.

An advanced degree is not essential for a job in project management, but it can be extremely helpful. Most project managers have a master's or Ph.D. degree, and some have M.B.A. or R.N. degrees. Qualifications depend in part on the requirements of a project. An early discovery project, for example, might require a Ph.D. degree, whereas an R.N. degree might be more useful for a clinical project manager. Project leaders are typically Ph.D.s or M.D.s.

Certificates in project management can be obtained from many universities and from the Project Management Institute. Although such certificates are desirable, exposure to the drug development process is far more valuable.

### Paths to Project Management

- Serve as a team member in one of the functional areas; this is by far the most common route to project management. Learn from a project manager who is willing to mentor you.
- Consider working as a project coordinator. This entry-level position leads to a project management role. Project coordinators assist project managers with tasks such as arranging meetings. By showing that you can facilitate productive meetings (a skill that can be learned), you have the chance to demonstrate leadership ability.
- Consider joining a clinical research organization (CRO) if you are interested in clinical project management. Working at a CRO provides excellent clinical research exposure and allows a transition to a job in a drug discovery company. CROs often hire Ph.D.s

with clinical experience to be group leaders or project managers. Those with undergraduate or nursing degrees can become clinical research associates (CRAs) or project coordinators; both of these jobs lead to career tracks in project management. If you are still in graduate school, try to gain some exposure to clinical trials.

- Attend project management society meetings and conferences; they are good places to network. Make as many contacts as you can. As with all careers, the more people you know, the more likely someone will open a door for you.
- Gain as much experience managing people as possible. Demonstrate the ability to be a level-headed, strategic thinker, and learn how to give lucid presentations.
- Obtain as much experience in drug discovery and development as you can. Become familiar with the internal and external factors that affect these processes, including business and product development issues. Take courses in drug discovery and development.

## RECOMMENDED TRAINING, PROFESSIONAL SOCIETIES, AND RESOURCES

### *Courses and Certificate Programs*

Project management certificates are offered in most local universities and can also be found by conducting on-line searches. Programs that are particularly recommended are offered by The Project Management Institute (www.pmi.org) and George Washington University. Barnett Educational Services (www.barnettinternational.com) offers classes that are designed specifically for people interested in clinical trials and drug development.

### *Project Management Societies and Resources*

Project Management Institute (PMI), www.pmi.org

Project Managers in Pharmaceuticals (San Francisco nonprofit), www.projmgr.org

Project Connections, www.projectconnections.com

### *Drug Discovery and Development Societies*

Association of Clinical Research Professionals, www.acrpnet.org

Drug Informational Association (DIA), www.diahome.org. The DIA has a strong project management group within the organization. Their annual meeting is a good place to make contacts and learn about project management in biotechnology and pharmaceutical companies.

### *Books and Magazines*

PMI offers several books about project management.

Covey S.R. 2004. *The 7 habits of highly effective people: Powerful lessons in personal change.* Free Press, New York.

### *Other Suggestions*

Consider taking classes in conflict resolution and public speaking. Become a member of Toastmasters International, www.toastmasters.org.

# 10

# Clinical Development

## Developing New Products to Benefit Human Health

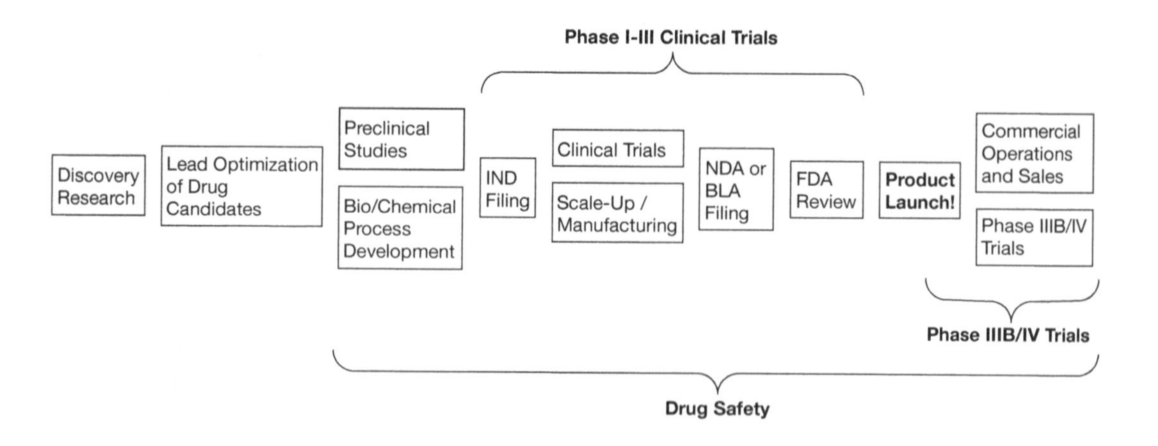


IF YOU ARE INTERESTED IN THE PRACTICAL SIDE OF SCIENCE and want to see the ultimate outcome of products that progress from basic research to approval for sale, clinical development may be the career for you. This is a career for those who enjoy the biological and medical sciences, are goal-oriented, and are highly motivated.

*Part of the appeal of working in clinical development is knowing that you can be instrumental in developing a drug that is having a positive impact on world health.*

Careers in clinical development can be highly gratifying. Because it encompasses multiple disciplines, there are many opportunities to learn new skills and technologies. An additional attraction is the potential for high salaries and exceptionally good job security.

## THE IMPORTANCE OF CLINICAL DEVELOPMENT IN BIOTECHNOLOGY AND DRUG DEVELOPMENT

Before a product can go to market, the company must first show that it is safe and effective by conducting clinical trials. After the trials are completed, a label is composed that defines the indication, dose, and preferred treatment population. It also includes characterizations of the product such as risk, safety issues, adverse events, and possible side effects.

## CAREER TRACKS: CLINICAL DEVELOPMENT AND OPERATIONS

Clinical development is a large department and includes many subspecialties (see Table 10-1). To generalize, there are those who work directly on the clinical studies (clinical development, operations, pharmacology, translational research, and project management) and those who provide support functions (biometrics and medical writers). Additional careers discussed in this chapter include those in drug safety, which is usually a separate department, and those in the U.S. Food and Drug Administration (FDA) (see also Chapter 12).

For simplicity's sake, and because these various roles and functions are so different, the clinical careers are combined into one section. The support functions, drug safety, and FDA positions are each described individually at the end of this chapter.

Some people make a distinction between the terms "clinical development" and "clinical operations." Clinical development, to some, is considered the medical and scientific side of development, whereas operations reflect the more tactical aspect of running trials. In large companies, these two roles are usually separated, but in most others they are one and the same. Typically, the higher-level medical personnel in clinical development are responsible for the oversight of the development plan. They design the clinical protocols, oversee drug safety, and are ultimately responsible for the information on the drug label. Clinical operations personnel are at the center of trial management, bringing the teams together, coordinating activities, and delegating roles and responsibilities.

**Table 10-1.** Clinical development subspecialties

| Branches in clinical development and operations | Biometrics | Clinical support | Working at the FDA | Drug safety |
|---|---|---|---|---|
| Translational medicine | Data management | Medical writers | Medical reviewers | Drug safety |
| Clinical pharmacology | IT quality | | | |
| Phase I studies | Biostatistics | | | |
| Phase II–III studies | Statistical programming | | | |
| Phase IIIb–IV studies | | | | |

The titles and roles are described below:

### *Chief Medical Officers*

The chief medical officer (usually an M.D.) is in charge of the company's development programs. He or she brings scientific expertise to the oversight of the development plan, strategy, budget, drug safety, and other high-level aspects of development.

### *Clinical Team Leaders or Clinical Program Directors*

Usually M.D.s or Ph.D.s, they are often responsible for writing and designing protocols and for leading the operations teams who execute the study.

### *Medical Monitors*

Usually M.D.s, medical monitors offer drug and patient safety oversight during the trials and provide the medical supervision to ensure that the study is conducted accurately.

*Medical monitors serve as the medical liaison between the company and the clinical investigators or CROs actually conducting the studies.*

### *Associate Directors, Directors, and VPs of Clinical Operations*

Directors have more scientific and strategic responsibilities and tend to lead programs in specific therapeutic areas.

### *Program Managers, Trial Managers*

Program managers supervise the operational side of clinical development and have project-management-type responsibilities. They oversee the execution of national or international trials, determine program budgets and timelines, manage contract research organizations, and help write regulatory submissions and safety documents.

### *Principal Clinical Research Scientists*

People in these positions might lead the protocol review and study design, oversee scientific aspects of the study, and make sure that trials adhere to regulatory requirements.

### *Clinical Research Associates (CRAs), Clinical Monitors, and Clinical Trial Coordinators*

Clinical research associates operate at the heart of clinical trials. They work with the nitty-gritty details of tracking patients and monitoring data at the hospital sites where the trials are run. They verify the accuracy of the data and ensure that the study is being conducted according to FDA guidelines.

### *Clinical Project Assistants and Contract Administrators*

A clinical project assistant is a person in training who is involved in tracking and assisting CRAs and program managers. Contract administrators deal with contracts and budgets.

### About clinical research associate careers

There are multiple levels of CRAs. They can be field-based, work in-house at a sponsor organization, or work for a contract research organization. Highly experienced CRAs are in demand and are well paid. They are hired for specific studies depending on their expertise. Many are nurses, science undergraduates, or doctors from foreign countries. Approximately 70–80% of a CRA's time is spent visiting study sites. Many CRAs eventually burn out from so much travel and move into more stable operational areas of clinical development, but it's a good point of entry and a great way to learn about drug development.

## Other Departments within Clinical Development

### Translational Medicine, Translational Research, or Translational Development

*Translational medicine intersects discovery research and clinical development.*

"Translational medicine" is a vague term with several definitions; the broadest of which is "taking research from 'idea to bedside.'" One aspect of translational medicine involves developing laboratory measurements of biological markers (biomarkers; i.e., protein and RNA levels, psychometric tests, etc.). Biomarkers can be used to select patients for clinical trials who would be more likely to respond to therapy, to measure whether the treatment is working in patients, for diagnostic purposes, and more. Biomarkers can also expedite studies by providing an additional screen during clinical trials to determine whether a compound has its intended effect (see Chapter 8).

### Clinical Pharmacology and Pharmacokinetics

Clinical pharmacology translates preclinical data into first-in-man Phase I studies. This includes testing dose-response and selecting doses for clinical trials. Clinical pharmacology studies evaluate pharmacokinetics (e.g., plasma drug levels over time), bioavailability and disposition studies (absorption, distribution, metabolism, and excretion of the drug, often termed "ADME"), safety, tolerability, and, if possible, pharmacodynamic effects of each compound (see Chapter 8).

### Clinical Project Managers

Project management is mainly an operations function in which the multidisciplinary aspects of programs are coordinated. Project managers generally do not work at the trial level. They oversee budgets and timelines, manage relationships with contract research organizations and other vendors, and have a variety of other responsibilities, such as coordinating and running project team meetings and providing oversight for programs (see Chapter 9).

***Careers in drug safety, medical writing, data management, biostatistics and statistical programming, and working at the FDA are discussed in more detail at the end of this chapter.***

### *Clinical Studies*

***Clinical protocols*** are documents that describe the study objectives, design, and methodology of trials, as well as the patient population and how data will be collected and analyzed. Protocols and Investigational New Drug Applications (INDs) are submitted to the FDA for review and non-objection before initiating Phase I human studies.

***Investigator sites*** are located in hospitals or private clinics where the clinical trials are run. Early investigational studies are often run in an academic institution by clinicians (the investigators) with a practice specializing in a particular therapeutic area. The physicians generate data by following the clinical protocol and recording the results.

***Contract research organizations (CROs)*** are vendors that conduct and/or manage studies for biotechnology and drug development companies (sponsors). Clinical trials and specialized functions (e.g., data management, registries) are often outsourced to CROs.

***Case report forms (CRFs)*** are used to record data collected on each patient enrolled in a clinical trial. This information is stored in databases and used to determine a drug's clinical effectiveness.

***Investigational New Drug Applications (INDs)*** are regulatory filings sent to the FDA before a company can enter Phase I clinical trials.

***New drug applications (NDAs)*** *(for small-molecule pharmaceuticals) and* ***biologic licensing applications (BLAs)*** *(for biologic therapeutics)* are large regulatory filings submitted to the FDA for product approval. These filings occur after the drug sponsor completes pivotal efficacy and safety studies, usually on large volunteer patient populations representative of the potential treatment group (see Chapter 12).

***Phase I studies*** test drug safety, pharmacokinetics, and clinical pharmacology. The studies are usually conducted on a small group of healthy volunteers. Dose-response curves are constructed and side effects are monitored to determine the optimum dose. The studies can test for a variety of qualities, such as food–drug or drug–drug interactions, pharmacokinetics, and pharmacodynamics. The generic drug industry conducts Phase I bioequivalence studies for products that are already on the market.

***Phase II studies*** are conducted on larger groups of subjects with the targeted condition or disease to test for proof of efficacy and to look for potential short-term side effects or risks associated with the drug.

***Phase III studies*** are large clinical trials that generally include hundreds or thousands of subjects. Studies are either controlled (randomized groups separated to receive investigational product, placebo, or approved drug therapy) or uncontrolled, and are intended to gather additional information about the effectiveness and safety of the drug and to evaluate the overall benefit–risk relationship. For drugs intended for chronic use, pivotal efficacy studies may last 90 days or longer, with a smaller "safety" cohort treated for a year or longer. After a successful outcome, the company can submit an NDA or BLA filing, and if this is approved, can request and initiate sales of the drug.

***Phase IIIb/IV studies*** are extensions of Phase III trials designed to study the drug in other disease states or in post-approval commitment trials requested by the FDA for additional surveillance. The company continues to learn more about the product, identify more disease indications, test the drug in special populations, identify possible drug–drug interactions, and more. Head-to-head programs may be run to show that the new product has specific advantages over its competition (see Chapter 11).

## CLINICAL DEVELOPMENT ROLES AND RESPONSIBILITIES

*Depending on one's position, those in clinical development may have some of the following roles and responsibilities:*

### *Development Strategy*

Development strategy encompasses all of research and development (R&D). Functional heads work in teams from marketing, preclinical research, and other areas to assess products and their stages of development. Together, they evaluate development programs, the cost, and the potential financial outcome if everything proceeds according to plan. This is translated into a budget that is aligned with corporate goals.

*In clinical development, implementation of strategy is largely an examination of the budget.*

Development strategy also involves making sure that the many moving parts in clinical development are properly functioning together as a unit. Clinical development is highly complicated, and many things need to take place in order to successfully develop a commercial product. Strategic decisions are made about the therapeutic approach, how to drive decisions, and how to balance authorities.

*R&D often has long-term horizons of at least ten to fifteen years.*

### *Designing Studies and Writing Clinical Protocols*

The clinical program directors, managers, or medical monitors are involved in the design, review, and evaluation of protocols. They lead teams to execute or implement the projects' goals.

### *Clinical Trial Management/Operations*

People in clinical operations are at the center of clinical trials. They review protocols and qualify, arrange contracts with, and manage CROs. They oversee field monitors and work with biostatisticians, medical writers, and the drug safety group.

Clinical research associates visit clinical sites to verify that the study is conducted per protocol and that the data and samples are being collected and processed accurately. They also ensure that the sites are complying with FDA guidelines. They collect the information and enter it into the database and generally manage processes. After each visit, the CRA documents findings. When studies are completed, they "close out" sites and write final reports.

### *Medical Monitoring and Drug Safety Assessment*

Once a study is up and running, medical monitors provide the oversight for patient safety and make sure that the trials are running smoothly. They review the clinical labs, and patient inclusion and exclusion criteria, and make sure that the appropriate patients are enrolled. They are the main contacts who field questions from investigators.

### *Data Management*

The information gathered from the clinical trials is entered into a database. Data from case report forms and collections of tissue samples, pharmacokinetics results, and more are stored in the database for the final analysis.

*The end result of clinical studies is data.*

### *Clinical Trials Analysis and Biostatistics*

After the clinical trials are completed and the last of the data is entered, the database is "locked," i.e., the information cannot be altered. Typically, the lead clinician and statisticians then work together to analyze the results and draw conclusions. They might ask new questions or sort the data according to different age or ethnic groups, for example. When FDA personnel review the data, they will run their own analyses and draw independent conclusions.

### *Interaction with Key Opinion Leaders*

Clinical development professionals spend time becoming acquainted with and interacting with renowned experts, known in industry as "key opinion leaders." These important individuals serve on FDA advisory boards that discuss the merits of particular drugs in specialized therapeutic areas, or they may serve as lead authors for the published results of the company's clinical studies.

### *Writing Standard Operating Procedures*

Standard operating procedures (SOPs) describing clinical trial execution and management are written to ensure consistency. The SOPs define processes step by step and also denote who is responsible for each step.

## A TYPICAL DAY IN CLINICAL DEVELOPMENT

*There are many roles in clinical development. Depending on your position, you may be involved in some of the following activities:*

- Attending various internal meetings for a long list of purposes, including but not limited to discussing timelines and milestones of clinical trials, status updates, budget and project reviews; determining where to publish data for clinical studies; writing or amending drug labels; and preparing for meetings with the FDA or key opinion leaders.
- Participating in meetings or teleconferences with clinical investigators, CROs, and other vendors. Contacting investigator sites if they are not performing well and offering assistance with patient recruitment.

- Reviewing and writing clinical protocols and a variety of other FDA submissions.
- Telephone conferencing with development partners, maintaining relationships, and sharing information.
- Contacting doctors who reported serious adverse events to inquire about specific details.
- Attending scientific advisory board panels, protocol reviews, and scientific meetings.
- Researching competition. Reviewing results of ongoing clinical trials from other companies.
- Discussing clinical data with statisticians.

## SALARY AND COMPENSATION

*Clinical development careers are lucrative and offer exceptionally good job security.*

Clinical development includes lucrative careers that tend to pay considerably better than those in most other areas of biopharma. The demand for qualified talent is high and the supply is low. M.D.s in this field can expect to earn higher salaries. Compensation for these jobs is comparable to that in regulatory affairs, business development, and marketing, and is better than in discovery or preclinical research, manufacturing, and quality. Those with M.D. degrees often earn more than those with other advanced degrees, but they carry a correspondingly greater level of responsibility. Doctors, nurses, and pharmacists may be compensated on a par with hospital work, depending on the position and therapeutic area. Those without advanced degrees are paid more in clinical development than in discovery research, and there is more opportunity for job advancement as well.

CROs tend to pay less than biopharma companies, but they provide more training for entry-level hires and are more likely to provide a broader exposure to the entire drug development process.

***How is success measured?***

Milestones and timelines are everything in this career. Milestones include patient enrollment ("first patient enrolled" and "last patient out"), the final database lock, completion of clinical study reports, whether the study has met the clinical endpoints (objectives of the study), and the ultimate measure of success—drug approval.

When a drug fails, it is sometimes difficult to determine whether it was because of poor development or because there was something inherently wrong with the compound or biologic. For this reason, people are evaluated according to their productivity and the quality of their work.

## PROS AND CONS OF THE JOB

### Positive Aspects of a Career in Clinical Development

*As a practicing doctor, you might touch the lives of 10,000 patients during your career. In companies, your work could touch the lives of millions of people worldwide.*

- It is immensely rewarding to know that you are contributing something beneficial to society and leaving your personal mark. Bringing safe and useful new drugs to market has a huge, positive impact on the world.
- Every day is different. Some bring interesting challenges and others bring frustrating nuisances.
- Original and applied clinical research is exciting. Outcomes are unknown until trials are completed, and each trial is unique. Your work is close to the market.
- There is a sense of accomplishment when milestones are reached and you can see progress based on your work. It is gratifying to be in an area where your work generates meaningful, publishable scientific data and contributes to the vast body of knowledge in medical research.

*In clinical development, you can stay up-to-date on the latest scientific advances without using a Pipetman.*

- Careers in clinical development currently offer exceptionally good job security. In addition, because so many functions exist within clinical development, there are many opportunities for professional growth and career development. This experience provides an easy transition into other areas, such as regulatory and medical affairs. There is also the opportunity to acquire on-the-job skills in areas such as accounting, biostatistics, and more.
- Clinical studies play a visible, pivotal role in companies. You will be contributing to the success of the company in a significant way.
- There is a tremendous amount of learning involved, especially when studying diseases in rapidly developing fields. It is exciting to be at the cutting edge of scientific discovery.
- This career provides an in-depth understanding of drug development and its various disciplines, from regulatory affairs to marketing. You can oversee the entire process of taking a compound from discovery all the way to post-approval.
- Clinical development careers present opportunities to meet many interesting and influential people. You could be interacting with clinical investigators and renowned experts, and internally, the clinical department tends to be composed of bright, considerate, and hard-working individuals.

### The Potentially Unpleasant Side of Clinical Development

- Depending on their roles, CRAs and clinical trial monitors can spend up to 80% of their time traveling! The extensive travel can be taxing.

- There is frequent pressure to meet constantly looming deadlines, and the objectives often seem to be "too much, too soon, with too little."

*There is always much to do and seemingly not enough time to do it.*

- The financial challenges of clinical development are immense. As a result, teams may be forced to cut corners, work their staff very hard, or discontinue product development due to insufficient funding.
- Clinical trials rarely go smoothly. There are many delays that impede progress, including legal, contractual, and patient privacy issues. This is particularly true for products and biologics that offer unmet medical needs, as the development path is unknown and can be full of unexpected bumps.
- Most clinical trials are not successful. (But the ones that are successful are worth the extra effort!)
- The amount of paperwork can be enormous. In big pharmaceutical companies, people can spend much of their time just taking care of bureaucratic details.
- It can be easy to become pigeonholed within your area of expertise.
- Some of the studies can be tedious and repetitive.
- Some investigators (academicians leading studies) have big egos and can be extremely arrogant. Working with them can be challenging.

## THE GREATEST CHALLENGES ON THE JOB

### *Just Making It Work! Getting Drugs Approved*

*You can have the most wonderful business plan in the world, but if the drug doesn't work in humans, then it just doesn't work.*

The process is extremely complicated, and most drug candidates fail in clinical development. Recent trends reveal that drug development costs have been increasing while at the same time there have been fewer successes. The list of challenges is long, but some of the most commonly encountered include the following:

- Designing the best possible studies with the cleanest clinical endpoints so that efficacy can be more precisely evaluated.
- Obtaining clinical trial results that show solid confirmation of clinical outcome.
- Determining the causes of unsuccessful studies. Was it because of something that you did? If so, can you change the clinical endpoint?
- Finding dosages that strike the right balance between maximal health benefits and minimal side effects.
- Finding the best indication for the appropriate population group. Many drugs fail in initial clinical studies, but may succeed later when applied to a different patient population. Thalidomide, for example, was originally prescribed as an antiemetic for preg-

nant women. Its use in that context was discontinued when its link to congenital malformations became clear, but it is now used as a cancer treatment.

- Knowing when to proceed and when to stop development. It is important to not give up on drug candidates too early, but you don't want to continue research on compounds that cause serious adverse side effects or seem ineffective.

#### *First-of-a-Kind Products to Treat Unmet Medical Needs*

*To go where no clinician has gone before...*

Developing therapies that are the "first of a kind" to treat a disease presents new challenges. These are unique trials that have not been conducted before and, as a consequence, unanticipated problems are likely to arise. Success with these novel products, however, will have a greater fundamental impact.

#### *Upholding Safety Standards*

The medical profession's command to "do no harm" is even more important in industry, because your decisions may affect hundreds of thousands of lives. Medical doctors in industry may be challenged with ethical situations in which they need to delay or terminate a study, actions that could directly affect the future of the company and its employees.

## TO EXCEL IN CLINICAL DEVELOPMENT...

#### *Agility in the Midst of Complexity*

*There is a science to trial design and an art to execution.*

Clinical trials are complex. Many factors, including the market and regulatory environments, need to be taken into account when preparing for potential study outcomes and making contingency plans for overcoming obstacles. It takes many years of experience to learn the tricks of the trade, and those who excel are able to maneuver within the complexity without overlooking the details.

## Are You a Good Candidate for Clinical Development?

*People who flourish in clinical development tend to have...*

*Hiring managers look for candidates who are highly motivated, outgoing, flexible, and detail-oriented.*

*A goal-oriented work ethic.* This is a vocation driven by timelines and budgets. You need to be efficient, resourceful, and willing to work long hours—in essence, you need to be a "driver." Self-starters who can work independently and possess a "can-do" attitude can excel.

*The ability to keep the big picture in perspective while simultaneously managing many details.* Drug development is complicated. Those who excel understand what the company ultimately needs to achieve but are still able to pay attention to the minutiae.

***Outstanding people skills and the ability to work well in teams.*** Clinical development is a team sport, so you need to be able to work smoothly with people from different departments and backgrounds. You need to be socially credible, competent, collaborative, fair, and considerate. You need to be receptive to criticism and constructive when you appraise others (see Chapter 2).

*The key to execution is good communication.*

***Exceptionally good communication skills.*** Because there are so many moving parts in clinical development, it is essential that you communicate well. These skills will be needed extensively for managing the many people and organizations involved in trials, such as CROs, clinical investigators, and clinical coordinators. You may spend a lot of time schmoozing, handling internal politics, entertaining vendors and clients, and serving as a liaison with collaborators. Those who have outgoing and gregarious personalities tend to be effective communicators who can diplomatically voice their opinions.

***Exceptionally good writing skills.*** The ability to write clearly, concisely, and quickly is valuable for drafting regulatory submissions and communicating with team members.

***Patience and perseverance.*** Clinical trials can take a long time, from one to five years, or sometimes longer. You need to be satisfied with celebrating small victories and mini milestones.

***A commitment to maintaining public safety and a strong sense of ethics.*** The nature of this work, treating patients with new, untried compounds, requires an empathetic concern for human safety.

***Compassion and sensitivity.*** For those working directly with patients, it is important to have sympathy for the desperate plight of people who enroll in clinical trials.

***Strong problem-solving skills, an ability to think creatively and independently, and flexibility.*** On any given day, issues will arise that possibly require a shift in priorities and focus. For example, the various components of trials and projects require constant reevaluation and risk mitigation. Many things do not go as planned.

***The capacity for strategic, critical, and analytical thinking.*** This job requires not just responding to current situations but also thinking ahead. You need to be able to make sound, science-based decisions and take carefully calculated risks when choosing how to design and manage trials and analyze results. Once a trial is initiated, there are limited chances to make changes.

***Good judgment.*** Those who excel have the creativity and persistence to see projects through, perhaps against all odds when the situation calls for it. It is equally important, however, to terminate programs that are unsafe, ineffective, or not meeting corporate objectives.

*You need to be able to juggle many balls in the air to be efficient in clinical development.*

***Extraordinarily precise attention to detail, an ability to multitask, and exceptional time management skills.*** The work load is heavy. There are many details, and they have to be monitored accurately.

*You may have the best clinical development background in the company, but if you can't motivate the team, you could fail miserably.*

***Leadership qualities (for senior management).*** Superb management skills are mandatory. Leaders should be focused, very knowledgeable about their field, good judges of people, motivational, and able to delegate efficiently.

***A broad understanding of drug development.*** This includes maintaining a firm grasp not only of the medical field, specific disease indications, your competitors, the clinical trial data, and pharmacokinetics, but also of finance, strategic planning, and more.

***A natural curiosity and the ability to learn quickly.*** Clinical development is an immense and dynamic field with a daily influx of new scientific information. Most clinical professionals are hired for their expertise in a specific medical area. To expand your usefulness into different therapeutic areas, you must be willing to continue educating yourself.

***An ability to work in a heavily regulated environment.*** One must be able to accept and work with regulations, processes, and procedures. Clinical data are collected and analyzed in accordance with strict FDA guidelines.

***Risk takers.*** Clinical development can be cutting-edge research. Most trials have never been conducted before. There is risk involved, but the ultimate rewards are well worth it.

***You should probably consider a career outside of clinical development if you are...***

- Unmotivated.
- One who cannot easily delegate work or is a micromanager.
- A person who prefers to work alone or cannot operate effectively as a team member.
- Dependent on short-term success and constant positive reinforcement.
- A person who is political or has an abrasive or temperamental personality.
- Not detail oriented enough or too detail oriented, a perfectionist.
- Unable to function well under deadlines or unable to understand the meaning of urgency.
- Inflexible and unable to handle reprioritizations.
- Negative, pessimistic, or a complainer.
- Unwilling to accept rules and regulations (consider medical affairs or marketing if you need a creative work environment).
- A medical doctor with a continued interest in clinical practice. Your work can be entirely reduced to paper, writing protocols, and reviewing data, although many companies allow their physicians to continue to practice on a limited basis.

## CLINICAL DEVELOPMENT CAREER POTENTIAL

Because clinical development comprises such a large assortment of vocational areas, there is unlimited room for career growth. One can advance to director or VP of clinical development, head of R&D, and ultimately, chief medical officer, chief operating officer, or CEO (see Fig. 10-1).

Experience in this field allows relatively easy transitions into other departments such as marketing, regulatory affairs, and medical affairs. It also positions one for jobs at the FDA, in drug safety, or in academia. It is not uncommon for senior clinical executives to become consultants, venture capitalists, or advisors to venture investment firms.

### Job Security and Future Trends

*The current demand for clinical development professionals far exceeds the supply.*

There is currently a tremendous demand for qualified talent in all of the various disciplines within clinical development. Job security depends in part on the financial situation of the company and its pipeline of products. Projects in clinical development can be discontinued suddenly due to efficacy or safety issues with the compound, lack of funding, and a multitude of external factors. If a product fails in clinical studies, the company may not survive, or it may be acquired or downsized. Small companies with one product in clinical studies generally offer less job security than the larger companies. People who are risk-adverse should consider larger biotechnology and pharmaceutical companies with multiple drug development programs. Be aware, however, that even large pharmaceutical companies can be bought up, merged, and reorganized. Regardless, if your company suffers one of these unfortunate fates, you can rest assured that you will be able to find another clinical position relatively easily.

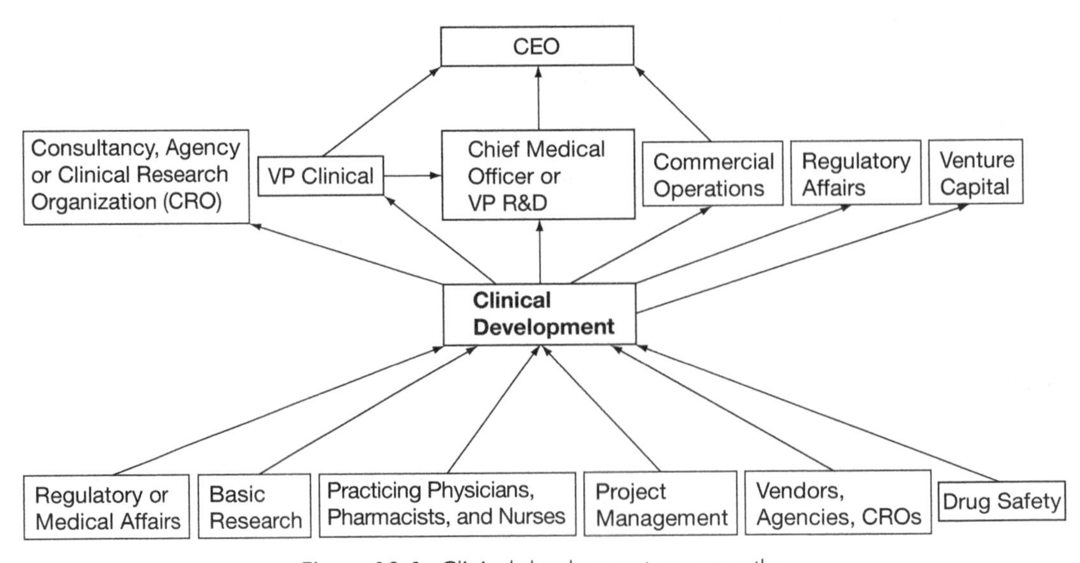


**Figure 10-1.** Clinical development career paths.

Demand for experienced clinical talent in biotechnology companies is expected to grow in the future. This is due in part to the increasing need to take compounds further into development before being partnered. Compounds in Phase II clinical stages bring in a higher price tag for the company and also pose less development risk for partners. Our large aging population, in combination with the sequencing of the human genome, will likely drive explosive growth in drug discovery and development.

The lure of far less expensive processes may drive significant parts of clinical development programs outside the United States. CRAs, medical monitors, and data management positions might be at risk of being outsourced. Data management is already frequently outsourced to vendors in the United States, and there is a trend to outsource those functions overseas. Clinical trial management, operations, and project management roles will likely not suffer the same fate, because those positions need to remain within the company to ensure direct and timely communication with senior management.

## LANDING A JOB IN CLINICAL DEVELOPMENT

### Experience and Educational Requirements

This is a field with multiple roles and functions, and as a consequence, a large variety of backgrounds are appropriate. A medical degree is generally required for positions such as chief medical officer and medical monitor, but most other positions do not require one. Those with medical backgrounds (M.D.s, Pharm.D.s, M.P.H.s, and R.N.s) tend to be the usual hires, but there are also many scientists with Ph.D.s, M.S., and B.S. degrees in the field. Computer database managers, lawyers, and epidemiologists can be found as well. Those with higher professional degrees, such as M.D.s and Ph.D.s, can often more easily move up the ranks to managers, directors, and VPs. In general, those with undergraduate and master's degrees tend to move up the operational ranks.

It is highly advantageous to have a medical background in this career. Physicians know how the medical system works and can speak with credentialed authority to the FDA about their clinical programs. Most large pharmaceutical companies prefer that their heads of development have medical degrees, but those with Ph.D.s are also frequently hired. M.D.s will usually be given more clinical or safety responsibilities and will focus their efforts more on clinical development and protocol writing as opposed to operations.

Most physicians join companies at the associate director level. They usually come from academic research centers, where they were industry-sponsored researchers or investigators. A renowned professor, however, can easily step in as a VP. Foreigners with medical degrees often join clinical development departments so they can apply their medical knowledge without having to become certified to practice medicine in the U.S. A common path for doctors is to become medical science liaisons first and then to transfer to clinical development (see Chapter 11).

Those with advanced science degrees (Ph.D. and master's) have acquired a vast body of information which may not be directly applicable to clinical development. As a general rule, biological scientists tend to migrate into clinical pharmacology, translational medicine, and other Phase I types of work, or project management and clinical scientist positions.

Nurses are ideally suited for CRA and clinical monitor roles, as their knowledge of medical charts, terminology, and abbreviations used in clinical practice are useful for development.

## Paths to Clinical Development

Even though there is high demand for candidates, it is not easy to enter a career in clinical development without prior experience.

- Consider entry-level positions such as CRAs, clinical study coordinators, or clinical trial specialists. These positions require little or no initial experience and serve as excellent starting points.
- Probably the easiest way to break into the clinical field is at a CRO. CROs are excellent places to learn the basics and gain exposure to multiple therapeutic areas and different types of drugs. They are well-versed in clinical execution, regulations, and guidelines, and the larger ones have outstanding educational and training programs.
- If you want to avoid the CRO route, consider a position as a project manager, study coordinator, or internal clinical research scientist. Your project management skills will later serve you well in clinical development. As a study coordinator, you will be coordinating various internal and external groups to advance programs. You will be responsible for oversights, budget tracking, and timeline management. You may also gain experience negotiating with CROs, scheduling meetings, and compiling information. Many clinical development professionals with Ph.D.s began as study coordinators, where their knowledge of biology and basic disease research could be readily applied.
- Consider obtaining a certificate or taking additional classes in clinical trial management. Although real experience is preferable, a certificate shows you have tremendous motivation and a desire to learn and work hard, and hiring managers will be more apt to take a risk on hiring you. There are literally dozens of different types of programs. Courses in auditing, clinical trials, project management, and medical writing, in addition to training in clinical trials, are available.
- If you are interested in a career in clinical pharmacology, consider taking courses in pharmacokinetics, formulation, and preclinical and clinical pharmacology.
- For those who are currently working in hospitals, obtain experience conducting clinical trials. Try to become a study coordinator.
- Apply to companies that are focused on the field of your specialty. People are often hired for their specific knowledge of a therapeutic area, even if they lack substantial clinical experience.

- Pharmacists and those with medical degrees from foreign countries might initially consider obtaining a position as a medical science liaison or work in drug safety before transferring into clinical development. These positions are good entry points and provide an introduction to the entire process of drug development (see Chapter 11).
- Consider joining the FDA (CBER, CDER, or CDHR divisions). Such experience will be invaluable for your career, will make you more marketable, and will provide a deeper understanding of the FDA's perspective on clinical studies. You will need to live near Washington, D.C. Apply for these positions at www.usphs.gov, www.fda.gov, or www.usajobs.gov. See "Career Snapshot: Working at the FDA" (page 120), and Chapter 12.
- If you are currently in academia, try to obtain clinical experience. Take on clinical projects or consider a postdoctoral fellowship in clinical research. If you are a doctor, select a fellowship with bench and clinical research; your published work will make your resume more attractive to companies.

## CAREER SNAPSHOT: DRUG SAFETY, PHARMACOVIGILANCE

Because of the limited cohort sizes inherent in Phase I–III clinical trials, it is difficult to determine whether rare adverse side effects will occur. But what happens after the product is approved and thousands of individuals are now taking the drug? Employees in the drug safety department monitor adverse events during clinical trials and continue to track safety-related events after the drug is approved ("pharmacovigilance").

*Drug safety is about "benefit–risk management": weighing the drug's benefits against the risk of adverse side effects. It is about making sure that drugs are as safe as possible and that patients are using them properly.*

The role of drug safety is to ensure that drugs are safe. Most drugs have some side effects, but usually the benefits outweigh the risks. The drug safety group identifies trends and any new risks that become obvious during clinical trials and after large-scale use. When new safety issues arise, this information is distributed to prescribers so that they can serve their patients better.

This department exists most often as an independent, separate entity, but in smaller companies it may be located in the clinical, regulatory, or medical affairs departments. In many larger companies, drug safety groups are separated into two departments: the preclinical department, where animal toxicology studies are reviewed (see Chapter 8), and human drug safety monitoring.

### *How Drug Safety Works*

The drug safety group is responsible for reviewing cases reported from prescribing doctors and medical monitors of clinical trials. Nurses, pharmacists, and doctors manage a call center to discuss, monitor, and process cases. Each case is tracked and documented in a drug safety database. Cases in which a serious adverse event is reported are recorded as "safety reports" which are summarized and submitted to the FDA on a periodic basis. Those in drug safety make recommendations as to whether the label should be amended or augmented.

*The drug safety group is the "eyes and ears" of the company.*

### *Drug Safety Roles and Responsibilities*

Depending on one's position, one might spend time:

- Reviewing cases and reading medical records (this is most of the work). For the case reports, making sure that the story is medically sound and the proper terminology is used. Contacting physicians for detailed questions about individual cases.
- Processing safety reports into a drug safety database. Information in the database can be searched and aggregated to determine whether a previously unrecognized safety issue with the drug has arisen.

*It is far better for the company to find a fault with a drug before the FDA does!*

- Reviewing the data in the drug safety database, and if new adverse events have been reported, meeting with the drug development teams to determine whether the label needs to be changed.

- Designing a safety plan for clinical protocols and for IND filings for each of the compounds being developed.
- Preparing the safety sections for NDA or BLA filings and periodic safety reports for the FDA.
- Reading new processes and guidelines, remaining updated on the latest medical issues.
- Teaching employees about drug safety and the proper procedures for interviewing patients and entering data into the database.

***In addition to the skills and traits of those who do well in clinical development, people who flourish in drug safety careers tend to have...***

- Good clinical judgment and expert knowledge of the product's pharmacology. If you know the product well, you will be able to discriminate between a serious adverse event caused by the drug and an event that is unrelated to the drug.
- The ability to distinguish the most relevant points from the many irrelevant ones. You need to be able to do the research, gather pieces of information, and put together an accurate story. A vast majority of drug safety cases are not product-related, and when they are, drug–drug reactions are often the culprit.

*Drug safety is like detective work—you gather the clues and put together a complete story.*

### *Positive Aspects of a Career in Drug Safety*

- A mixture of activities provides job variety.
- If quality of life is important to you, this can be a great career. It can be an 8-to-5 job and requires little travel. You are not on call, and when you leave at the end of the day, your work stays at the office.
- The job fills an important need of society. You are contributing to the safe use of products and affecting public health on a grand scale.
- There is much autonomy, and these positions carry decision-making authority.
- Those with drug safety experience are in high demand, and there is exceptionally good job security. And the demand is expected to grow!
- This position pays well. Those in drug safety are generally paid more than those in hospitals or pharmacies with a comparable level of experience.
- This position can serve as a launching pad to careers in medical affairs (medical science liaisons in particular), clinical development, regulatory affairs, marketing, and project management.

### *The Potentially Unpleasant Side of Drug Safety*

- The work load can be heavy, and the work environment can be fast-paced.

- There is considerable paperwork, and much time is spent reviewing cases in front of a computer.
- With time, the work can become repetitive.
- Work can become a nightmare if a drug is recalled and has many case reports of serious events.
- Cases often involve doctors who prescribe the wrong drugs, poorly diagnose the patients, or do not check for potential drug–drug interactions.

#### *Drug Safety Career Potential*

There is a high demand for qualified individuals, and that demand is expected to grow. Depending on the company and level of its sophistication, those with M.D., Pharm.D., R.N., and M.P.H. degrees, as well as college-educated employees, can apply. Epidemiologists are needed for identifying trends in the data. Entry-level positions are available for entering cases into the database, interviewing patients, writing case narratives, and handling details.

*Drug safety professionals are in high demand! For doctors, pharmacists, and nurses, this occupation can be an excellent entry into biopharma.*

#### *Tips for Obtaining a Position in Drug Safety*

- Clinical experience and pharmacological knowledge are needed.
- Experience in medical information (see Chapter 11) may facilitate an easier first entry.
- Educate yourself about regulatory requirements; review the safety guidelines at the FDA Web site and international ICH standards (ICH= International Conference on Harmonisation of Technical Requirements for Registration of Pharmaceuticals for Human Use).
- Read the journal *Pharmacoepidemiology and Drug Safety.*
- Join or attend meetings at the International Society for Pharmacoepidemiology (www.pharmacoepi.org). Job postings are listed at the site.

## CAREER SNAPSHOT: MEDICAL WRITING

If you enjoy writing and interpreting data, have an interest in medical studies, and possess strong analytical skills, then you might be interested in a career in medical writing. Medical writers collaborate with authors to write and/or edit various documents and manuscripts. The writer's function depends on the nature of the project, the work styles of the team members involved, and the time available. Medical writers summarize information from various sources to compile registration dossiers used in clinical brochures, expert reports, NDAs, and INDs. Medical writers follow regulatory guidelines and in-house style guides to prepare clear, concise, and accurate documents that will help facilitate drug approvals. It is important to present data that support the key conclusions the company has developed based on the results of the studies.

Medical writing is an excellent career option for scientists, because many of the same skills used in research are applied, such as analytical thinking, data manipulation, and evaluating and reviewing information. If you aspire to be a famous author, however, carefully consider your alternatives, because this is a highly regulated industry, and there is no room for artistic flair or creative input. The content is dictated by data, and there are rigid language requirements.

*Sadly, you will not become famous for your work as a medical writer.*

There is a common misconception that this vocation involves working in a room by oneself. For some positions, particularly for freelance writers and consultants, this might hold true, but in fact, much time is spent interacting with content experts and team members to determine the key messages and to drive the review process.

***There are many different types of medical writers. Depending on your job and the department you work in, you might be working on:***

- Dossiers for clinical, regulatory, or chemistry manufacturing and controls (CMC) sections for IND or NDA filings.
- Grants and manuscripts, articles intended for publication, material for conferences and continuing medical education (CME) events.
- Basic research reports (e.g., animal studies) written for journal submissions, CMC sections, or other studies.
- Safety documents.
- Software documentation.
- Marketing, promotional material, brochures.
- Clinical protocols, investigator brochures, and clinical study reports.
- Brochures for advisory board meetings.
- Drug package inserts.

Depending on their expertise, medical writers can be found in various departments, including regulatory affairs, clinical development, or medical affairs.

### *The Workload*

From start to finish, putting together a typical study report can take at least eight weeks. Sometimes the amount of data to review and summarize can be daunting. Often, there are mountains of information, perhaps three to five feet high, which need to be reviewed and summarized into 200–500 pages of text. After writing the first draft, documents might be reviewed by as many as 15 people and are eventually signed off by an executive VP.

*Clinical study reports are like a drug's mini-history.*

### The Writing Process

In some companies, the content experts collect the data and clean it up, statisticians analyze it and add tables and figures, and writers distill it and put the content onto paper so that it is clear, concise, accurate, and properly formatted according to regulatory standards. In other companies, medical writers analyze the raw data (frequently after biostatisticians have reviewed it) and draw conclusions which are then shared with the team.

After the data from the clinical study are collected, the team meets to discuss the key findings. The medical writer often creates the first draft of the document, coordinates subsequent reviews, and finalizes the text. Usually the content experts write their own sections, such as the CMC and nonclinical sections. Medical writers, however, are often charged with driving the process and obtaining input from clinicians and statisticians to produce a final document ready for submission.

### Salary and Compensation

Medical writers tend to be paid slightly more than those in discovery research.

### A Typical Day as a Medical Writer...

- Most of one's time is spent writing, analyzing raw data, and organizing information.
- The rest of the time is spent in meetings with team members, talking about the data, planning, clarifying issues, and scheduling projects.
- A little time is devoted to reading background information.

### In addition to the skills and traits of those who do well in clinical development, people who flourish as medical writers tend to have...

- A strong, broad science background.
- Exceptionally good technical writing skills.
- The ability to quickly process information and organize large quantities of data.
- The ability to drive the entire process of reviewing and finalizing documents in a diplomatic, team-oriented way.
- A flexible and adaptable personality. A willingness to hear criticism.
- A high tolerance for frustration.
- Somewhat introverted personalities. When driving projects, however, one needs to be outgoing because of the communication demands.
- Particularly strong attention to detail.

### Positive Aspects of a Career in Medical Writing

- Medical writers are needed in many different facets of drug discovery and development.

Experience in this field can offer the opportunity to expand your knowledge in other areas for career growth and continued learning.

- With experience, you can find lucrative positions as a freelance writer and consultant, which offers the advantages of working from home on your own schedule and avoiding company politics.
- This is a marketable skill, and there is an exceedingly high demand for scientists and medical professionals with a talent for writing.
- It can be intellectually challenging and enjoyable to analyze and bring together an honest and accurate story of how a drug performed in the clinic.
- Medical writing is important to companies. Your documents may be part of regulatory filings for drug approvals.
- In some companies, medical writers can telecommute.
- It is a great way to learn about the minute details of drug discovery and development and how conclusions are derived from data. It is exciting to be involved in the process of seeing a whole story emerge.
- There is constant learning and exposure to new science.
- It is a very satisfying feeling when projects are completed.
- There is no or little travel, depending on the project (time is mostly spent working on the computer).

### *The Potentially Unpleasant Side of Medical Writing*

- Because there are varying degrees of responsibility and educational backgrounds in this career, there can be misconceptions about the roles and responsibilities of medical writers (some bosses believe that they are glorified secretaries, for instance).
- There is constant deadline pressure.
- Data presentation can be challenging. There are millions of data points that need to be summarized, and often it is difficult to arrive at a group consensus on the best way to explain the results.
- Medical writing requires long periods of time working in front of a computer.

### *How is Success Measured?*

The ultimate measure of success as a group and company is FDA approval of products. Personal measurements of success include meeting timelines and keeping customers happy.

### *Medical Writing Career Potential*

Those who are interested in upward mobility can rise to VP of Medical Writing in some companies. Those who are interested in pursuing alternative areas can relatively easily parlay their skills into project management, regulatory affairs, clinical development, medical affairs (medical communications), corporate communications, and marketing. Those who excel typically become freelance writers.

### *Experience and Educational Requirements for Medical Writing*

There is great variety in the educational requirements for a career in medical writing. Some companies require only an English degree, whereas others require a science or medical background. In large pharmaceutical companies, the vast majority have Ph.D. or Pharm.D. degrees. In addition, it is advantageous to be familiar with desktop publishing tools for making charts, graphs, etc.

As for background, most people evolve into medical writing from clinical development, regulatory affairs, and discovery research.

### *Tips for Entry into Medical Writing*

- When applying for positions, assemble a portfolio of your written work. Write articles for newsletters to boost your portfolio.
- Hiring managers look for publications and projects of which you have been in charge.
- Knowledge of FDA regulations, ICH guidelines, and the entire drug discovery and development process is optimal. Gain experience writing regulatory submissions or familiarize yourself with the guidelines. Classes on these topics are offered.
- You can obtain a certificate or take classes in medical writing. The Drug Information Association (DIA; www.diahome.org), the American Medical Writers Association (www.amwa.org), and select university extension programs offer courses and professional certifications. Also consider reviewing information from the Board of Editors in the Life Sciences (www.bels.org). Classes in statistics are beneficial.
- As for books, the "bible" in medical writing is *American Medical Association Manual of Style: A Guide for Authors and Editors*, by Cheryl Iverson, et al., Ninth Edition, published in 1998 by Lippincott Williams & Wilkins (Philadelphia). Another recommended book is *Writing and Publishing in Medicine*, by Edward J. Huth, published in 1999 by Lippincott Williams & Wilkins (Philadelphia).

## CAREER SNAPSHOT: DATA MANAGEMENT

*Data management is needed to ensure the quality and completeness of data.*

If you have a dual interest in computer and medical sciences, data management can be an exciting avenue to explore. Data management is needed to capture data and ensure its quality and completeness. Those in clinical data management may have the larger responsibility of determining the ultimate end use of data and how to translate it into reports and trial results. Data management professionals are also needed for capturing information from drug safety and preclinical research (e.g., pharmacology and toxicology).

There are several subspecialties within data management:

- Data entry personnel for clinical trials.
- Programmers who support data management needs.
- The Information Technology quality group, which runs automated quality checks on the data (see "Computer Validation" in Chapter 13).
- Data coordinators: people who liaise with the CROs who provide the data.
- Coders: Coders take reports written by doctors in free text and basically translate them into common, standardized terminology to make analyses possible.

### *Data Management Roles and Responsibilities*

- Designing data collection instruments. These classically have been paper case report forms (CRFs), but are increasingly taking the form of electronic CRFs, which directly capture and store data in a standardized format.
- Designing databases to store the data.
- Entering data from case report forms during clinical trials. Typically performed by the data management group, this is commonly an entry-level function or is outsourced to CROs.
- Defining and designing data checks and writing quality validation programs. Programs are written to survey the data for errors and inconsistencies and to clean the data.
- Running electronic checks against data to identify potential errors. Query documents are written and sent to doctors running trials for suggested corrections.
- "Freezing" or "locking" the database after the data from the study are complete and of good quality. After this time the data cannot be altered.
- The locked database is then submitted to the biostatistics department for analysis. This group organizes and draws conclusions from the analyses of the data and writes programs to display the tables, listings, and graphs for regulatory submissions.

### *How is Success Measured?*

The classic measurement is the time needed to go from the "last patient out" until the database is locked. Another yardstick for success is the number of errors in the data processing.

### *Salary and Compensation*

Salary and compensation in these careers are somewhere near the middle of what those in clinical operations receive.

*Positive Aspects of a Career in Clinical Data Management*

You can have a larger positive social impact in clinical data management than in other high-tech careers.

- Clinical data management can be more interesting and satisfying than other information technology (IT) careers, and it is easier to feel that you are having a positive social impact.
- It provides an opportunity to interact with interesting people with diverse backgrounds. The entire process of clinical development can be very interesting.
- It is exciting to be the first person to review the data, identify trends, and see the final outcome of the trials.
- You can be at the forefront of cutting-edge IT technology.
- There is some opportunity to telecommute and some flexibility on work hours, depending on the company.
- Expect a fair amount of job security. Clinical data management is important to companies because clinical trial data are a requirement for FDA approval. Biopharma can be a more stable industry than high tech.
- There is a high demand for qualified and experienced individuals.
- This is a wonderful opportunity for people who lack clinical backgrounds but want to enter the field.
- This can be an 8-to-5 job and requires little travel.

*The Potentially Unpleasant Side of Data Management*

- This is a regulated industry and, as such, is heavily focused on documentation.
- Data management is not a well-understood profession, and you may need to explain to other employees what you do and why people should pay attention to you.
- Data management is going through a tremendous and slow-moving transformation to "electronic data capture." As a consequence, data entry analysts may lose their jobs or be required to learn new skills.
- Work is episodic and depends on periodic clinical urgencies.
- There is limited potential for career growth, but it does allow advancement into other areas.
- Data entry is tedious, and there is constant turnover of analysts. When the bright, capable, or adept analysts become knowledgeable, they quickly advance. New employees must continually be found and trained.
- An adversarial relationship often develops between clinical development and data management members. Data managers are, in a way, spot-checking the clinical monitor's work (e.g., exposing points that are missing, finding errors in data, etc.).

- It can be tempting to compromise on quality in order to meet aggressive timelines—you must sustain a balance between the two.

***In addition to the skills and traits of those who do well in clinical development, people who flourish in data management tend to have...***

- Strong analytical, mathematical, and investigative skills.
- A certain level of comfort and aptitude with technical systems. Computer programming skills are required.
- Knowledge and understanding of the clinical trial process. Good clinical data management professionals tend to be conversant with the right medical terminology and can distinguish data points that are relevant from those that are not.
- The ability to design data entry forms so that it will be easy to clean the data after the trial, and an understanding of why that's important.

*Data management is like detective work—you review data from patients and help others understand whether the facts are related to the disease or product.*

*Data Management Career Potential*

There are several career options for those in data management. One can stay in this field and move up the ranks from manager to director to vice president. Many go into clinical operations or join vendors who specialize in data management. It is also possible to eventually become the chief information officer (CIO).

*Experience and Educational Requirements for Data Management*

A diverse set of backgrounds qualify for data management careers. There are many with B.S. or computer science degrees. Some start in programming, whereas others focus on building databases. People commonly enter data management from nursing, IT, and data entry positions. The most frequent direction is from the clinical department. Nurses who are familiar with medical terminology and clinical procedures, and CRAs with monitoring experience, have ideal backgrounds. As mentioned above, an intimate understanding of the clinical trial process and a science or medical background are advantageous.

*Tips for Entry into Data Management*

- Develop your computer programming and database-building skills.
- Most of the industry uses Oracle's database platform. Enhance your Oracle and SQL (Structure Query Language) skills. Specify your SAS, DB, SQL, and Oracle database knowledge in your resume.
- Learn medical terminology and concepts.
- Join a CRO or a niche provider that offers clinical data management services.

- For those in academia, ask whether the school is conducting clinical research and inquire about obtaining a part-time student data entry position or other clinical trial-related job.
- Consider an entry-level data entry position. Advancement is quick for fast learners.
- Consider obtaining a certificate in clinical data management offered by select universities. The Society for Clinical Data Management (www.scdm.org) offers a certification program. The Drug Information Association (www.diahome.org) offers conferences.
- Subscribe to *Applied Clinical Trials*, a free magazine, at www.actmagazine.com/appliedclinicaltrials.

## CAREER SNAPSHOT: BIOSTATISTICS AND STATISTICAL PROGRAMMING

*Statisticians: the number mavens.*

Statisticians manage and analyze the ultimate end result: data. They convert numbers into meaningful comparisons, often between a placebo and various drug treatment groups, which indicate either success ("$p<0.05$") or failure of clinical trials or experiments. They work on projects that span everything from discovery research to preclinical and clinical studies and even extend into marketing and post-marketing efforts. The results of their analyses form the evidence used to show a drug's efficacy, for example, and may be published in journals or in regulatory filings.

Statistical programmers serve a similar role. They neatly package data for viewing in the form of tables and listings for publications and regulatory submissions.

### *Statistics Roles and Responsibilities*

- Running statistical analyses on data, analyzing data, summarizing conclusions, and generating reports.
- Providing input into the development program for products in the pipeline.
- Serving as the statistical liaison with the FDA.
- Working with the clinical and regulatory teams to design studies—defining how the studies are structured and specifics about how data will be collected and analyzed.

*Statisticians are part detective and part number junkie.*

***In addition to the skills and traits of those who do well in clinical development, people who flourish in statistics careers tend to have...***

- Strong statistical, mathematical, and analytical skills.
- A good grasp of how to distill a set of data into a form that can be analyzed and easily understood.
- A particularly strong ability to pay attention to detail.

### *Salary and Compensation*

Biostatisticians are well compensated. They are paid similarly to those in clinical operations.

### *Positive Aspects to a Career in Statistics*

- It is exciting and interesting to be involved in clinical trials and to help design appropriate measures that will succinctly answer questions.
- It is mostly a 9-to-5 job, except during submission preparations.
- Telecommuting is an option in most companies.
- There is high demand for qualified individuals.
- Some career advancement and development is possible. There is also much potential to enter related areas of clinical development, such as regulatory affairs, project management, operations, and quality.

### *The Potentially Unpleasant Side of Statistics*

- There is intense deadline pressure.
- Sometimes there are difficult personalities in the clinical development department with whom you must work.
- If the study is not designed well and data measurements are not taken into consideration before the trial begins, the data analyzed after a trial can often be messy and incomplete, making it difficult to draw conclusions.

### *Careers in Statistics*

Statisticians specialize in therapeutic areas. Each indication has its own unique endpoints and measurement tools. For example, Alzheimer's clinical trials use psychological measurement tools, whereas cancer trials measure clinical outcome by completely different parameters. There are "bench statisticians" whose primary role is to analyze data. Others specialize in the design of studies or manage other statisticians.

### *Experience and Educational Requirements*

Most statisticians and programmers have math, computer science, or statistics degrees, at the masters or Ph.D. level.

### *Tips for Entry into a Career in Statistics*

- The easiest entry into statistics is via data management or by earning a degree in statistics. Alternatively, you can obtain a position as a bioanalyst, a programmer who works with the statistics team to help analyze data.

- Certificate programs are available at select universities and associations. Relevant societies include a SAS users' group called Pharmasug (www.pharmasug.org), the American Statistical Association (ASA, www.amstat.org), and the Association of Clinical Research Professionals (ACRP, www.acrpnet.org). Both the ASA and the ACRP have subgroups focused on biopharma statistics.

## CAREER SNAPSHOT: WORKING AT THE FDA

Working at the FDA is a wonderful way to learn the process of clinical development and to have a significant and global impact on drug safety. Although many of the FDA positions have more of a regulatory affairs role, there is also a group that evaluates the clinical protocols. They tend to be medical professionals (medical reviewers).

*Clinical trials must not expose patients to unreasonable risks.*

The FDA is divided into eight centers and offices. The Center for Drug Evaluation and Research (CDER) and the Center for Biologics Evaluation and Research (CBER) are focused on pharmaceutical and biological products, respectively. There is also a Center for Devices and Radiological Health (CDRH). For additional information, see www.fda.gov. See also "Working at the FDA" in Chapter 12.

### *A Typical Day as a Medical Reviewer*

Most of the time is spent reviewing applications (mostly IND filings or updated clinical trial results) for clinical trials conducted by academic research centers or pharmaceutical/biotechnology companies. This includes reviewing the design of the trials, ensuring that the study's answers will be meaningful, and ensuring that the approach is safe.

*Other time is spent...*

- Analyzing clinical trial data with the help of statisticians.
- Conferring with fellow FDA or advisory members to reach a consensus after data are reviewed.
- Writing reviews and reports. Depending on the level and stage of the trials, medical reviewers grant approval for a proposed clinical trial to proceed or, if it is deemed unsafe, place a hold status on it until the company can satisfy the FDA's comments or recommendations.
- Presenting data to advisory committees composed of experts outside of the FDA.
- Attending many meetings, for specific centers or the entire FDA.
- Attending annual society conferences and meeting with patient advocacy groups.
- Giving presentations to the public about what the FDA does.

*Positive Aspects of Working at the FDA*

- Careers in the FDA are intellectually stimulating, and the work is varied. There is tremendous exposure to the latest scientific and medical advancements on a broad range of topics—you will be reviewing clinical protocols from a variety of institutions.
- Working at the FDA provides a great opportunity to meet the leading experts and bright scientists developing the latest therapies.
- You can have a larger impact on global health than by helping patients individually. This is a very important protective role for society as a whole.
- You can work flexible hours and telecommute some.
- Many FDA employees do not work weekends or nights.
- The benefits are good: paid vacations, sick leave, retirement benefits, and life and health care insurance.
- There is no specific retirement age, so you can continue to work after turning 70.
- The FDA tends to have a congenial work atmosphere.

*The Potentially Unpleasant Side of Working at the FDA*

- It does not pay as well as industry positions and is at the lower end of the pay scale when compared to private practice.
- Most positions are located near Washington, D.C., so there are geographic constraints.
- For clinicians in private practice, there may be considerable adjustment to working in a hierarchical structure where you are no longer your own boss. At the FDA and in industry, you will often enter at the lower end of a totem pole and most of the time will not be the ultimate decision maker.
- The FDA has limited staff and budget, resulting in heavy workloads.
- There is constant shifting in the organizational structure and processes, which tends to be inherent in most behemoth organizations.

*In addition to the skills and traits of those who do well in clinical development, people who flourish at the FDA tend to have...*

- The ability to apply regulations fairly and evenhandedly, without bias or prejudice.
- A willingness to serve the public and to work long hours.
- A willingness to divest yourself of any conflict of interest. For example, you must be willing to refuse gifts from sponsors because it might suggest that you are developing a bias in favor of those organizations.
- A careful and thoughtful approach when speaking to reporters, and the sensitivity to keep some things confidential.

### *Experience and Educational Requirements for Working at the FDA*

The FDA is composed of an experienced cadre of individuals. Because the medical officer role involves reviewing and making clinical protocol decisions, an M.D. degree is often required. Some pharmacists and advanced nurses also serve in medical positions. There are many other positions besides medical reviewers.

The FDA attracts doctors from private practice, academia, and the military who would like to be involved in regulatory processes and want to maintain a connection with governmental service.

Many start as clinical or medical reviewers, move up to team leader, and then up the ladder to several director-level positions, which eventually report to the United States FDA Commissioner. The Commissioner reports to the Secretary of Health and Human Services, who reports to the President of the United States of America!

### *Tips for Entry into the FDA*

- Gain as much experience as possible writing and designing clinical protocols and running trials.
- Remain up to date with the latest medical research and review processes at the FDA.
- Serve on editorial boards for journals; serve on advisory committees for the FDA.
- The FDA prefers to hire individuals who specialize in particular therapeutic areas, so choose a field and become an expert.
- To apply for jobs, visit www.usajobs.com, www.fda.gov, and www.usphs.gov.

## RECOMMENDED TRAINING, PROFESSIONAL SOCIETIES, AND RESOURCES

### *Courses and Certificate Programs*

Certificates are offered at Association of Clinical Research Professionals (www.acrpnet.org)

Certificates in clinical trial design and management are offered at select universities and extensions

Classes in biostatistics

Classes in pharmacology and pharmacokinetics

### *Societies and Resources*

The Drug Information Association (DIA; www.diahome.org) is a commonly recommended, broad, and comprehensive organization that covers everything related to drug development. It holds conferences and provides career development, networking opportunities, and job postings.

Association of Clinical Research Professionals (www.acrpnet.org): Probably the best organization for CRAs.

CenterWatch (www.centerwatch.com): Broad information about ongoing clinical studies, job postings, lists of recruiters, and many other resources.

ClinicalTrials.gov (www.clinicaltrials.gov): Lists ongoing clinical trials.

American Society for Clinical Pharmacology and Therapeutics (www.ascpt.org): The oldest (ca. 1900) and one of the largest professional organizations that is interested in promoting and advancing human therapeutics; largely made up of M.D.s, Ph.D.s, and others in academia, pharma, and the FDA.

Clinical Focus at BioSpace (www.biospace.com): Lists clinical job opportunities and the latest clinical trial studies. Subscribe to the free BioSpace newsletter and receive a daily E-mail about the latest clinical trials and jobs.

The FDA Web site (www.fda.gov) has a tremendous amount of information.

### *Recommended Books and Magazines*

Spilker B. 1991. *Guide to clinical trials.* Lippincott Williams & Wilkins, Philadelphia.
*This is the "bible" for clinical research.*

*Goodman & Gilman's the pharmacological basis of therapeutics,* ed. L. Brunton, J. Lazo, and K. Parker. McGraw-Hill Professional, 11th edition 2005.
*This is the "bible" for clinical pharmacology.*

Fletcher A., Edwards L., Fox A., and Stonier P. 2002. *Principles and practice of pharmaceutical medicine.* John Wiley & Sons, New York.
*This book provides a good overview of clinical trials.*

*Applied Clinical Trials* (www.actmagazine.com/appliedclinicaltrials)
*This is a free magazine.*

# 11

# Medical Affairs

## Working in the Post-Approval World

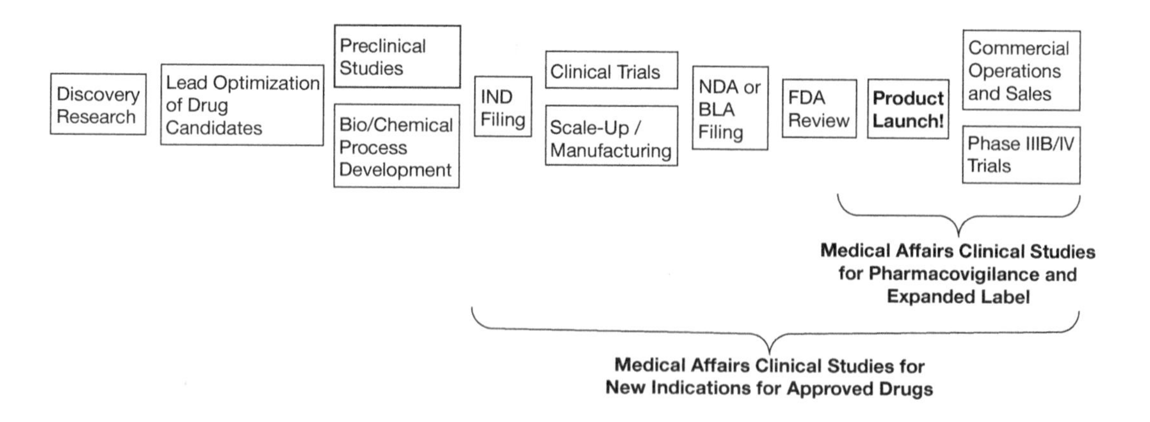


AFTER A PRODUCT HAS BEEN APPROVED and is ready for sales, one may think that product development has ended, but it is really just the beginning of the next phase: medical affairs. The medical affairs department provides medical support for a company's marketing efforts and serves as a bridge between R&D, sales, and marketing. The main objective in medical affairs is to generate additional information about the approved drug and to supply that information to the world, with the goal of enhancing health care providers' understanding of the product's attributes and utility.

*In medical affairs, you need to know more than just the technical details about a product—you also need to understand the market and how the drug is performing in the "real world."*

*Medical affairs brings together the product, the science behind the product, and the product's impact on the marketplace.*

### THE IMPORTANCE OF MEDICAL AFFAIRS IN BIOTECHNOLOGY AND DRUG DEVELOPMENT

Companies define medical affairs in various ways, but it is typically considered the culmination of the integrated activities involved in post-drug-approval efforts (although some activities, such as Phase IIIB trials, are conducted after regulatory filings and before mar-

Table 11-1. Common medical affairs subspecialties

| Clinical research | Medical communications | Miscellaneous areas |
|---|---|---|
| Investigator-sponsored trials | Library sciences | Medical education (CME) |
| Company-sponsored trials | Medical information | Medical science liaisons |
| Clinical support (data management, biostatistics, etc.) | Scientific publications<br>Competitive intelligence | Drug safety |

ket approval). After a drug has been approved, the medical affairs department provides an array of important services (see Table 11-1). It continues drug safety surveillance, provides a call center for customers, conducts clinical trials for efficacy and label expansion, publishes results from those trials and updates customers, and more. The department sometimes resides in clinical development or commercial operations, but is most often an independent entity.

## CAREER TRACKS IN MEDICAL AFFAIRS

### *Medical Affairs Clinical Research and Operations*

In a role similar to clinical development, teams design and manage Phase IIIB and IV clinical trials. Phase IIIB studies are often undertaken to analyze or resolve any outstanding safety or efficacy questions after Phase III trials and after a regulatory dossier has been submitted. Phase IV trials are conducted to further assess and define the risks, benefits, and new therapeutic applications of an approved drug.

*Medical affairs is just a tiny step across the street from clinical operations.*

### *Medical Communications and Information*

In general, this department researches, analyzes, aggregates, and communicates scientific and medical data to customers and the company. The medical communications department is generally composed of four different groups: medical information, competitive intelligence, scientific publications, and library sciences.

*The people in medical affairs have to walk a fine line between providing objective information about a drug and using promotional language.*

- *Medical information.* This is the company's medical call center—providing technical support for doctors and other health care professionals, as well as patients and caregivers. Nurses, pharmacists, and doctors ("drug product specialists") manage phone lines, answer medical inquiries, and provide information about drugs. Examples of questions they might field include those about off-label uses of a drug or allergic reactions. This group also collects information for the medical affairs and marketing departments about the questions it receives and the ways in which doctors are using the products.

- *Competitive intelligence.* There is often a small group within medical communications that provides various forms of competitive information. It gathers and analyzes data about comparable products in other companies' pipelines.
- *Scientific publications.* One important function of medical affairs is to report clinical trial results to doctors and key opinion leaders. The scientific publications department helps authors of Phase III and IV studies plan, write, and submit publications. This group also makes certain that the information is presented at key meetings and continuing medical education (CME) events through abstracts, posters, booklets, and reprints.
- *Library sciences.* Some medical affairs departments manage medical libraries. They maintain institutional journal subscriptions, assist with literature searches, and order books and articles for the library.

### *Continuing Medical Education, Medical Education, or Medical Marketing*

Sometimes as part of marketing, this department's goal is to educate doctors about the product in continuing medical education (CME) programs. In strict accordance with FDA-mandated policies, medical affairs professionals are not allowed to promote drugs, so CME events are usually run through independent agencies. The programs are reviewed by academic bodies to ensure that the information being presented is unbiased and scientifically rigorous.

*Medical marketing connects the product's medical focus with its commercial aspects.*

### *Medical Science Liaisons*

Medical science liaisons (MSLs) are also known as regional medical scientists, clinical liaisons, scientific liaisons, or scientific affairs personnel. Stationed around the country (or around the world) and working closely with sales reps, MSLs provide expertise about products and diseases. They inform customers about the latest clinical findings and provide information about off-label use of their company's products. They answer questions, interact with industry experts, and find potential investigators for Phase III and IV trials.

*Medical science liaisons are the field application scientists of drug companies.*

#### *The importance of MSLs*

A drug's "label" refers to the written information about the drug that appears on or with its package, which is the insert provided with a drug purchased at a store or pharmacy. Drug labels list the FDA-approved uses of the drug, as well as directions for use and warnings about side effects and contraindications. Sales reps are only allowed to discuss information that is "on-label," even if that information is two or more years out of date.

MSLs, on the other hand, are permitted to discuss the latest clinical, safety, and other "off-label" information (not yet approved by the FDA), but they are not allowed to promote or sell drugs. Their ability to inform physicians of relevant new data is invaluable support for the sales effort.

### Drug Safety, Pharmacovigilance

Pharmacovigilance involves closely monitoring drugs for adverse effects (drug surveillance) during clinical trials and after drug approvals, as well as reporting drug safety information. Nurses, pharmacists, and doctors manage phone lines to discuss and process adverse event cases. To learn more, see Chapter 10.

### Data Management and Biostatistics

Data managers and biostatisticians handle and analyze the large volumes of information generated from Phase I–IV clinical trials, and they spot trends in adverse events. To learn more about careers in data management, see Chapter 10.

## MEDICAL AFFAIRS ROLES AND RESPONSIBILITIES

*Medical affairs comprises a hodgepodge of different functions. The following is a generalized summary and is by no means comprehensive.*

### Strategic Guidance

Medical affairs professionals provide strategic advice for clinical development programs, for potential business development deals, and for sales and marketing efforts. They work in multifunctional teams to determine, for example, whether the market dictates that resources be devoted to brand extensions or to further efficacy studies.

### Clinical Studies

Clinical operations in medical affairs involve the review, approval, and management of Phase IIIB and IV clinical trials. Medical affairs personnel provide leadership for the trials, which includes working with commercial project management teams, overseeing reporting requirements, and filing documents.

***There are two types of Phase IV trials in medical affairs:***

1. ***Investigator-sponsored studies.*** These studies are typically run by clinicians in academic or government institutions. The "sponsoring company" (the biopharma company with the approved drug) supplies them with a grant or drug supplies. Investigators submit a formal proposal to the company, which undergoes a fairly rigorous and tiered scientific review process based on scientific merit. Most investigator-sponsored studies involve specific drug–drug combinations, special populations, and retrospective studies examining specific drug safety parameters. Medical affairs professionals (typically medical affairs liaisons) generate proposals for studies, maintain ongoing contact with the investigators, make sure that their results are valid, and help to disseminate their results in the form of publications or abstracts.

2. ***Company-sponsored studies.*** These studies are usually run by the in-house clinical group or are outsourced to vendors (contract research organizations [CROs]). Medical affairs professionals may be involved in managing studies, sites, and data.

#### *Registries*

Drug registries are databases of case histories from patients who have either used the medication or have a particular disease for which the product has been approved. By maintaining registries, companies hope to uncover empirical evidence showing beneficial responses to a drug. This method is a cost-effective way to capture data in real treatment settings without paying for exorbitantly expensive clinical trials.

Using the information gathered from the registry database, the biometrics group can run sophisticated analyses that eventually provide the raw material for articles publicizing conclusions such as a drug's superior efficacy or fewer side effects. Registries are maintained in house or by CROs or specialized vendors.

#### *Information Dissemination*

MSLs and those in medical marketing and communications oversee the important role of informing the world of the company's latest clinical trial results. This group disseminates newly generated information about the benefits, use, and safety of approved drugs to doctors and key opinion leaders. They plan publications, give scientific presentations, run symposia, and provide medical information about the products to the sales force. In addition, they work with continuing medical education agencies and other educational forums and manage advisory boards that include key opinion leaders.

#### *Consultancy Programs*

Medical affairs professionals run consultancy programs, which are a form of market research. Working with marketing or sales professionals, MSLs invite doctors to meetings and solicit their opinions and advice about a product or a drug. This information is shared with the marketing and medical affairs departments to develop strategic ways to improve the product or increase visibility.

## A TYPICAL DAY IN MEDICAL AFFAIRS

*A person in medical affairs might be involved in some of the following activities:*

- Overseeing and supervising the many people and CROs running clinical research.
- Maintaining a database to track investigator-sponsored study commitments.
- Attending internal meetings in which budgets, strategy, status updates, etc., are discussed.
- Traveling to medical or educational conferences and society meetings.

- Speaking to doctors at hospitals, answering their questions, and discussing off-label use of drugs.
- Providing clinical and scientific input for the marketing team.
- Evaluating and critiquing business development opportunities.
- Managing CME agencies.
- Presenting new data at international and national conferences and arranging meetings to answer questions after conferences.
- Contacting investigators about the possibility of conducting Phase IV studies.
- Reviewing promotional pieces for proper messaging and regulatory compliance.

## SALARY AND COMPENSATION

Salaries are commensurate with those for other clinical and marketing positions. For M.D.s, salaries are generally comparable to what most doctors earn in clinical practice, depending on the specialty.

***How is success measured?***

In general, successful medical affairs professionals efficiently provide the information needed to support a drug while it is on the market. Because medical affairs is a department composed of many disciplines, the measurements of success depend on one's specific role. Examples of success include completed submission of NDA or IND filings, effective oversight of drug safety, or production of scientific rationale that leads to Phase IV clinical research. For MSLs in particular, it can be difficult to effectively track individual contributions to the sales effort, although rather subjective metrics can be used, such as the number of client visits and customer feedback.

## PROS AND CONS OF THE JOB

### Positive Aspects of a Career in Medical Affairs

- The job is dynamic. Even though the management of clinical studies can be routine, the market is constantly changing, and each therapeutic area has its unique challenges.
- The work is intellectually stimulating. You must stay abreast of industry news and promising new developments as well as continue to educate yourself when managing trials in new therapeutic areas. You need not only to keep a sharp scientific focus, but also to be able to draw expertise from marketing and clinical research. The many roles that are part of medical affairs allow individuals to learn new fields and gain a broader understanding of drug discovery and development.

- Ultimately, you may be able to help more patients by working in medical affairs than you ever could by working in a clinical practice or hospital.
- A career in medical affairs provides the opportunity to interact with and build relationships with interesting, dynamic, and intelligent people, both inside and outside the company. The collegial, team-oriented environment in biopharmaceutical companies can be very appealing.

*It is immensely gratifying to know that you have educated doctors about your product so that they can better serve their patients.*

- It can be rewarding to be an expert on a specific drug and its therapeutic area and to serve as a knowledgeable resource for colleagues and customers.
- This career allows a flexible schedule and much independence. In general, you can arrange your own calendar, work on weekends, and plan to attend those conferences that interest you.

### The Potentially Unpleasant Side of Medical Affairs

- For MSLs in particular, about 30–50% of the time can be spent traveling, which can quickly become tedious and can cause a strain on family life.
- The role that MSLs play is often not clearly defined in companies. You may not be directly accountable for much of what you have produced.
- If you are working on only one particular disease, the work can become tedious.
- Some people prefer clinical research to medical affairs. They prefer the purely applied science of clinical studies.
- If you are a practicing physician, the loss of autonomy may require considerable personal adjustment.

## THE GREATEST CHALLENGES ON THE JOB

### *"Herding Cats"*

As with business development, getting functional heads aligned on strategic issues can be challenging. Senior management has its own priorities, as do the marketing, clinical, and research teams. They often have competing objectives, and it can be difficult to arrive at common agreements.

### *Taking into Account "Real-World" and Market Conditions*

*Working in medical affairs is like trying to aim at a moving target.*

Understanding how a drug is behaving in the "real world" presents unique challenges. The drug's performance, patient compliance with dosage regimens, doctors' prescribing practices, the performances of competitors' products, and knowledge of the disease must all be taken into account. As the market changes and more data are collected, the targets and goals of medical affairs evolve. Whereas clinical development involves a more cookbook approach, the processes and goals in medical affairs are more ambiguous.

## TO EXCEL IN MEDICAL AFFAIRS...

*Knowledge is King*

Many of these positions involve close interaction with physicians and key opinion leaders, so a comprehensive understanding of the product and marketplace and a high level of sophistication and expertise are needed.

*Flexibility and Comfort with Ambiguity*

To excel in medical affairs, one needs the ability to work well with the multifunctional and ambiguous nature of medical affairs. The goals are clear for Phase I–III clinical trials—to get the drug approved. After a product is approved, however, solid justification is needed for continued studies. Those who excel can think strategically in terms of marketing and remain very cognizant of the limitations imposed by regulatory requirements.

### Are You a Good Candidate for Medical Affairs?

*People who flourish in medical affairs careers tend to have...*

*Competency and knowledge.* It is important to have a solid understanding of the technical material, to be able to serve as an expert resource, and to be well prepared.

*Outstanding communication skills.* Exceptionally good speaking, writing, and reading skills are a must. Being able to communicate with individuals and in small or large groups is required.

*Flexibility with prioritizations and scheduling.* Most days in medical affairs do not go according to plan, particularly for MSLs. You need to be adaptable, able to quickly re-prioritize, and willing to drop your current task for a new demand.

*A "team player" attitude.* People in medical affairs need to be able to work in a multidisciplinary and collaborative environment. It is particularly important in this career to be a friendly, supportive, and responsible team member (see Chapter 2).

*A creative and flexible mind.* Strong problem-solving skills and the ability to think strategically are required. It helps to be able to freely associate new ideas, to be open-minded about adapting variations on themes, and to be willing to accept new approaches.

*The ability to work independently.* Most of the positions in medical affairs require initiative and the ability to work without direct supervision.

*Good judgment.* Part of working independently involves knowing when to ask questions, when to solve problems, and how to make decisions on one's own.

***Gregarious, outgoing personalities.*** Many in medical affairs tend to be outgoing because of the high level of contact with customers and because of the communication demands of working with many individuals.

***Excellent time, organizational, and project management skills, and an ability to multitask.*** Most people in medical affairs work on many projects simultaneously, and many of the positions require the ability to manage external contractors.

***An interest in science and desire to learn new technologies.*** This career involves exposure to new technology and science and requires continuous advanced learning.

***You should probably consider a career outside of medical affairs if you are...***

- Someone who enjoys a rigid schedule and is unable to adapt well to last-minute changes (except in medical information and drug safety).
- Most comfortable in a highly structured and closely supervised work environment (except for medical information and drug safety).
- Incapable of making independent decisions or unable to determine when those decisions need to be made.
- Not very detail-oriented, not good at time management, or unable to multitask.
- Not interested in working in an atmosphere where "the customer is king" (for medical communications).
- A person who wants to be the primary decision maker or is used to always being right.
- Someone who is self-important.
- Not self-disciplined (especially for those who work from home).

## MEDICAL AFFAIRS CAREER POTENTIAL

There is a wide variety of positions within medical affairs to explore. Those who wish to advance to other areas can move into marketing, business development, clinical affairs, sales training, managed care at an HMO, or public affairs (see Fig. 11-1). Other possibilities include joining one of the many vendors or agencies that provide a service to the medical affairs department, moving into pharmacoeconomics, or pursuing other management roles.

### Job Security and Future Trends

In general, positions in medical affairs tend to offer excellent job security. This is because medical affairs departments are needed whenever a company has a product on the market. Jobs in industry are never guaranteed, however, and if the product is revoked by the FDA, you may quickly find yourself back on the job market.

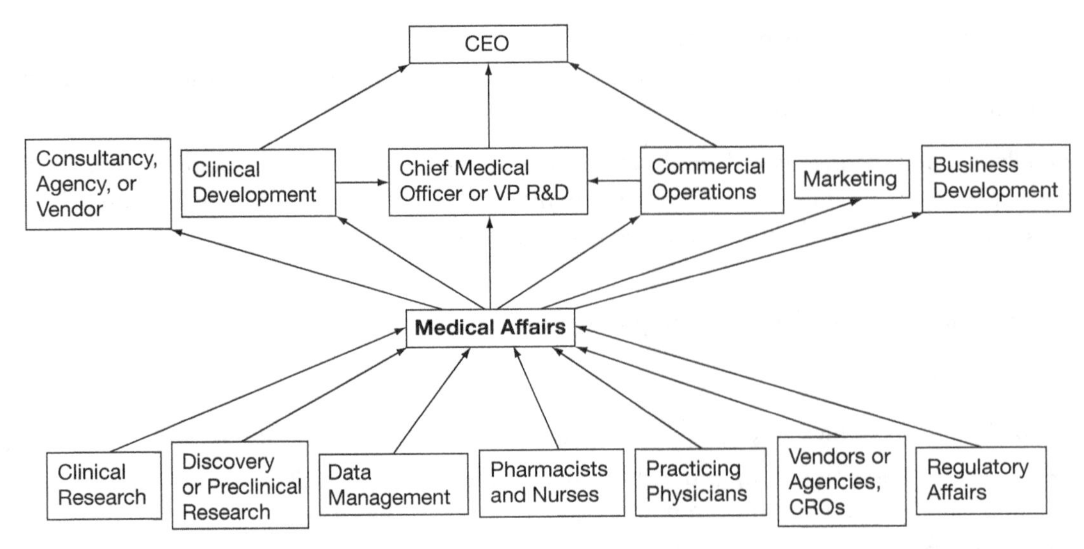


**Figure 11-1.** Common medical affairs career paths.

There is already a high demand in biotechnology for people with clinical development experience. The ability to work in both the clinical and medical affairs departments provides you with even more marketability, career versatility, and job security.

## LANDING A JOB IN MEDICAL AFFAIRS

### Experience and Educational Requirements

Careers in medical affairs combine marketing with clinical development and regulatory affairs. As a consequence, a variety of backgrounds are useful, including those in medicine, science, library or computer sciences, and biostatistics.

Most people need some clinical research experience to qualify for medical affairs operations positions. This group's requirements are the same as those for clinical operations, except that you need to cultivate an additional interest in marketing and a better understanding of regulatory issues, for which you need superior problem-solving skills. Any technical, scientific, or clinical degree will suffice (e.g., Ph.D., M.P.H., R.N., and M.D.). Promotions (particularly to director-level positions) are more easily obtained by those with advanced medical or scientific backgrounds.

The educational backgrounds of those in medical communications are highly varied. People in this department include those with Pharm.D., M.D., Ph.D., and M.S. degrees, as well as those with backgrounds in library science and administration. Support positions in this group require a B.S. degree in any area, but more advanced degrees are often a prerequisite for promotion. Most of the Ph.D.s in this department tend to migrate into medical writing and competitive intelligence. Nurses and pharmacists with clinical experience typically have the best backgrounds for the medical information department, and people in the medical education group commonly have an M.D., Ph.D., Pharm.D., R.N., or M.B.A. degree.

Customers expect MSLs to know a lot about drugs, how a disease is treated, and all of the pharmacologic options. In the general biopharmaceutical industry, approximately 80% of liaisons are Pharm.D.s and nurses, and 20% are Ph.D.s and M.D.s, but these numbers vary with each company and its products. Because it can be more difficult to explain the mechanism of action of biologics, many biotechnology companies prefer to employ MSLs who have the medical training to explain data in greater detail and in scientific terms.

## Paths to Medical Affairs

- If you are interested in medical affairs operations, consider working in clinical development first. Develop an intimate understanding of disease states. Learn about pharmacology, currently available therapies, and the needs of the health care field.
- Apply to companies with products already on the market. Companies with only one product in Phase III trials are risky—if the drug is not approved, you may soon be looking for another job.
- Consider joining a CRO. CROs prefer to hire people with scientific and medical backgrounds, and they can be easier entry points than biopharmaceutical companies, which typically require prior experience. CROs provide training and more exposure to multiple therapeutic areas, and many more types of products.
- Consider joining one of the many agencies that provide a service to the medical affairs department. For example, there are many CME and medical communication agencies that seek specialists to run their programs and interact with companies and medical professionals. This is another way to bypass the biopharmaceutical companies' initial requirements for industry experience.
- If you are considering an MSL position, talk to the liaisons and sales reps that you know. Most liaisons were originally customers who expressed an interest in such work to sales reps and were subsequently recruited or referred by them.
- If you are in clinical practice, become involved in investigator-sponsored trials. Joining a sponsoring company is much easier after relationships have been established.
- If you are still in school, apply for fellowships or residencies in a biopharmaceutical company or serve as an advisor for CME programs. Do what you can to be perceived as a leader in your field—speak at forums, for example. Companies prefer to hire perceived experts.

## RECOMMENDED TRAINING, PROFESSIONAL SOCIETIES, AND RESOURCES

*Societies and Resources*

Center for Business Intelligence (www.cbinet.com) offers seminars on medical affairs

Drug Information Association (www.diahome.org)

Academy of Pharmaceutical Physicians and Investigators (appinet.org)

American Medical Writers Association (www.amwa.org)

American Society of Health-System Pharmacists (www.ashp.org)

*Books and Magazines*

Fletcher A.J., Edwards L.D., Fox A.W., and Stonier P. 2002. *Principles and practice of pharmaceutical medicine.* John Wiley & Sons, New York.

# 12

# Regulatory Affairs

## The Final Challenge: Passing the FDA Test

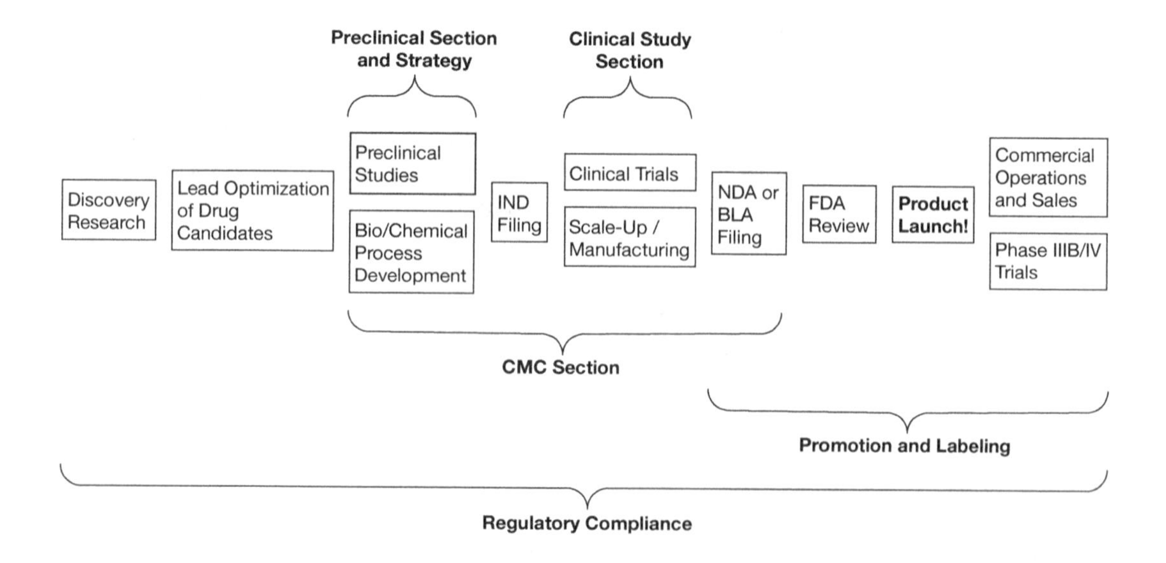


If you are detail-oriented and enjoy writing and working on teams, regulatory affairs can be an attractive career area that pays well and offers remarkably good job security. These careers allow you to exercise your scientific, medical, or clinical knowledge while working with cutting-edge science, technology, and policy-defining laws. People in regulatory affairs are at the center of the inner workings of drug discovery and development. They have an essential role in overseeing the multidisciplinary process of obtaining product approval by health agencies such as the U.S. Food and Drug Administration (FDA).

*Regulatory affairs is an exciting career path that offers exceptionally good job security.*

## THE IMPORTANCE OF REGULATORY AFFAIRS IN BIOTECHNOLOGY AND DRUG DEVELOPMENT

*The goal of regulatory affairs is to move the product and filings through the FDA review and approval process as quickly and as smoothly as possible.*

Regulatory affairs personnel play a crucial role in managing the regulatory approval process in companies. They provide strategic advice for the company's therapeutic and business development programs, they oversee the regulatory submissions process, and they serve as the primary liaisons between companies and the health authorities.

Regulatory affairs is usually, but not always, a separate organizational unit within companies. To manage the heavy workload and the many people involved in major submissions, most regulatory departments are subdivided into core teams, depending on company philosophy and size. A common organizational approach is to divide the work according to a particular research program or project. Other approaches include divisions based on whether the drug candidates are in development or on the market, or according to disease focus or technical areas. There is typically a separate group dedicated to publishing documents and archiving information.

## CAREER TRACKS IN REGULATORY AFFAIRS AND WORKING AT THE FDA

### *United States Regulatory Affairs Liaisons*

The regulatory affairs liaison is the most common type of position (see Fig. 12-1). They facilitate communication between the company and the FDA, and they represent the company to the FDA during the submission process. Within the company, they help prepare the large, complex regulatory submissions and other documents. They interact extensively with content authors (the people who produced the data) to make sure that the content is accurate, well-presented, and correctly formatted. They also offer strategic input before and during clinical development to expedite product approval. They are responsible for ensuring that regulatory requirements within the company are being fulfilled.

Various sections of the filings (e.g., preclinical, clinical, and chemistry manufacturing controls [CMC]) may be covered by separate groups of liaisons. There may also be a group responsible for labeling and advertising; this group submits filings to the FDA's Division of Drug Marketing, Advertising, and Communications (DDMAC).

### *International Regulatory Affairs Liaisons*

International regulatory affairs liaisons are usually based in or near the territories for which they have responsibility. They specialize in international registration and market clearance requirements. Regulatory affairs liaisons based in the U.S. are likely to focus on interactions with the FDA and, if there is no company presence in Europe, the European Agency for the Evaluation of Medicines (EMEA). These organizations represent the two major

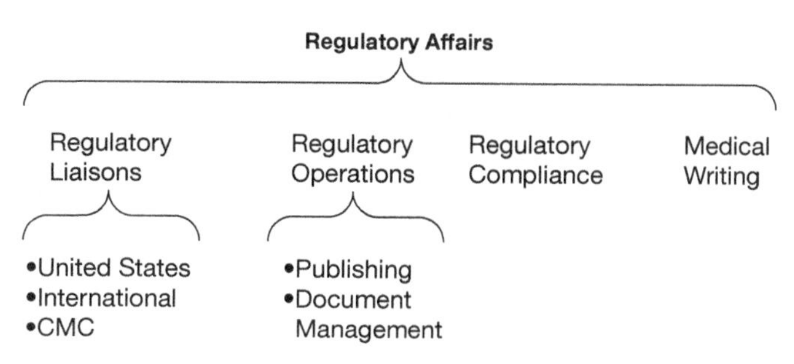


**Figure 12-1.** Common regulatory affairs specializations.

worldwide markets, but approval is generally sought in the rest of the world as well. Other international regulatory agencies include Health Canada and the International Conference of Harmonization (ICH).

### *Regulatory Operations: Publishing*

Previously, companies owned large photocopiers to print the massive amount of material, but now entire publishing systems are maintained on computers. Technical publishers oversee and generate highly complex hard-copy and electronic filings for submission to the regulatory authorities. Like editors of 100-volume books, they tie everything together and make sure that the table of contents, tabs, and other contents are functional and accurate. For electronic filings, they make sure that the files can be opened and are in the appropriate hypertext.

### *Regulatory Operations: Document Management and Archiving*

Documents and communications to and from the regulatory agencies are archived and maintained at a central facility so the information is organized and can be easily retrieved.

### *Medical Writers*

*Medical writers convert data into text.*

Medical writers draft summary documents of large, detailed reports, including regulatory submissions, clinical protocols, clinical study reports, and more. They typically produce a draft of the text from raw data, and send the information to the content authors for several rounds of reviews and revisions. This is a specialized art, for which degrees are available. To learn more about medical writing, see Chapter 10.

### *Government Affairs*

Some regulatory affairs professionals serve as conduits between the company, the FDA, and Congress in Washington, D.C. Because the regulations are constantly evolving, government affairs specialists work to improve or develop new policies, often by lobbying government representatives.

### Regulatory Compliance

Although the main role for regulatory affairs is to manage the process of regulatory review, regulatory compliance specialists ensure that internal systems and procedures maintain product quality and meet regulations. Regulatory compliance is discussed in more detail in Chapter 13.

*To learn about regulatory careers at the FDA, see Career Snapshot: Working at the FDA on page 153.*

### Regulatory Submissions

In the United States, there are a variety of regulatory filings, including relatively short, routine responses to FDA questions and updates, as well as the more voluminous filings such as Investigational New Drug Applications (INDs), New Drug Applications (NDAs), and Biologics Licensing Applications (BLAs).

An IND is the preclinical data submitted before companies are permitted to initiate clinical trials in humans. NDAs or BLAs are very large filings requesting market approval after the pivotal Phase III studies (see Chapter 10). To offer a sense of the magnitude of information submitted in a typical NDA filing, the paperwork can fill as many as three 18-wheelers! The good news is that the FDA is now accepting electronic submissions. It takes about 10 months for the FDA to review the content, but this can be shortened to a 6-month priority review for products that address unmet medical needs (see next paragraph). During this time, the company's manufacturing facilities also must pass preapproval inspections by health authorities. Review periods of an NDA or BLA are very active times for the teams supporting a product because FDA requests for information can require extensive, time-sensitive responses. This exchange of information culminates in careful negotiations on the final wording of the product label (described in Chapter 11).

"Unmet medical needs" is the catchphrase that refers to new drugs intended to treat severely debilitating illnesses such as Parkinson's or Alzheimer's disease, for which approved treatment options are limited or nonexistent. New products often have novel mechanisms of action, as opposed to the "me-too" or "copycat" drugs.

## REGULATORY AFFAIRS ROLES AND RESPONSIBILITIES

*Depending on the level of responsibility and type or size of the company, a regulatory affairs professional may have some of the following roles:*

### Regulatory Intelligence

Regulatory intelligence is needed for making important clinical or business development decisions. The regulatory professional provides information to the project teams about regulatory and development issues previously encountered with specific drugs and therapies, as well as the most recent FDA guidelines and regulations.

*Guidance documents (Guidelines) are written by the FDA and other regulatory authorities and represent the agency's current thinking on specific topics.*

### Regulatory Strategy

Regulatory professionals often assist in the strategic planning of clinical development programs. They analyze potential pitfalls where products might fail in the FDA approval process and help assess the different available development options. For example, any activities or applications that can shorten drug development times and lengthen marketing exclusivity, such as orphan drug status or accelerated approval, can be part of a company's strategy.

### Regulatory Submissions

*Regulatory affairs liaisons compile information from the different disciplines and package it for the FDA to review.*

Regulatory affairs plays a pivotal role in interacting with every department that contributes to FDA submissions, including discovery research, preclinical, clinical, bio/pharmaceutical product development, and manufacturing. The data are collected from each department, analyzed, and edited to make sure that they support the conclusions well enough to obtain approval before being submitted to the FDA. Regulatory professionals help the project heads assemble their arguments and clearly express the rationales for their sections. The regulatory liaisons might not necessarily provide technical expertise, but they are able to assess the application from a regulatory perspective.

### Documentation and Publishing

Regulatory professionals oversee the technical aspects of publishing and archiving the large volume of documentation so that all submissions conform to regulatory formats and the FDA has easy access to the information.

### Meetings and Communications with the FDA and Other Health Authorities

Regulatory professionals meet and communicate with the FDA and other health authorities during various stages of product development to ensure that the company's development plans are in line with FDA expectations. One important regulatory role is crafting appropriate and persuasive responses to FDA queries in order to expedite the approval process.

### Labeling, Promotion, and Post-Marketing Vigilance Reporting

After a product is approved for marketing by the FDA, regulatory liaisons continue to seek approval for the exact wording of the drug label information when new, pertinent information becomes available. (See Chapter 11 for a definition of "label." In this context, it is the information about the medicine in the drug packet.)

Some regulatory professionals provide oversight of product advertising and promotion. They determine whether claims made about products include the appropriate safety information and whether they are in strict accordance with regulatory requirements.

### Project Management

Regulatory affairs work frequently involves many project management-like responsibilities, particularly in smaller companies. Responsibilities include managing contract research organizations (CROs), setting up meetings, organizing and/or supervising teams, setting timelines, and identifying the key people who need to be involved in projects.

#### *Business Development*

It is not uncommon for regulatory affairs professionals to be asked to explain the regulatory hurdles that may need to be surmounted to obtain approval for products being considered for business development. They may also participate in due diligence to assess the regulatory strengths and weaknesses of the company's products or the products of potential partners or acquisitions.

#### *Writing Standard Operating Procedures (SOPs)*

When putting new programs in place, SOPs are written to ensure that internal department processes are well defined and that workers are fully informed and well trained so that the guidelines are understood and followed.

#### *Regulatory Compliance*

Regulatory compliance professionals provide general oversight of programs and make sure that team members comply with regulations. They are involved in many activities, such as qualifying the sites and vendors for clinical trials and lab tests; managing SOPs to make sure that they are followed according to guidelines and that records are maintained; validating electronic systems used for clinical trials, safety reporting, electronic publishing tools, and computers; and overseeing and hosting inspections from health authorities.

## A TYPICAL DAY IN REGULATORY AFFAIRS

*Depending on your level and role in the organization, on a typical day in regulatory affairs you could be:*

- Attending project team meetings, which may include technical, labeling, core, or clinical teams, or subcontractors, to discuss project status updates or strategic planning issues.
- Communicating and negotiating with health agencies; setting up and preparing for meetings or responding to the FDA's questions.
- E-mailing documents, i.e., circulating documents for review, approval, and sign-off.
- Reviewing promotional material or the label to make sure that the claims are appropriate and that the proper amount of safety information is included.
- Reviewing and preparing documents for filings for health authorities, such as INDs, NDAs, or BLAs, or preparing documents for orphan or fast-track designation.
- Researching the history of certain drugs and gathering information about new regulatory guidelines. Calling the FDA for input.

- Dealing with crises and "putting out fires." Assessing the situation, solving the problem, and getting everyone back on track.
- Writing SOPs. Ensuring that internal departmental processes are well defined and that people are trained and following the SOPs.
- Hosting inspections by health authorities. Alerting functional area leaders that an inspection is imminent and then meeting the inspectors and taking notes. Making sure that staff is available to answer questions and following up.
- Scrutinizing the clinical data with the statistics department.

## SALARY AND COMPENSATION

There is a shortage of qualified regulatory affairs professionals, because it takes many years to learn the nuances of the trade, the training is highly specialized, and there is not enough awareness about the field in academia or industry. Therefore, the salaries are high. Regulatory professionals may earn more than their counterparts in quality, discovery research, toxicology, manufacturing, project management, and process sciences. For people without advanced degrees, the pay in regulatory affairs can be 10% to 30% more than for comparable positions in research.

*Because of high demand and the specialized expertise required, regulatory affairs positions pay well.*

***How is success measured?***

Although the ultimate sign of success for a company is when it receives drug approvals from the FDA, product filings involve so many people that personal credit cannot be easily assessed. There is a general tendency for regulatory affairs professionals to learn to content themselves with smaller victories like completing individual steps in a program or meeting deadlines along the way to the completion of regulatory submissions. Other measures of success are based on how long it takes to get the submission through the FDA, the number of rounds of questions, and how swiftly questions are answered.

## PROS AND CONS OF THE JOB

### Positive Aspects of a Career in Regulatory Affairs

- It is exciting to be part of a team that helps develop a product that will benefit humanity. Completing submissions and sending them to health authorities are tangible results due, in part, to your efforts as a team member.
- Regulatory affairs is in the center of drug development. You will learn the ABCs of how drugs are developed: the big picture, the minutiae, and how it all ties together.
- Challenging, multifaceted issues constantly arise, and there is little daily routine.

- Regulatory affairs professionals have crucial roles, particularly in small organizations whose survival may rely entirely on one drug approval.
- There is excellent job security.
- This is a good career for people without advanced degrees. Depending on the company, you can progress to the level of vice president without an advanced degree, which is unusual for most other departments.
- Regulatory strategy is intellectually stimulating and personally rewarding. Predicting how the FDA will respond to submissions and preparing accordingly is strategically challenging.
- Negotiations can be fun and challenging—convincing health authorities to agree to your approach can be exciting.
- You will be constantly learning new technology, processes, and science, and there may be an opportunity to interact with people in other departments.
- Regulatory affairs professionals tend to travel less than their coworkers in other departments.

*You'll learn an enormous amount in regulatory affairs—about as much as you would learn in a Ph.D. or M.D. program.*

### The Potentially Unpleasant Side of Regulatory Affairs

- Because regulatory affairs is the last step before submitting materials to the FDA, you have to deliver on time, even if coworkers working upstream of you are late. As a result, regulatory professionals frequently work overtime leading up to deadlines. And with the continual flow of information and filings, the deadlines keep coming.
- Much time is spent putting out fires and dealing with crises. There is an endless list of situations that can cause emergencies, such as the discovery of a new impurity in your drug, a serious adverse effect on a patient in a clinical trial, procedures that have been done incorrectly by clinical monitors, and new filing deadlines.
- Regulatory strategy is complex and rather ambiguous, often requiring guesswork. For new, unfamiliar, or unstudied areas of drug development, including most biologics, there are no regulatory "recipes" to follow; every compound is unique. As an added complication, people's illnesses do not always behave the way they are described in textbooks. Because of this complexity, some FDA rules must be strictly followed, whereas others are subject to interpretation. Determining the difference can be like trying to read the FDA reviewers' minds. Sometimes even the rules that don't make sense need to be followed exactly, and even then, approval is not assured. In short, development requires a customized and intrepid approach, with no guarantee of success (see page 146, "To Excel in Regulatory Affairs...").

*The last remaining domino in product development is regulatory affairs.*

*There are frequently no single best answers in regulatory affairs.*

- Regulatory compliance professionals are perceived by some as policemen or nuisances. This is because when you want compliance with the rules and regulations, you frequently have to tell researchers to perform experiments that they don't want to do or else inform them that they can't do something that they want to do. On some days, employees might be grateful for your advice, but on other days, they will resent it.

## THE GREATEST CHALLENGES ON THE JOB

### *Balancing FDA Requirements and Internal Constraints*

*There is an internal conflict between what the FDA requests and what the senior management wants.*

There are certain situations in which a regulatory professional is personally liable, and it is important to guard against any situation that would compromise your career and legal standing. For a common example, upper management may want to expend the minimal effort to receive marketing approval, but at the same time, it is important to not submit poor-quality or insubstantial data to health authorities. The consequences can be dire: The FDA can impose fines or shut the company down if it is found to be noncompliant. Not just regulatory affairs individuals, but any company representatives who were aware of fraudulent data, can be personally liable for statutory crimes with civil penalties.

### *Rallying the Troops, "Management without Authority," and Keeping up with the Heavy Workload*

There is a strong project management component to regulatory affairs, and as such, the same issues dealing with "management without authority" apply (see Chapter 9 for a more detailed description). Team members may feel overwhelmed with the workload, and it can be challenging to keep them happy and focused.

### *Staying Current with the Regulatory Guidelines*

It is challenging to stay current with continually changing regulatory guidelines. This can be particularly daunting when taking the worldwide perspective into consideration, because each of the international regulatory authorities has its own constantly changing requirements.

### *Stakeholders Can Lose Objectivity and Take Things Personally*

*Telling a researcher that his experiments need modification can be like telling someone that her child has faults.*

On occasion, inventors or content authors can become so enamored of their product that they lose their objectivity, and any suggestion that a set of experiments might not satisfy a regulator can be incorrectly taken as a personal affront.

## TO EXCEL IN REGULATORY AFFAIRS...

### *Great Regulatory Aptitude and Strategic Thinking Capabilities*

There are many who can unerringly put together submissions, but to be really great, one must be able to approach product development plans and submissions with a strategic appreciation of the corporate goals and regulatory requirements. A successful strategic approach requires high-level thinking that involves understanding the subtleties of the political environment, the market, recent issues with the FDA, and many other considerations that take into account the big picture.

*Strategic thinking is about deciding which wars to fight and how best to win, not where to place each soldier.*

### *Years of Experience Working with the FDA*

Those who excel in this field have a keen intuition and judgment usually derived from years of experience working with the FDA. Armed with comprehensive information about a specific product, they assess the regulatory climate and develop insights that help them plan the most convincing presentations for the FDA.

### *Understanding the Drug Development and Regulatory Process*

*Those who excel do not overreact, but supply only a reasonable amount of resources to fix situations.*

Those who excel are able to understand the complex issues in drug development and to grasp their multidimensionality. Exceptional regulatory professionals can determine the most expedient and inexpensive approach to market approval with the minimum amount of studies. They know how to interpret the FDA guidelines to decide which studies are needed and, just as importantly, which are not. Those with less experience tend to be overly conservative with guidelines and follow the rules to the letter, whereas experienced personnel understand that certain guidelines might not apply.

### *A Strong Science Background*

*Regulatory affairs requires a comprehension of every step of drug development, from beginning to end.*

A scientific background is essential for success in regulatory affairs. In order to be very knowledgeable about the products and how they are made and tested, it is important to understand the tremendous amount of science in manufacturing, toxicology, pharmacology studies, clinical data, feasibility studies, and the mechanism of action for products described in the regulatory filings.

### *Team Management*

The regulatory submissions are extensive and cumbersome and require input from content experts in different technical fields. There is an incredible amount of work to do and many people to manage under intense deadline pressure. The leaders who are able to motivate and guide the teams efficiently are the ones who develop a track record of success.

***Building Relationships and Establishing Trust with the Health Authorities***

The FDA is the ultimate gatekeeper to market approval in the United States. Those who excel in regulatory affairs have developed relationships with FDA reviewers built on trust and credibility. Consistent honesty with reviewers and thorough, well-composed analyses can result in a greater willingness on the part of health authorities to negotiate and identify creative alternatives during the drug approval process.

## Are You a Good Candidate for Regulatory Affairs?

***People who flourish in regulatory affairs careers tend to have...***

*You need to have the eyes of a hawk: the ability to see both the big picture and the minute details at the same time.*

***The ability to focus on details and the big picture simultaneously.*** Accuracy and attention to detail are required.

***The ability to work in a highly regulated environment with many rules and processes.*** Although creative input is also useful, particularly for problem solving, there are clear limits in regulatory affairs.

***Outstanding oral communications skills.*** In regulatory affairs, you will interact with people in almost every other function in the company; therefore, strong oral communication skills are essential.

***Exceptionally good reading and writing skills.*** Much of what you do is communicated in written form, which needs to be articulate and easily understood by the FDA. It is important to be able to express yourself clearly and concisely. You need to be able to write persuasively so that the health authorities understand and agree with the information that you have provided. There is often a considerable amount of reading required and tomes of information to gather and review.

*To be successful, you need to be personable and able to work well with people.*

***Excellent interpersonal skills and the ability to work with many personality types.*** Being a team player is a necessity. Regulatory affairs is a highly matrixed team function; you rely on others and they depend on you. You need to be responsible for your own efforts and to be able to delegate work to colleagues. (For a more detailed description, see Chapter 2.)

*You need to be somewhat assertive and not mind debating on a regular basis.*

***Leadership skills.*** VP-level professionals tend to be self-confident, speak up at meetings, and trust their own judgment. They have the forthrightness to contact FDA reviewers for input or consultation when questions arise.

***Exceptionally good listening skills.*** Due to the constant communication demands, being a perceptive listener, especially when talking to the FDA and to team members, is important. Information garnered from keen listening skills allows you to assess other people's opinions so that you can construct the optimal strategies.

***The ability to think strategically about complex issues.*** In regulatory affairs, you need to think at least two steps ahead as you move forward. Bringing all the pieces together to compose a consistent story for health authorities requires a strategic mind.

*A strong work ethic, stamina, and endurance.* Boundless energy and the desire to work hard are required for preparing submissions. Some projects may drag out for years! It takes enduring energy and persistence to chase minute details. Most organizations are insufficiently staffed to handle the extra work leading up to filing deadlines, and some people even camp out in the office to get the work done.

*A strong science background.* Without a science background, it may be difficult to understand what the sections in the filings are about and whether the data support the conclusions.

*Intellectual curiosity.* Regulatory affairs provides an in-depth exposure to an abundance of new science. It is important to have an interest in extending the breadth of your knowledge into many fields so that you can interact smoothly with team members. In order to be effective, you should keep up with the current scientific literature.

*Strong analytical, objective, and critical thinking capacity.* You need to be able to accurately and objectively evaluate data and make independent judgments. This includes the ability to review information with a critical eye and to determine whether the data support the conclusions.

*The ability to quickly re-prioritize.* You need to be intellectually nimble enough to make reasonable decisions and quickly change course upon receipt of new data. It is also important to be politically sensitive to the ramifications of selecting one project rather than another.

*Excellent organizational and multitasking skills.* Many projects are often simultaneously in progress and may require different types of tasks.

*A high tolerance for pressure and the ability to manage daily crises.* Regulatory affairs professionals need to be able to cope with the intense stress of looming corporate deadlines for filings. Because crises are often constant, some people say that you need to be an "adrenalin junkie" to cope.

*A thoughtful and contemplative approach.* Because you represent the company to health authorities, it is important to think before you speak. You need to be careful about how you say things so that you later don't have to retract or modify your statements.

*The capacity for creative and flexible thinking.* Regulatory professionals tend to be relatively inventive. They can review the data and "spin" it in such a way that the FDA might be more likely to accept it. They tend to be flexible and willing to compromise while remaining scientific, honest, and objective.

*Superb diplomatic and negotiation skills.* Regulatory affairs professionals should be able to negotiate effectively and diplomatically and build win–win scenarios with FDA officials. Within the company, they need to be self-confident enough to state and defend their positions but diplomatic enough to express difficult opinions and have them accepted in a cooperative spirit. It is also important to be able to moderate and facilitate meetings and help teams resolve conflicts.

*A moderate level of perfectionism.* Wanting everything to be perfect is a good trait up to a point, but it may become a hindrance at higher levels. Regulatory filings will never be perfect, and eventually deadlines force submission. Strict perfectionists tend to overwork themselves and suffer burn-out.

***A work attitude that is goal-oriented but not pushy.*** To complete the heavy workload and meet deadlines, it is helpful to have a goal-oriented work ethic. To be effective, you must be able to delegate with high expectations while remaining professional and respectful of your colleagues.

***You should probably consider a career outside of regulatory affairs if you are...***

- A "free spirit" who would have a difficult time working in a highly regulated environment.
- Reckless or biased; if you are judgmental or make snap decisions.
- Not interested in scientific detail, sloppy, too lazy to tackle difficult and complicated problems.
- Not willing to ask questions, listen, and gather information.
- One who prefers to work alone in your office and does not enjoy interacting with others.
- One who enjoys perfect order. Regulatory affairs can be chaotic. Interruptions and changing priorities are to be expected.
- In need of receiving constant recognition for your work, very self-important.
- A naysayer or a devil's advocate.
- Inflexible or doctrinaire.
- Too soft-spoken and meek.
- Unable to perform well under adversarial conditions or likely to worry about the risk of losing coworkers' allegiance and affection.

## REGULATORY AFFAIRS CAREER POTENTIAL

*Most people "fall into" regulatory affairs. Once in, they usually don't leave.*

After being trained in regulatory affairs, people usually don't leave. This is an enormous field, and there is an abundance of new areas where you can perfect your skills and learn. There are opportunities to work in different therapeutic areas, expand globally, or take on compliance or other levels of responsibility. It is easy to move up in the ranks. You can start as a document or regulatory specialist and move from manager to director to vice president. Regulatory affairs professionals with experience also commonly choose to go into consulting (Fig. 12-2).

Because regulatory affairs is part of every aspect of the business, it prepares individuals for general management and for career tracks in other disciplines. Many professionals go into program or project management, where, as in regulatory affairs, the role is to drive complicated programs and oversee the work of many people in a multidisciplinary capacity. Those with regulatory affairs training can also eventually transfer into careers in clin-

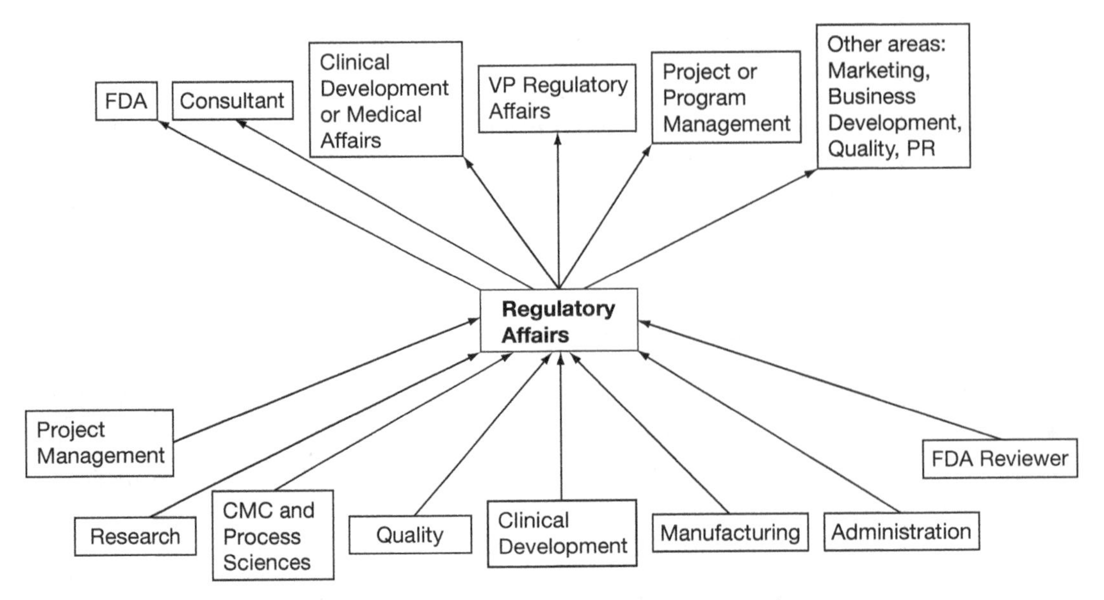


**Figure 12-2.** Common regulatory career paths.

ical and medical affairs, marketing, quality, business development, governmental affairs, public relations, or advertising (see Fig. 12-2).

## Job Security and Future Trends

*Regulatory affairs requires specialized expertise that is very marketable and in high demand.*

Regulatory affairs positions are among the most difficult to fill, due to the limited supply of qualified individuals. Consequently, there is excellent job security. Those who are proficient and well known will be constantly sought by recruiters, and those who have experience working at the FDA are in even greater demand (see "Career Snapshot: Working at the FDA" on page 153).

Demand for regulatory affairs professionals is predicted to increase, because the industry is becoming even more regulated. The rules are becoming more complicated, and they are constantly changing. With the introduction of electronic submissions, the landscape will continue to change for years to come.

As long as a company has products in development and on the market, a regulatory department is mandatory, but if something goes wrong, regulatory positions can be easily sacrificed. Regulatory VPs are often blamed for failed drug approvals, even if they weren't responsible. Regulatory managers and directors, however, can usually avoid the axe.

This career is unlikely to be outsourced overseas. Although some of the work can be accomplished remotely, one still must interact in person with others in the company and be present for face-to-face meetings.

## LANDING A JOB IN REGULATORY AFFAIRS

### Experience and Educational Requirements

Most people find their way into regulatory affairs by working as a member of a project team and becoming involved in the documentation of a product submission. Others come from operational or publishing backgrounds. In general, people migrate from many disciplines into regulatory affairs. Preclinical, pharmacology, discovery research, clinical research, chemistry, manufacturing, quality, administrative, or documentation backgrounds are good. Those with clinical trial experience as physicians, pharmacists, and nurses in hospitals also do well.

An advanced degree is not mandatory for regulatory affairs in most companies. Drug discovery and development experience, basic intelligence, and the ability to think analytically are generally held in higher esteem than are educational backgrounds. Individuals who do possess advanced degrees, however, tend to start higher and progress further and faster. Ph.D.s, Pharm.D.s, M.P.H.s, and M.B.A.s are commonly found, and people with J.D. degrees are particularly in favor due to their ability to negotiate with and present convincing cases to the FDA. Larger pharmaceutical and biotechnology companies tend to require professionals with advanced degrees for director- and vice president-level positions, whereas a master's degree often suffices for smaller biotechnology and medical device companies.

It has finally dawned on the school system that regulatory affairs is an excellent career with great potential and that there is a healthy demand for qualified individuals, so new university programs are sprouting up which provide degrees or certificates in regulatory affairs or sciences.

### Paths to Regulatory Affairs

*It is difficult to obtain an initial position in regulatory affairs without prior experience. Here are some tips for entering the field:*

*Training in drug discovery and development is highly valued in regulatory affairs.*

- If you are already in a biotechnology company, try to gain experience by working on a section of a regulatory submission. If you are doing bench work, join a project team so that you will have an opportunity to work on a section filing.
- Consider obtaining project management experience first. The skills garnered as a project manager provide excellent preparation for a career in regulatory affairs. In addition, project management will expose you to the entire clinical and drug development process (see Chapter 9).
- Consider working at a contract research organization (CRO), which offers considerable exposure to a greater number of scientific disciplines and therapeutic areas. It is generally easier to obtain jobs at CROs than at biopharma companies, which often require prior experience.

- When considering a company to join, conduct due diligence on how the management teams operate. Do they appreciate working in a regulated environment? Will they support their head of regulatory affairs? Will they follow the FDA rules when times are tough, or will they try to take shortcuts?
- Regulatory affairs is an emerging field, and many universities are starting to develop preparatory vocational programs. Select universities offer specialized certificates or even full master's degrees in regulatory sciences. The Regulatory Affairs Professional Society (RAPS) offers a Regulatory Affairs Certificate program (RAC), but you need at least two or three years of full-time experience on the job to qualify for the exam.
- Consider applying for a position as a reviewer at the FDA. As a reviewer, you will see the inner workings of the FDA and the entire process of drug approvals, and you will gain tremendous exposure to many different therapeutic areas and classes of drugs (see Career Snapshot: Working at the FDA below).

## CAREER SNAPSHOT: WORKING AT THE FDA

If you seek total immersion in regulatory affairs, probably the best way to really understand how the product review process functions is to work at the FDA as a product reviewer. It is a tremendous learning opportunity and can be a highly secure and rewarding job.

*The FDA's primary mandate is to protect the public health.*

Drug product reviewers assess applications for new drug programs and help companies develop safe and effective products. The FDA's mission is to protect public health by making sure health care products are safe while simultaneously bringing new therapies to the market efficiently.

### *Depending on the Level and Position, a Job at the FDA Might Entail...*

- Reviewing filings such as INDs, NDAs, and BLAs.
- Developing policies to expedite product approvals or increase safety.
- Attending meetings with academic or company sponsors.
- Developing guidance documents.
- Attending advisory committees composed of leading experts who provide their opinions about drugs.

*Having a good idea for a drug doesn't necessarily mean that it's a good idea to put the drug in people.*

### *Positive Aspects of Working at the FDA*

- Safeguarding the public is a highly important role. The legal and health consequences of your decisions may affect thousands of people.

- There is tremendous exposure to the latest cutting-edge technologies, a wide swath of products, evolving policies, and public issues.
- Benefits such as pension plans, health care, and retirement packages are provided.
- Excellent job prospects await you after your FDA experience. Companies will want to hire you.

### *The Potentially Unpleasant Side of Working at the FDA*

- Salaries are considerably lower compared to industry positions, perhaps by as much as 30%.
- The job is very demanding and stressful. The FDA is constantly underresourced and overburdened.
- The FDA is a large bureaucratic organization and consequently, things move slowly.
- Most positions are located in or near Washington, D.C.

### *Careers at the FDA*

There are distinct disciplines for product reviewers, such as preclinical, chemistry, clinical, statistics, and manufacturing. M.D.s tend to become medical reviewers (see Chapter 10). After you gain exposure to the regulatory process, you can specialize in a particular product area, such as gene therapy. After serving as a product reviewer, the next career moves are to become a branch chief for a therapeutic area and then deputy division director. Depending on your ambition and interest, you can become involved in policy development, education, or outreach programs to represent the FDA perspective at home and abroad.

Besides product reviewers, a variety of other positions are available, such as research scientists, fellows, or project managers, who interface with academic and industry investigators, organize and run meetings, and submit amendments to filings.

Inspectors usually have a master's or bachelor's degree and operate across the country. There are many laboratory and administrative positions, international liaisons, legal and policy experts, epidemiologists, engineers, and more. There are also careers in the device section of the FDA. The Center for Devices and Radiological Health (CDRH) typically seeks engineers, as well as other experts described above, who can review the medical device filings.

### *Experience and Educational Requirements*

There is almost always great demand for talented individuals with technical backgrounds at the FDA. To obtain a position at the FDA, a minimum of a B.S. degree in science is required. The majority of candidates with medical or science backgrounds enter the agency with postdoctoral or industry experience. The FDA values experience in health care, biotechnology, or regulatory affairs in addition to academic credentials. A broad industry perspective and diversity of technical experience is appreciated. Epidemiologists and those with statistical backgrounds are also needed at the FDA for tracking trends in adverse event reports for marketed drugs.

#### *Paths to a Career in the FDA*

- For more information, visit the FDA Web site at www.fda.gov and click on "job opportunities," or visit the U.S. Public Health Service Commissioned Corps at www.usphs.gov or www.usajobs.com. Look for information about opportunities for physicians, scientists, and other health-related professionals at the FDA, the Centers for Disease Control (CDC), and the National Institutes of Health (NIH). There are other centers to consider as well: the Center for Biologics Evaluation and Research (CBER), the Center for Devices and Radiological Health (CDRH), the Center for Veterinary Medicine (CVM), and the Center for Food Safety and Applied Nutrition (CFSAN).
- If a reviewer position is unobtainable at the Center for Drug Evaluation and Research (CDER) or if you would like to continue to conduct laboratory research, consider obtaining a regulatory affairs fellowship or research position at the FDA. People who transfer from academia to research positions at the FDA will generally conduct applied research. Once in a research position, you can eventually move up the ranks to become a reviewer.
- The FDA prefers to hire experts in new and emerging fields, so consider becoming a specialist in the latest therapeutic approaches. They also seek to hire people with a wide range of experience, so include a comprehensive list of your technical skill sets and expertise in specific therapeutic areas on your application.
- If you apply for positions at the FDA, keep in mind that acceptance is a formal process that can take up to six months or longer. You must be a U.S. citizen with no criminal record. The FDA will require that you divest any stock that you own in drug discovery companies. You may submit your resume every six months.

## RECOMMENDED TRAINING, PROFESSIONAL SOCIETIES, AND RESOURCES

#### *Courses and Certificate Programs*

The Regulatory Affairs Professionals Society (RAPS, www.raps.org) offers Regulatory Affairs Certificates.

The Food and Drug Law Institute (www.fdli.org) is a private organization of lawyers interested in the FDA. They provide classes and offer legal internships.

### *Societies and Resources*

The Regulatory Affairs Professionals Society (www.raps.org) is the best for those interested in a career in regulatory affairs; its Web site also has job postings.

The Parenteral Drug Association (www.pda.org)

The American Society of Quality (www.asq.org)

Pharmaceutical Education & Research Institute (www.peri.org)

The Food and Drug Law Institute (www.fdli.org)

The Organisation for Professionals in Regulatory Affairs (TOPRA, in Europe)

The Food and Drug Administration (www.fda.gov)

The American Medical Writers Association (www.amwa.org)

The U.S. Public Health Service Commissioned Corps (www.usphs.gov) provides opportunities for physicians, scientists, and other health-related professions at the FDA, the CDC, the NIH, etc.

The Institute of Validation Technology (www.ivthome.com) offers a self-learning program about regulatory affairs.

### *Biotechnology and Pharmaceutical Societies with a Broader Audience*

The Drug Information Association (www.diahome.org/DIAHome/) is an umbrella organization that covers many disciplines in drug discovery and development. They have regulatory sections and job postings.

Biotechnology Industry Organization (www.bio.org)

The Pharmaceutical Research and Manufacturers of America (www.phrma.org)

### *Medical Device Societies*

Advanced Medical Technology Association (www.advamed.org)

National Emergency Medicine Association (www.nemahealth.org)

Association for the Advancement of Medical Instrumentation (www.aami.org)

# 13

# Quality

## Consistently Making Good Products

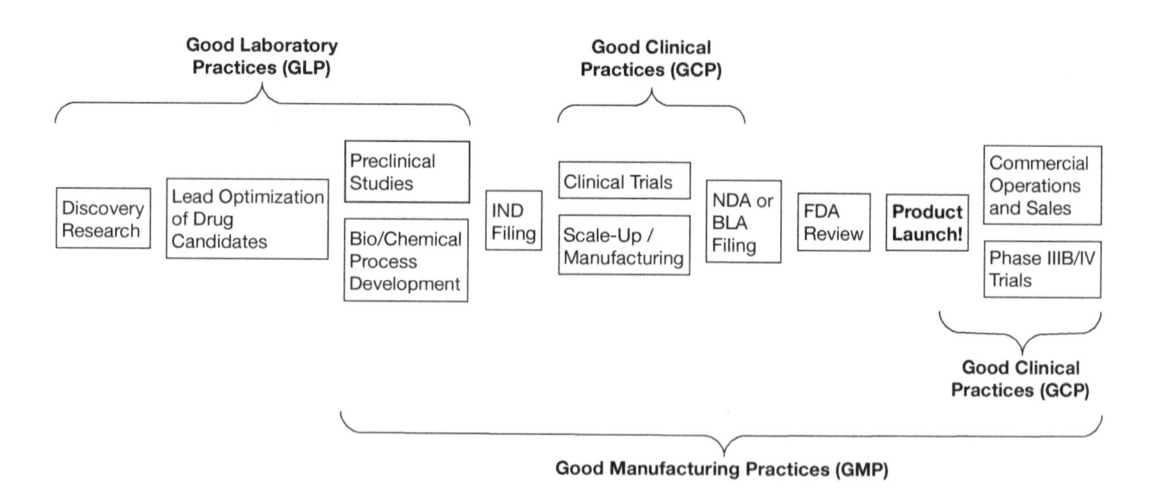


FOR THOSE WHO LIKE TO TAKE DETAILED NOTES AND HAVE an interest in developing procedures so that processes can run more efficiently, a career in quality might be appropriate. Quality, which includes quality control (QC), quality assurance (QA), regulatory compliance, and quality systems, ensures that products are consistent and that processes and systems comply with FDA standards. Quality departments offer diverse vocational areas with many different specialties and encompass jobs that range from routine and predictable to challenging and full of surprises.

### THE IMPORTANCE OF QUALITY IN BIOTECHNOLOGY

Quality ensures that products are consistent and that company procedures comply with FDA regulations. For therapeutic companies, quality also ensures that products are pure so that they are safe for human consumption.

*Unless a product goes through a regulated quality process, it cannot be used to diagnose a disease or treat a patient.*

*The importance of quality is illustrated in the following examples:*

*Quality professionals are akin to chefs working at fancy restaurants: They write detailed instructions so that novice cooks can make the same consistently tasty dishes every day.*

Most of the data in the early stages of product development are derived from discovery research. When production is transferred to the manufacturing division, quality professionals translate the protocols developed in discovery research into standard operating procedures (SOPs).

When an IND or NDA filing is submitted to the FDA, the company is making a promise to manufacture and test a product using specified processes. Quality professionals independently verify that this commitment is being met. They set up SOPs so that information is recorded properly and the company can prove that the processes have been carried out in the agreed-upon manner.

Quality provides specific instructions so that all information is tracked, archived, and linked, and can be easily retrieved and reviewed. This faithful paper trail can be used during audits by the regulatory authorities.

**What does it mean to be compliant?**

"Compliant" means that the company is following specific internal auditing procedures in accordance with the FDA regulations. The compliance group acts like an internal version of the FDA; reviewing systems, processes, and procedures to make sure that they are meeting regulations and guidelines and that coworkers are following SOPs.

## CAREER TRACKS: QUALITY CONTROL, QUALITY ASSURANCE, REGULATORY COMPLIANCE, AND QUALITY SYSTEMS

As shown in Table 13-1, there are several different specializations in quality. These represent vastly different work functions and skill sets.

As a general rule, the more technically inclined chemists and the "lab rats" tend to flock to quality control, the biologists prefer quality assurance and regulatory compliance, and computer scientists prefer quality systems. In small biotechnology companies, quality assurance and quality control may be parts of the same entity, but in most organizations, these two functions are in separate and independent departments. Regulatory compliance can be part of the regulatory affairs or quality assurance departments or it can be a distinct entity.

**The difference between quality control and quality assurance**

The quality control group provides *measurements* to demonstrate that the job is being done correctly and that the products meet specifications. The quality assurance group provides the *documentation* that the job is being done correctly and that systems for monitoring quality are in place.

Table 13-1. Common quality specializations

| Quality control | Quality assurance and regulatory compliance | Quality systems |
|---|---|---|
| Chemistry lab | Document and review | Validation (computer validation/IT quality, and more) |
| Microbiology lab | Auditing group | Training |
| Cell biology lab | Prescription drug compliance | Document management |
| Animal group | Regulatory compliance (sometimes in QA, regulatory affairs, or a separate department) | Corrective and preventive action |
| Environmental monitoring | | Quality engineering |
| Metrology | | |

## Quality Control

The quality control department tests products and verifies that manufactured products meet specifications.

*In most companies, Quality Control includes:*

### *Chemistry Labs*

QC chemists verify that drug products are pure. They apply many different types of biochemical and inorganic chemical analyses to check for impurities and verify the composition of the final product.

### *Microbiology Labs*

QC microbiologists conduct sterility testing and environmental monitoring. Good Manufacturing Practice (GMP) regulations dictate that even when sterility is not required, it is still necessary to show that certain objectionable organisms, such as strains of *E. coli* or other disease-causing organisms, are not present (see box on page 162, "The 'GXP' regulations").

### *Cell Biology Labs*

QC cell biologists ensure purity and verify the composition of the final products. Tests using PCR and biological cell culture assays are conducted to check for purity and quality.

### *Animal Group*

Animal handlers maintain animals for preclinical toxicology studies. This department may also be part of the preclinical division.

### *Environmental Monitoring*

Employees in this department properly dispose of trash. They make certain that the material that goes down the drain is compliant with Environmental Protection Agency regulations.

### *Metrology*

Metrologists calibrate the machinery and instruments.

## Quality Assurance

There are typically two functions in quality assurance: QA professionals work closely with all departments to develop and implement systems and procedures to assure compliance with regulations, and they create and archive records to show that procedures are being done correctly. Quality assurance is more paper-oriented than quality control.

*Quality guarantees that the company is conforming to the agreements made with the regulatory authorities.*

*In most companies, quality assurance has three main functions:*

### *Documentation Review*

This group handles and reviews the paperwork from manufacturing and laboratories. In some companies, they also maintain the clinical documents.

### *Auditing*

It is extremely important to monitor products, procedures, and systems. The usual source of problems is bad raw material provided for drug products. Therefore, it is important to visit and audit raw material suppliers and contractors and verify that everything is being done properly.

### *Prescription Drug Marketing Act (PDMA) Compliance*

There is a group which ensures that the company is in compliance with PDMA. This law was enacted to ensure that prescription drug products purchased by consumers are safe and effective and that counterfeit, adulterated, misbranded, subpotent, or expired drugs are not sold to the American public.

## Regulatory Compliance

Whereas the main role of regulatory affairs is to manage the process of submitting documents to the health agencies, the regulatory compliance group conducts surveys within an organization to ensure that systems and procedures have accounted for quality and that they meet regulations. Regulatory compliance is usually located either in the quality assurance or regulatory affairs departments (see Chapter 12).

*The regulatory compliance group is like an internal FDA.*

Regulatory compliance professionals train employees to ensure that they are familiar with FDA regulations, guidelines, and internal procedures and processes. They audit various providers involved with the development and validation of electronic systems for clinical trials, safety reporting, electronic publishing tools, and computer validation.

## Quality Systems

Large companies have a separate group called "quality systems" that is typically divided into five subgroups: validation, training, document management, corrective action, and quality engineering.

*Validation*

*"Validation" means that you have tested the system and that it is doing what it is supposed to do.*

Validation can be defined as the means of establishing documented evidence that a specific process or system will consistently produce a product meeting its predetermined specifications and quality attributes. Validation can be broken down into many different categories, including the validation of computer systems, various equipment, processes, analytical methods, and facilities.

*Computer Validation*

Validation of computer systems is required if they are used for FDA-regulated purposes. Such activities may include the testing of the software, hardware, databases and their back-up and recovery, business continuity, and disaster recovery procedures.

This field is very complicated. Laws assert that data collected for product development efforts, such as preclinical or clinical information, need to be controlled. Procedures and controls need to be in place to ensure the authenticity, integrity, and the confidentiality of the data.

*Information Technology (IT) Quality*

*Like a master plumber, IT quality checks for leaks and ascertains that the pipes are made of the correct materials.*

This department reviews the overall quality and integrity of the IT infrastructure and protects the data. It puts certain controls in place to reduce the risks associated with the use of computer systems. Risks include hardware failures, network security threats, viruses, communication errors, accidental erasure of data, and natural disasters.

*Training*

GXP regulations require that personnel be trained to meet their assigned job requirements. Most organizations achieve this through a training program.

*Document Management*

*Some professionals in quality assert that "if it was not documented, then it was not done!"*

In all regulated areas, documentation plays an extremely important role in satisfying regulations. Regulations require written procedures, batch records, test specifications, test results and summary reports, validation documentation, training records, etc. These documents must be maintained and controlled.

*Corrective and Preventive Action*

Programs are implemented to investigate problems, to correct those issues, and to make provisions that will prevent problems from recurring in the future (i.e., quality improvement).

### *Quality Engineering*

Quality engineers test systems and applications. They define metrics and run programs to identify trends or problems in the product development process. These procedures provide step-by-step descriptions of exactly what the systems are expected to do.

#### *The "GXP" regulations*

The "GXPs" are FDA regulations that govern the development, testing, and manufacturing of drugs, medical devices, and biologics. GXPs are the law and are published in the Code of Federal Regulations, under title 21. Organizations regulated by the FDA must comply with the GXPs.

GXPs include three main types of regulations:

- Good Laboratory Practices (GLPs) regulate discovery research and preclinical studies in the evaluation of new drugs, medical devices, or biologic products.
- Good Manufacturing Practices (GMPs) regulate the manufacture, testing, and distribution of new drugs, medical devices, or biologics for human or veterinary use.
- Good Clinical Practices (GCPs) regulate the design, conduct, performance, monitoring, auditing, recording, analysis, and reporting of clinical trials.

#### *What does it mean to work under GXP regulations?*

*There are always two pairs of eyes to look things over in quality.*

GXP regulations outline, among other things, how to specify test results. For example, if there is a failure in manufacturing, the GMPs outline the proper procedures and requirements for conducting and documenting investigations. In the lab, bench scientists are required to record results in ink and sign their laboratory notebooks (see 21CFR58.130(e), below). At the end of the day, or after a batch run, the supervisor checks the notebooks to make sure that they have been properly filled out. A quality assurance member will ultimately review and double-check these records.

#### *The Code of Federal Regulations, 21CFR58.130(e):*

*"All data generated during the conduct of a nonclinical laboratory study, except those that are generated by automated data collection systems, shall be recorded directly, promptly, and legibly in ink. All data entries shall be dated on the date of entry and signed or initialed by the person entering the data."*

## QUALITY ROLES AND RESPONSIBILITIES

*Because quality covers such diverse vocational areas, many different functions are performed within an organization. The following are some of the roles and responsibilities:*

### *Laboratory Testing*

Employees in QC run various analyses on samples from product batches. The final results are reviewed and signed off by a supervisor and sent to QA.

### Validating Systems

Employees in quality systems develop and test systems based on their intended use. Although the users define the system requirements, the validation managers ensure that those requirements are being met.

### Providing Compliance Oversight for Contract Service Providers

QA is responsible for compliance oversight and for managing vendors. They might develop business relationships with service providers, write and execute quality agreements and contracts, or manage work processes.

### Laboratory Compliance

Laboratory compliance professionals review protocols and vendors, and ensure that the work done at the bench is compliant. For example, they may review the in vitro and animal experiments completed in support of regulatory submissions.

### Conducting Vendor Audits

> You cannot outsource quality—you can only outsource the work.

The sponsor (the company developing the product) is ultimately responsible for the device or drug, even if it is produced by vendors or contract service providers. It is extremely important, therefore, that a sponsor confirms that the services or products provided by a third-party vendor comply with regulations. To do this, sponsors typically conduct vendor audits. Depending on the type of audit, an auditor will assess whether the vendor complies with national and local regulations and meets the sponsor's requirements. Auditors typically review systems, processes, procedures, and supporting documentation such as training records, and they interview the staff.

### Conducting Internal Audits

Compliance professionals ensure that their own organization is prepared for FDA audits by conducting internal audits.

### Being Audited and Hosting Inspections by Health Authorities

Before and after a new drug approval, FDA regulators may inspect manufacturing facilities and clinical investigator sites or may visit the sponsor organization. Regulatory compliance professionals prepare for and manage these inspections. They ensure that all documentation under review is complete, accurate, and accessible. They serve as the primary point of contact for the inspectors, gather and provide the requested documentation, and arrange interviews with employees and inspectors. In addition, they are typically responsible for coordinating whatever follow-up activities are requested by the inspectors.

### Document Review for Product Release

QA professionals spend time reviewing relevant documentation, determining whether activities related to the release process were performed appropriately and in accordance with standard operating procedures, and making assessments about whether the product is suitable for release. Documentation associated with the release process is extremely important, because this information is used in the event of a product recall.

#### Good Documentation Practices

Good recordkeeping is very important because the FDA audits companies by reviewing their documentation. It is important to retain and be able to retrieve information long after employees have left the company. If data is well organized, legible, clear, and easy to retrieve, it makes the life of an FDA inspector easier.

*All documentation associated with development, testing, and manufacturing of a drug, device, or biological product must be archived and retained in accordance with the applicable regulations.*

#### Document Control, Writing and Managing SOPs and Policies

QA professionals draft a variety of different documents, such as standard operating procedures and policies. Policies are high-level documents that define how regulations will be put into practice. SOPs, on the other hand, are the detailed work instructions.

#### Training Employees

Quality professionals train employees on new SOPs, policies, and other types of work instructions. They make sure that all training is documented.

#### Regulatory Submissions Work

Individuals in quality are involved in the review of documents that go into FDA regulatory filings. They confirm that the data are consistent with what was collected in the clinical trials and that the conclusions are reasonable and based on the data.

#### Handling Complaints

After a product is approved, a group is responsible for handling quality-related product complaints and recalls. When there is a complaint about a product, they document, investigate, and resolve the problem.

## A TYPICAL DAY IN QUALITY

*There are many different types of roles in quality. Depending on the position, a typical day might include some of the following activities:*

- In QC, running laboratory analyses, analyzing the data, and recording the final results, which are signed by a supervisor and sent to QA.
- Reviewing documents, such as reports, protocols, or FDA submissions, to ensure that applicable regulatory requirements have been observed.
- Ensuring compliance with applicable external regulations and guidelines as well as internal policies and procedures.

- Auditing systems, processes, procedures, vendors, and contractors to make sure that applicable regulations are being followed. Writing audit reports and determining that "observations of nonconformities" are adequately addressed.
- Developing and implementing training programs to ensure that the staff is up to date on new or revised policies and procedures and ensuring that all training is documented.
- Writing quality agreements and acting as a liaison with vendors and contractors.
- Participating in projects that relate to the organization's infrastructure; for example, installing new global or document management systems and training employees to use them.
- Attending meetings. As a general rule, supervisors and managers attend more interdepartmental meetings than do individual contributors.
- Keeping abreast of FDA activities. This includes reviewing FDA warning letters to other companies in order to gain insight into FDA trends.

## SALARY AND COMPENSATION

In general, people in quality careers are compensated a little less than those in R&D, but about the same as those in manufacturing. There are specific areas in quality where demand for qualified talent is so high that compensation is very good. In addition, for most areas, expertise in a particular area can lead to a lucrative consulting business.

***How is success measured?***

Gaining product approvals and passing regulatory inspections are the ultimate measures of success.

*Quality is about consistently doing things right.*

There are many key performance indicators that are specific to the individual areas of quality, such as:

- Having a good compliance history.
- Passing regulatory inspections without major problems.
- Routinely making good quality products with no issues from the FDA.
- Testing and demonstrating that new systems meet the acceptance criteria.

*If you can improve processes for coworkers and make their jobs easier, you have done your job well.*

As a department, measures of success include:

- Being seen as a resource or partner working with others to improve processes within the organization.
- Being able to implement changes and identifying mechanisms that will reduce, mitigate, or eliminate regulatory risk.
- Increasing efficiency, saving coworkers' time, or making their work processes easier.

## PROS AND CONS OF THE JOB

### Positive Aspects of a Career in Quality

- Quality plays an important role in society and in companies. Your work could potentially affect patients who are being treated with drugs.
- There tends to be good camaraderie in the quality department. The routine nature of much of the work allows people to have fun on the job. People in quality tend to be friendly, team-oriented, and respectful.
- Some roles require the ability to understand a combination of perspectives in other technical areas such as regulatory affairs or manufacturing. These jobs are interesting because they involve many functional areas and allow you to keep up to date with the latest industrial microbiological and QC techniques.
- Quality provides a good balance of administrative and laboratory work.
- If you enjoy working in a highly prescribed environment where few challenges will threaten to upend your day, a job as a QC analyst is ideal. Most of the work is fairly routine, but there is some variety.
- If you prefer the intellectual challenge of solving problems, managerial roles are a good fit. There is endless potential for problems, and lots of opportunities to think through technically challenging solutions.
- If you improve the efficiency of coworkers, your efforts can have a large positive impact on an operation. The automation of review processes, for example, saves colleagues time.
- In general, this is a 9-to-5 job. You can do your job and return home without taking your work with you. Supervisors may need to work extra hours to resolve problems when things go awry.
- There are abundant opportunities to learn new skills, and the career allows professional growth and development. A person with a bachelor's degree has more room for advancement in quality than in other vocational areas of biotechnology, such as in discovery research.
- You will learn new things every day. Some areas of quality, particularly in computer systems, can change quickly, requiring continual learning of new regulations and the use of new tools and systems.
- Quality provides the opportunity to understand the overall picture and many nuances of product development and manufacture. You will be exposed to a wide range of business processes, government regulations, new technologies, and more.
- For auditors, the job can have an international scope. Whether you work in the United States or elsewhere, the chances are good that you will be involved in an audit involving another country.

- Quality positions, except for auditors, tend to require less travel compared to many other careers.

### The Potentially Unpleasant Side of Quality

- Work can be highly repetitive and tedious. Many positions, particularly QC analysts, require that the same tasks be repeated over and over again.
- Auditors travel extensively, as much as five months out of the year. For those in senior roles, even more traveling may be needed for FDA meetings and other events.

*In quality assurance, you will either travel extensively or not at all.*

- A sudden increase in production can lead to long hours and increased stress, and it may be difficult to maintain high quality standards. For some poorly managed companies, the pressure is continuous and increases the likelihood that mistakes will be made, possibly with disastrous consequences.
- Upper management frequently does not comprehend or appreciate how much time it takes to set up and manage quality systems. Inadequate planning or unexpected system problems can threaten the ability to meet corporate deadlines.
- The quality assurance and compliance groups are sometimes viewed as a "necessary evil" or "the police" who won't let others do what they want to do (see "Greatest Challenges"). It is difficult to be effective in an organization when coworkers have a negative attitude about your group.

## THE GREATEST CHALLENGES ON THE JOB

#### *Being Proactive about Product Quality*

A quality department should be independent and empowered to make important decisions, even if halting manufacture or recalling a product is required. A good quality department will have the foresight to anticipate problems and will advise senior management of the necessary mitigation steps.

#### *Staying Current with the Constantly Changing Regulatory Environment*

The regulations do not change much, but the interpretations and ways to meet them do. As FDA regulators run inspections and review submissions, they identify new trends and change their priorities accordingly. These new considerations must be continually identified and adopted by your organization.

*In most cases, the FDA regulations do not spell out how to specifically do things.*

#### *Diplomacy*

Diplomacy is needed at all levels. To keep quality high, you must tactfully inform fellow coworkers if their work does not meet FDA requirements or if deficiencies have been found as a result of an audit.

## TO EXCEL IN QUALITY...

### Optimizing Quality and Improving the Bottom Line

> Quality is the quickest way to improve the business's bottom line—it is simply good business practice.

Those who excel are able to incorporate quality so that it improves the business's bottom line. They consider quality in the context of the organization's business goals and balance the need for excellence with an awareness of costs. The FDA regulations are purposely written vaguely so that they can be customized and adapted to the needs of the company. Quality professionals who do well are able to interpret regulations and follow them as needed without exceeding them unnecessarily.

### Resilience and Diplomacy

Problems will be encountered within the company, and there will be changes in the strategic direction of an organization. People who excel understand the importance of quality in the overall context of discovery and development, and they continually promote quality in the organization. They sense the urgent need to protect the public and insist on working to resolve problems until acceptance or cooperation is gained.

## Are You a Good Candidate for Quality?

Because this field encompasses many different types of roles, it can accommodate a wide variety of personality styles. For example, introverts may prefer QC analyst roles, whereas extroverts may prefer supervisory or QA positions.

### People who flourish in quality careers tend to have...

***Meticulous attention to detail.*** An eye for detail is critical when reviewing batch records and conducting audits. Supervisors need to be able to see the details of processes and procedures within the context of the larger business perspective.

***A systematic, methodical, and organized approach.*** It is important to be able to record comprehensive and organized notes and to have a systematic approach.

> You must be aware of the guidelines and operate within them.

***The ability to remain focused on the rules.*** You need to follow the rules and to implement and communicate them to the organization. The FDA regulations tell you *what* you must do, but not *how* you should do it. Flexibility and creativity can be assets when interpreting rules, particularly if you are in upper management.

***The desire to work in a prescribed work environment.*** For QC analysts (not supervisors), most of the work is routine and predictable.

***Exceptionally good problem-solving skills and the ability to handle sudden emergencies and changes.*** For supervisors, problem solving can be a constant necessity. Changes in procedures, odd samples, and broken instruments all challenge your ability to keep processes compliant.

*Good writing skills.* Descriptive writing skills are needed for translating written regulations clearly into SOPs and drafting a variety of documents. Good communication skills are needed for promotion to supervisory levels.

*Good presentation and verbal communication skills.* There is considerable interaction between coworkers. QA professionals, in particular, need to be able to clearly explain the rules and regulations and how to use new systems.

*Diplomacy and tact.* For managers and supervisors, the ability to skillfully negotiate is needed for a myriad of reasons, from trying to help people understand what the requirements are and why they are important, to defending why you are asking people to do what they think is unnecessary, to being tactful when auditing investigator sites (especially if the lead clinician has a big ego).

*Excellent interpersonal skills.* The senior-level quality professionals tend to be outgoing and have good people skills because they have to interact daily with so many people. Interacting effectively with a variety of personalities means dealing with people openly and honestly and being able to handle adversarial situations (see Chapter 2).

*A team-oriented attitude.* People in quality tend to be collegial and respectful of others. You have to be able to work well with other people.

*The ability to stick to your values.* You may have to defend your principles when you know that something is wrong. Consumer safety is of paramount importance in quality.

*Time management skills.* It is important to prioritize tasks to make the most efficient use of your time. It sometimes only takes one unexpected problem to throw off your schedule. This is particularly true for supervisors.

*Good listening skills.* To gain credibility and acceptance, it is important to be sympathetic and understanding and to allow people to explain their problems. If you demonstrate that you really understand, it tends to be the beginning of cooperation, acceptance, and the resolution of problems.

*Service or customer orientation.* In quality systems or QA, you are providing a service for coworkers in the form of training or forwarding knowledge of the rules and regulations and making sure that procedures and systems are compliant. You are there to help people figure out how to apply the regulations to their work processes.

*Information-gathering skills.* In QA, it is important to seek information from the other groups with whom you interact so that processes and procedures will run more smoothly and so that you can be prepared for auditing and overseeing compliance.

*Resilience and tenacity.* Some of the work can be tedious and routine. Tenacity and resilience are needed to continue to work efficiently and effectively.

*General knowledge.* For many quality positions, particularly in QC, a general knowledge base is better than a specialized one. Procedural demands vary greatly. Those with specific knowledge in a single area should consider working in large organizations that can afford specialists.

***Knowledge of the drug discovery and development process and regulations.*** It is imperative to understand the basic processes of drug discovery and development in order to put processes and regulations in context.

***The ability to be a supportive coach for the success of the organization.*** Your role in quality should be more as a coach than a police officer. Coworkers should be able to learn from their mistakes and improve in a supportive environment, rather than by fear of punishment.

***You should probably consider a career outside of quality if you are...***

- Unable to perform well in adversarial situations or if it is important for you to be liked by everyone.
- A prickly, brilliant scientist type.
- Disorganized, not detail-oriented, illogical, or unsystematic in your approach.
- Too creative or a nonconformist.
- Too rigid, unable to cooperate, or have a "my way or the highway" attitude.
- Willing to take shortcuts, regardless of the rules or ethical considerations.
- One who can't appreciate and accept the FDA regulations.
- Unable to maintain the courage of your own convictions.
- One who takes pleasure in identifying errors and administering justice.
- More interested in solving fundamental biological problems (consider discovery research instead).

*The prickly, brilliant scientist might be able to survive in R&D, but not in quality.*

## QUALITY CAREER POTENTIAL

Because quality encompasses numerous careers, there are many areas to explore. You can find a rewarding career as an eventual VP of Quality or Chief Compliance Officer (see Fig. 13-1), or you can move into operations and higher-level management. For those who are interested in career possibilities other than quality, experience in this field can be applied to regulatory affairs, manufacturing, project management, marketing, business development, information management (bio-IT), drug safety, medical affairs, clinical development, medical writing, and other areas.

As in discovery research, there tend to be two general career tracks in quality: managerial or "individual contributor." Managers can move up to the VP ranks, whereas individual contributors add significantly to their department, but do not seek supervisory responsibilities.

People usually do not return to discovery research after they have been in quality. This is because the speed of technical advances and the sheer volume of literature in discovery research are difficult to keep up with, especially if you are in another field.

*Those who enter quality rarely return to discovery research positions.*

**Figure 13-1.** Common quality career paths.

### Job Security and Future Trends

Careers in quality tend to be fairly secure. As long as your company has products that have been approved, and as long as the regulations keep changing, there will always be a need for quality departments. The current high demand is expected to continue for most quality areas. In particular, there is a strong demand for specialized quality functions like computer validation and, if you have that rare combination of exceptionally good supervisory skills and appropriate technical training, for middle management. The news is not as good for lower-level personnel: New FDA initiatives that support the increased implementation of automated testing may result in fewer jobs for external quality inspectors and lower-level laboratory technicians.

Outsourcing overseas will not be an issue for many positions in quality. These responsibilities cannot be outsourced, because the company remains ultimately responsible for the product.

## LANDING A JOB IN QUALITY

### Experience and Educational Requirements

A broad technical base and understanding of the regulations are required for this work. Because the amount of technical knowledge needed is immense, it is generally easier to teach a scientist about quality principles than to teach science to someone with a quality background. Quality is not taught in the educational system, so experience is highly valued. Career advancement for those with a general science bachelor's degree is more likely in quality compared to other vocational areas in biopharma. An advanced degree is an advantage, however, and Ph.D.s in technical fields are common at the supervisory levels and above.

The most common path to quality careers is from discovery, preclinical research, or chemical development, but it depends on the subspecialty. For example, people in computer validation need computer knowledge, and people in regulatory compliance tend to have clinical experience.

A general biology background is good for QA, whereas QC departments are composed of a mixture of individuals with various degrees, mostly those with a bachelor's, master's, or Ph.D. in analytical or clinical chemistry, microbiology, or medical technology. Generalists, rather than QC specialists, are needed in small biotechnology companies, whereas large pharmaceutical companies can afford to hire specialists. An advanced degree by itself is insufficient preparation for the broad range of skills and procedures called for to run a QC lab in a small company.

A wide range of experienced people can enter compliance positions, although it is common to have a clinical background. For laboratory compliance, preclinical or discovery research and analytical laboratory experience is more appropriate. In manufacturing, backgrounds in the pharmaceutical industry, chemistry, and engineering are common. For the quality systems roles, degrees in computer science or engineering are common.

## Paths to Quality

- For a career in QC, consider gaining as much experience with a wide variety of laboratory procedures as possible. Generalized technical experience is good preparation.
- Learn the GXPs and regulations. This information is available on the FDA Web site at www.fda.gov.
- It may be easier to secure your first job with a vendor that provides a service to biopharma companies. There are a plethora of service providers, and working with them will provide training, a broader exposure to multiple therapeutic areas and drug classes, and an introduction to the product development process. Once you have become experienced, gaining employment in a biopharma company may be easier.
- For research scientists, a role in QC might be easier at first than one in QA. You can immediately apply your bench skills in QC and later switch to QA.

*It is easier to go from discovery research to QC and then to QA.*

- If you are in academia, making the transition into quality might be easier if you first work in discovery or preclinical research. This will allow you to familiarize yourself with how the drug development process works and to gain insight on how discovery research and quality fit into the grand scheme of things.
- Consider joining management consulting firms as a validation specialist.
- Obtain a position at the FDA as an auditor. For more information about working at the FDA, review Chapter 12.

## RECOMMENDED TRAINING, PROFESSIONAL SOCIETIES, AND RESOURCES

### Societies and Resources

Society of Quality Assurance (www.sqa.org)

American Society for Quality (www.asq.org)

Regulatory Affairs Professionals Society (www.raps.org)

Drug Information Association (www.diahome.org)

British Association of Research Quality Assurance (www.barqa.com)

Association of Clinical Research Professionals (www.acrpnet.org)

The Institute of Validation Technology (www.ivthome.com)

International Quality & Productivity Center (www.Iqpc.com)

Parenteral Drug Association (www.pda.org)

International Society of Pharmaceutical Engineering (www.ispe.org)

American Association of Pharmaceutical Sciences (www.aaps.org)

### Courses and Certificate Programs

Center for Professional Advancement (www.cfpa.com) provides training in regulations and computer validation and offers quality system classes.

Introductory classes to data management, bioinformatics, cheminformatics, and computer validation.

University extensions and select universities offer classes or masters programs in quality, compliance, and regulations.

American Society of Quality offers certifications and classes in auditing and statistics.

Regulatory Affairs Professionals Society (www.raps.org) provides classes in regulatory compliance and certificates.

### Books and Magazines

*BioProcess International* (www.bioprocessintl.com): a free magazine.

*BioPharm International* (www.biopharminternational.com): a free magazine.

Lieberman H.A., Rieger M.M., and Banker G.S. *Pharmaceutical dosage forms: Disperse systems* (vol. 3). Marcel Dekker, New York.

Prince R., ed. 2004. *Pharmaceutical quality.* DHI/PDA, River Grove, Illinois.

*This book can be purchased through the Web site of the Parenteral Drug Association (www.pda.org).*

Vesper J. 1997. *Quality and GMP auditing: Clear and simple.* CRC Press, Boca Raton, Florida.

# 14

# Operations

## Ensuring that Processes Run Smoothly and Efficiently

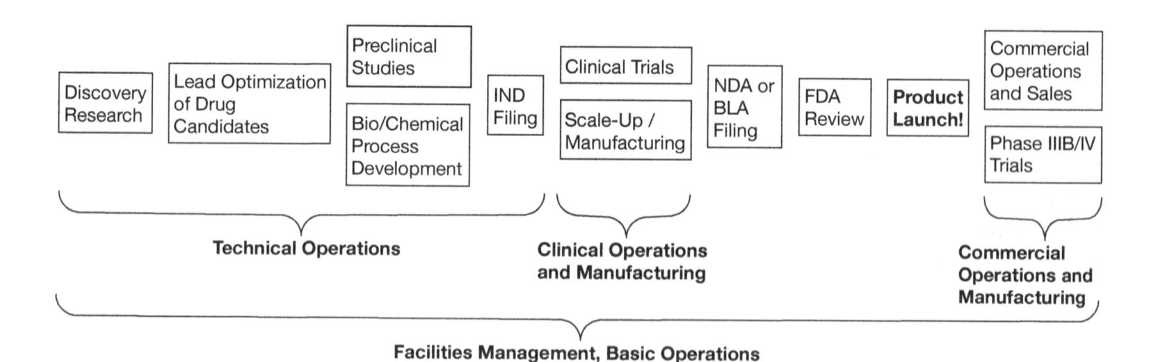


*People in operations gain satisfaction from producing and marketing good-quality products of which the whole company can be proud.*

FOR A CAREER THAT COMBINES SCIENCE, business, finance, and organizational management while offering broad exposure to the nuts and bolts of biotechnology and drug development, try operations. Those who enjoy making their workplace more efficient and productive can thrive in this department, particularly when they can see the fruits of their labor in the form of drugs and products in the clinic. Self-motivated, go-getter types who enjoy problem solving, delivering on expectations, and communicating with coworkers in all parts of the corporation are well suited for a career in operations. Opportunities for career development are excellent, and advancement tends to be rapid. Constant deadline pressure can make the work quite stressful, but at the end of each day, you can be proud of your contribution to the health of the company and of the consumer.

*A career in operations allows you to "wear many hats."*

## THE IMPORTANCE OF OPERATIONS IN BIOTECHNOLOGY AND DRUG DEVELOPMENT

Operations plays an influential and strategic role in business and is intimately tied to sales, marketing, technical support, and product development. At its core, operations is about manufacturing and distributing products to customers at the highest level of quality for the lowest cost. To accomplish this and its many other tasks, an operations group typically consists of a manufacturing department with all of its attendant functions. These include supply chain (raw materials), logistics, facilities management, shipping/receiving, purchasing, procurement, quality control, manufacturing engineering, and even more, depending on the type of company.

Although this chapter focuses on drug development, careers in operations are not limited to companies that develop drugs. There are many equivalent careers involving the development and production of instruments, reagents, diagnostics, medical devices, platform technologies, services, environmental waste management services, and chemical and agricultural applications (see Chapter 6).

*Because the operations environment requires intensive teamwork and involves a matrix with other departments, it provides an excellent opportunity for managers and employees to exercise their talents and enjoy their shared workplace.*

## CAREER TRACKS IN OPERATIONS

Operations is responsible for the day-to-day running of the business and includes a hodgepodge of activities. There are three major areas in operations:

### *Technical Operations*

Technical operations can be part of discovery or product research, or it can be a subcategory of operations. For early-stage companies without commercial products, this area oversees the day-to-day management of the company.

### *Clinical Operations*

Companies often have a clinical operations group to manage and run clinical trials (see Chapter 10). Clinical manufacturing and production can either be a research and development function or an operations function, depending on the structure of the organization.

### *Commercial Operations*

Commercial operations is more of a business function. A commercial operations team oversees the production of marketed products and manages the budget, sales, company infrastructure, international operations, etc. For more information, see Chapters 18, 19, and 20.

## Divisions in Operations

### *Manufacturing*

The manufacturing division is concerned with quality, production, packaging, and the distribution of products. After experimentation and refinement have been completed, those in manufacturing institute, oversee, and further optimize the efficiency of the production process.

*There are two general types of manufacturing:*

*Clinical manufacturing.* In large companies, there may be a separate group, usually called process development, that prepares drug candidates or medical devices on a pilot scale for testing in clinical trials. Clinical production is generally less strictly controlled than commercial manufacturing. The work tends to be more varied and scientifically challenging compared to commercial manufacturing because it consists of refining processes that will be scaled up later for commercial manufacturing. For more information, see Chapter 15.

*Commercial manufacturing.* Commercial manufacturing takes place after products have been approved or refined and are ready for large-scale production. Therapeutic products are made under strict "current Good Manufacturing Practices" guidelines, which were developed by the U.S. Food and Drug Administration (FDA) in the 1980s to ensure that the U.S. drug supply is reliable and safe (see Chapter 13). The types of tasks that managers in commercial manufacturing deal with are cycle time reduction, process improvements, and cost reduction. In small companies, commercial manufacturing is usually outsourced to vendors. This department once typically employed large numbers of lower-level workers, many with minimal educational backgrounds. Now, it increasingly hires more educated and technically savvy workers, mostly college-educated science graduates.

### *Engineering and Programming*

Engineering expertise is required for all sorts of activities that are needed for running companies and manufacturing plants. Engineers manage a large variety of tasks, including fluid dynamics, steam sterilization, remote control or large-scale continuous types of operations, plant design, assessments, overseeing of equipment, etc. Software programmers are needed for controlling and monitoring automated systems and processes.

### *Validation and Technical Services*

Those with engineering and scientific backgrounds write protocols and design studies to oversee the proper operation of equipment, facilities, and processes. Validation requirements are described in Chapters 13 and 16.

### Procurement

Procurement handles the purchasing of the raw materials used to make products. If a company is big enough, there may be procurement or purchasing managers or a supply chain group.

### Supply Chain Management

Supply chain management is used to economically manage the large number of raw materials used in manufacturing. Supply chain management defines the specifications and processes used to monitor and control the movement of materials from the many vendors so that items are received in the correct sequence, time, and place.

### Logistics

Those in logistics plan and schedule manufacturing operations (including production) to ensure that the steps in manufacturing are monitored and running smoothly.

### Shipping and Receiving

This group oversees the shipping of products to customers and the management of contracts with deliverers, such as United Parcel Service and Federal Express. They monitor the receipt of raw materials and other general goods and services, which can involve customs clearances, regulatory interactions, and more.

### Quality Control

Quality control personnel test products or batches to ensure that they meet regulatory and customer requirements. Some companies have their own quality departments, whereas in others it is part of operations or regulatory affairs. To learn more, see Chapter 13.

### Quality Assurance

Quality assurance workers document the quality test data and provide regulatory guidance.

### Facilities and IT Infrastructure

Many subdivisions are found in facilities and information technology (IT) infrastructure, including security, facility leases, IT infrastructure, energy-saving initiatives, pilot plant operations, air handling, systems controls, refrigeration, equipment maintenance, and other elements required to keep production running smoothly. Other IT-related careers are discussed in Chapter 16.

### Process Development

Potential therapeutic products are initially synthesized in small quantities for clinical trials. After successful clinical trials, process development engineers devise methods for large-scale manufacturing to help reduce costs and improve efficiency. The early steps of process development for therapeutics are described in more detail in Chapter 15.

#### *Project Management*

Companies assign manufacturing or operations project managers to a variety of cross-functional programs. Some monitor large improvements, such as building expansions, in addition to general product development management. The key goal of project management is to make sure that the right people know the right information at the right time. For more information, see Chapter 9.

#### *Finance*

Large companies typically assemble a separate finance group in operations that is responsible for budgetary controls. For example, if a company is interested in making significant changes to a plant or adopting new processes, those in finance assess the costs and estimate the potential return on investment.

#### *Human Resource Management*

As many as 200–5000 people can be employed at a plant that operates 24 hours a day, 7 days a week. Due to the large number of employees in manufacturing, many companies establish a separate, manufacturing-specific human resources department in addition to the company's general human resources department.

## OPERATIONS ROLES AND RESPONSIBILITIES

*Because operations represents a hodgepodge of functions, the following is a general summary of some basic roles and responsibilities. The specifics depend on the position and type of company.*

#### *Manufacturing, Product Monitoring, Testing, and Release*

Products are produced, packaged, and distributed based on demand. Multiple steps are involved in testing products before they are released for sale.

#### *Hiring and Training New Employees and Managing People*

A tremendous amount of time is spent hiring and training employees. In a large department, there can be additional human resources-related functions to perform, such as disciplinary actions, performance reviews, and more.

#### *Capacity Utilization*

The biggest challenge in operations is preparing enough products for the future needs of the research and development departments or the marketplace. Another important task is to make sure that the manufacturing plant is running at maximum capacity. Most plants operate full-time, day and night.

### *Troubleshooting*

Managers do an extensive amount of troubleshooting, particularly in process development.

### *Documentation*

Some roles, most prominently in quality assurance, require extensive writing, evaluation, and review of documents.

### *Vendor and Alliance Management*

A key aspect of operations is the management of vendors, contract manufacturing organizations, and partners. This role may include obtaining and negotiating agreements pertaining to confidentiality, research collaborations, and consulting.

### *Ensuring Total Quality Systems*

"Total quality systems management" refers to the maintenance of good business practices across the entire organization. Managers may be responsible for overseeing or contributing to routine technical evaluations of all company departments to ensure overall quality.

### *Facilities Management*

One day of work lost at a facility due to technical difficulties might result in the loss of millions of dollars. Facilities management includes maintenance and improvement of equipment, oversight of renovations, introduction of new instrumentation, and more.

### *Budget Management*

Those in operations might forecast and manage budgets for scientific projects, renovations, and other expenses to ensure that resources are properly spent and processes are optimally supported.

### *Strategy and Portfolio Management*

Those in operations may implement major decisions and higher-level strategic projects as designated by the executive team. Examples include portfolio management, cost-savings initiatives, resource allocation, and much more. This work entails providing the requisite documentation to support specific efforts and helping to manage and implement processes.

### *Safety*

Operations personnel in environmental health and safety establish liaisons with various governmental and advocacy groups to make sure that rules are followed, documents are signed, and yearly safety inspections are passed. Operations safety managers are responsible for internal audits within their company. They prepare and manage safety manuals and ensure that employees wear appropriate personal protective equipment, such as proper shoes. The environmental health and safety group cooperates with outside organizations such as the FDA, the Nuclear Regulatory Commission (NRC), the United States Department of Agriculture (USDA), the Animal Care and Use Commission (ACUC), the Drug Enforcement Administration (DEA), various environmental groups, and more.

### *Communications*

Announcements from upper management publicly recognize outstanding employees, inform employees of company progress, project the value of the company, and, when necessary, release bad news. Companies sometimes make these announcements verbally. They also may distribute newsletters and maintain an internal Web site or intranet.

## A TYPICAL DAY IN OPERATIONS

*Depending on your role and rank, you can expect some of the following managerial or project management activities in a typical day's work:*

*There are so many types of responsibilities in operations that no two days are the same.*

- Attending meetings, especially if you are a manager. Meetings might entail assessing data and production processes, prioritizing the next set of tasks, discussing the latest customer or company problem, or discussing the manufacture of existing or new products.
- Solving urgent manufacturing or operational problems.
- Reviewing analyses to determine the best business practices for making new products.
- Presenting data to various types of audiences.
- Implementing or helping sites adopt new systems or processes.
- Ensuring the safety of the workers and conducting risk management for the plant.
- Completing work orders and documentation and advancing products to the next step in development or manufacture.
- Determining the number of employees and components needed to reach manufacturing and shipping goals.
- Tracking the number of orders placed and the number of products shipped on time, and comparing with numbers from previous months.
- Negotiating contracts or revising terms with vendors.
- Developing financial forecasts.
- Recruiting and interviewing prospective employees.

- Mentoring employees who report directly to you and delegating responsibility to them.
- Training employees, writing performance reviews, and taking disciplinary action.
- Reviewing quality assurance and manufacturing documents.

## SALARY AND COMPENSATION

In general, salaries for employees in operations tend to start lower but rise higher than for those in discovery research with an equivalent amount of experience. This is because there are more types of managerial positions and advancement is quicker in operations. Salaries are generally lower in manufacturing.

***How is success measured?***

In general, the goals are to consistently manufacture and deliver quality products on schedule and within budgetary allowances. If these processes run smoothly, continuously, and robustly, the company's bottom line is improved. Your personal success depends in part on having positive interactions with your team and meeting productivity demands.

Other signs of success include:

- Keeping the cost of goods flat.
- Maintaining a high safety standard.
- Managing capital projects and product launches.
- Maintaining high productivity index measurements.
- Having few or no customer complaints.
- Having no product failures on the enterprise resource planning system (ERP, a business management system).
- Having a minimum number of back orders and products in inventory.
- Having the right number of employees, and training them so they perform well.
- Passing safety tests, and not having Occupational Safety and Health Administration (OSHA)-recordable incidents.
- Minimizing waste problem indicators like scrap material.

## PROS AND CONS OF THE JOB

### Positive Aspects of a Career in Operations

- Jobs in operations are fast paced. Every day is different, there are many types of responsibilities, and your work requires constant multitasking. This may not be true, however, for basic manufacturing positions.

*People enjoy the fast-paced interactive environment found in operations.*

- Your personal work effort can make an important and noticeable difference in the organization, and it is immensely rewarding to be responsible for successfully making products. It is satisfying to see the tangible, finished products and rewarding to see increased efficiency reflected in dollar amounts.

- There is a sense of purpose created by knowing that you help make quality products that will improve or save patients' lives (perhaps even those of your friends and family).

*You can take pride in the fact that you are producing health care products as opposed to widgets.*

- Your job can be intellectually and technically challenging and can provide an opportunity to apply your scientific, business, or engineering acumen and analytical skills. These skills are constantly put to the test, especially when developing new products or technologies.

- Product development is an essential function in companies and can be a highly visible responsibility. It is enlightening to discuss a potential technology with customers and collaborators and to brainstorm about the best use of it in the marketplace.

- Operations requires both science and business prowess. You can learn new skills and become a generalist who can then move into other areas.

- Operations provides an opportunity for you to spend much time training and mentoring employees.

- There is more room for career growth in operations than in discovery research, and you can climb higher up the management ladder, even without an advanced degree. You won't be pigeonholed in an area that is limited by your specific technical knowledge.

- Through your extensive work forming external alliances and partnerships and handling contract vendors, you will gain valuable negotiation, business development, and alliance management experience.

- These careers offer opportunities to meet new people and provide exposure to different business areas. People in operations often interact with individuals at all levels in the legal, business development, clinical development, facilities management, IT infrastructure, marketing, and sales departments.

- Many nonmanagerial positions require little travel and have 8-hour days. Some teams have flexible schedules.

- Some organizations allow those in operations to publish papers. There may be an opportunity to issue patents, particularly those related to the process chemistry sciences.

- Job security in operations is slightly higher than in discovery research.

### The Potentially Unpleasant Side of Operations

- Repetition of tasks can make the work tedious, although this mainly applies to positions that do not require scientific backgrounds. The level of monotony depends on the company. If the company is developing several diverse products, work usually remains interesting.
- While the more routine jobs are from 9 to 5, work hours for managerial positions can be up to 12 hours per day. If you are managing a commercial facility, you will likely be on call 24/7 and may be awakened at any hour of the night to answer questions. It can be difficult to follow a fixed schedule or to plan vacations.

*For some positions, be prepared to be attached to a cell phone, Blackberry, or pager at all times.*

- Jobs in operations can be high pressure. There is little room for error, and a lot of money at risk. A failed production run can significantly affect the budget of the entire company.
- The executive management team often has unrealistic expectations about how long it takes to develop products and processes. At the same time, operational managers are expected to be omniscient.

*It often seems like everything is needed yesterday.*

- There is a limit to how much the company can improve processes, and any new changes can be cumbersome to introduce. To improve efficiency, you first need to validate possible improvements, and there is the tendency not to bother.
- Depending on the position, there can be a lot of paperwork. Every activity is documented, so that drug products, supply chain management, equipment, and inventory can be monitored and tracked for quality.
- Sometimes when products are poorly designed or ill conceived, the blame is incorrectly placed on manufacturing.
- You may soon find yourself separated from basic research, as science and technology move very quickly. Even so, there is much new science to learn in operations.

## THE GREATEST CHALLENGES ON THE JOB

#### *Planning Product Supplies*

The challenge to deliver on time never ends, because the demand for products constantly changes. Unexpected spikes in demand can throw off production schedules. Most companies engage in "just-in-time manufacturing," but when something goes wrong, that approach can escalate and confound problems. It seems that time, resources, and employees are often in short supply.

*Overstock or back orders? It is important to not have too many or too few products in inventory.*

*People Management*

Much time is spent on people issues, such as awarding promotions, providing solace for unhappy employees, training, staffing, and more. Because production requirements vary greatly, overseeing the hiring, firing, and training of large numbers of employees involves constant, delicate, and time-consuming attention.

*"Herding Cats"*

When working in a matrixed environment, it can be time-consuming and difficult to obtain agreement about goals from the various other departments, some of whom may have different and sometimes competing agendas. This is termed "herding cats" and is described in more detail in Chapter 17.

*Changing Demands from Research*

Requests for various amounts of drug substances for animal studies or clinical trials often change, making it challenging to maintain the right supply.

## TO EXCEL IN OPERATIONS...

*Decisiveness and the Ability to Finish on Time with Minimal Costs*

Technical improvements and cost reduction measures translate into real savings. Those who excel have extensive experience and the knowledge required to find and design innovative solutions. They also have an overall, in-depth understanding of the processes and products.

*The Ability to Think Strategically and Align Different Departments*

Those who excel have a thorough understanding of the organization and of the role played by each of its divisions. Because these groups are tightly intertwined, decisions must be made in such a way that the effects and implications for all of the various interests are considered, favoritism is avoided, and the organization's members are aligned with the same goals and priorities.

*People who are highly motivated and are astute, logical thinkers will do well in operations.*

*The Ability to Communicate and Motivate Effectively*

An essential skill in operations is the ability to communicate clearly and effectively and to motivate coworkers and employees.

### Are You a Good Candidate for Operations?

*People who flourish in operations careers tend to have...*

*Superb organizational skills.* This area requires being hyper-vigilant on the deliverables and being accountable and responsive. Excellent organizational and systematization skills are required.

*An outstanding ability to multitask.* You need the ability to handle short-term crises and long-term planning issues simultaneously and to be able to prioritize. There may be times when your multitasking skills will be put to extreme tests.

*You must be able to rapidly solve immediate problems and simultaneously plan for the future.*

*A driven, self-motivated, go-getter personality.* You must be willing to take the initiative and not be afraid to tackle difficult problems. You should be ready to pitch in with the task at hand even if it's not your responsibility or you are higher on the managerial ladder.

*An ability to work under pressure and with limited timelines.* The environment is constantly changing: Spikes in demand and back orders are common, as are impatient calls from sales reps urgently requesting the manufacture of products for their customers. Production problems and other vexing issues are common, so it is important to remain calm and continue focusing on the overall goals and to quickly resolve pressing issues.

*The sociability to deal with a wide variety of people.* Operations involves extensive teamwork and the ability to work well with people from various cultural and educational backgrounds. It is advantageous to be outgoing, congenial, considerate of others, and able to see other people's points of view. It is important to treat people with respect, regardless of differences in background or position in the company hierarchy.

*If you offend the janitors, they may not take away your trash.*

*Excellent attention to detail.* Some positions require much documentation and the ability to pay attention to detail.

*An ability to negotiate.* Strong negotiation skills are needed externally with vendors and contractors and internally within research and development to get the product out within the given time frame. Products coming out of research are usually not completely developed to the point that they can be easily manufactured. You may need to stand your ground and be ready for challenges when negotiating.

*Over-promising can be a perilous flaw in operations. If you agree to ship something, it had better be delivered.*

*A flexible attitude.* There are many situations in which a flexible attitude will mitigate problems, because a given solution may work well for one problem but not for another. It is important to be able to listen carefully to other people's opinions and suggestions, to consider investigating what other companies have done, and to think about alternate approaches when you are looking for answers.

*Excellent, proactive problem-solving skills.* Strong technical capabilities combined with creative troubleshooting skills are assets for prioritizing and solving problems.

*Exceptionally good communication and presentation skills.* Due to the extensive interactions with people, the ability to clearly communicate about deliverables is important. Those in operations should be able to give lucid speeches to large audiences.

*An ability to delegate.* To handle many responsibilities, you may need to judiciously delegate tasks to others and offer guidance and confidence.

***The ability to work in a highly regulated environment.*** Although there is room for some creativity, especially when problem solving or embarking on new projects, operations is best suited for people who are able to adhere to agreed-upon methods and processes. Constant attempts at improvement can waste time and can even be counterproductive.

***You should probably not consider a career in operations if you are...***

- Shy and prefer to work alone.
- Not willing to pay attention to details.
- Not willing to adhere to the accepted rules and regulations.
- Someone who is slow or unable to work under constant pressure and fixed timelines.
- Expecting to be an inventor or product champion.
- One who prefers not to work with people who are below you on the corporate ladder.

## CAREER POTENTIAL IN OPERATIONS

A career in operations offers a path up the managerial ladder to plant manager, vice president of operations, and, if you have more business experience, general manager or chief operating officer and CEO (see Fig. 14-1). The chief commercial officers and chief operating officers are generally considered next in line for CEO stewardships.

One career benefit of operations is that you can move horizontally into many disciplines. The extensive interaction with so many different disciplines makes transfers into other areas relatively easy. In fact, it is probably a good strategic move to work in several departments before becoming a senior-level executive. For those who seek a career in oper-

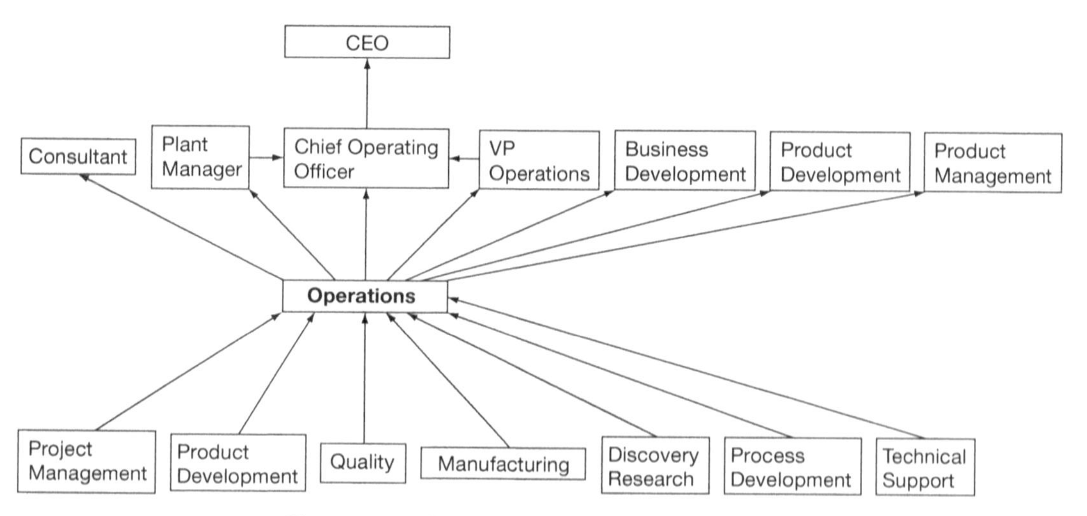


**Figure 14-1.** Common career paths in operations.

ations as a launching pad into other areas, the common destination is product development. Other areas include customer support and technical service, quality, marketing, business development, corporate communications, clinical operations, and consulting.

For those in entry-level manufacturing, there is a plethora of areas to explore, including inventory control, purchasing, distribution, quality, planning, and bioproduction. Many people obtain M.B.A.s and then move into product and project management or marketing roles.

### Job Security and Future Trends

There is a shortage of people with operations experience in biotechnology, because it takes a long time to gain experience in the various disciplines. Many new plants are being built, and there are not enough people to operate and run them. Within operations, senior-level executives possessing outstanding technical and managerial attributes and workers with a B.S. degree and more than five years' manufacturing experience remain in constant demand.

There can be greater job security in operations than in discovery research for a variety of reasons. For companies with marketed products, operating manufacturing facilities are essential for revenue, whereas after a product moves into clinical development, many small biopharma companies can't afford to continue discovery research. Skills gained in operations are readily transferable from one biopharma company to another; you are not limited to a specialized product area as you are in discovery research.

Manufacturing positions are relatively secure as long as the department is running well, but these departments are continually subject to cost-cutting measures. One current trend, for example, is to automate processes using robotics, which will affect some lower-level manufacturing positions. Another trend is to reduce labor costs by outsourcing manufacturing overseas. Due to the highly sophisticated techniques required, biologics will be more difficult to outsource than pharmaceutical products, so expect a delay in biotherapeutics outsourcing.

As the biopharma industry outsources more jobs, those with operations and manufacturing experience will be able to provide their expertise in contract research organizations or small start-ups. Knowledge of international regulations will likely become a new industry standard.

## LANDING A JOB IN OPERATIONS

### Experience and Educational Requirements

Most people do not initially set out to go into operations—it usually happens serendipitously. People can enter the vocation from just about any area, although they commonly do so from manufacturing, quality, or process development. Other areas include buying and planning, project management, and discov-

*Operations is a career into which you can migrate after having had industrial experience in other departments.*

ery research. They can also work their way up from the various subdivisions in manufacturing and operations (see Fig. 14-1).

Because operations is becoming more science- and business-oriented, a technical degree is now required more often, and an M.B.A. is desirable. A B.S. degree with a science background is considered the minimum, and advanced degrees are required for managerial levels in some companies. Most employees have a technical degree in either engineering or science. Depending on the type of company, there are also more highly specialized positions for those with quality or software engineering backgrounds. For lower-level manufacturing positions, a high school degree is a minimum requirement, depending on the organization.

Advanced degrees are generally not required for senior-level operations and manufacturing positions, but they are beneficial. Directors typically have B.S., M.S., Ph.D., or M.B.A. degrees. Those with advanced degrees will need some management experience before being promoted. Those with Ph.D. degrees tend to migrate primarily into process development, project management, process improvement, and program management.

## Paths to Operations

- To obtain initial operations positions, there are two main avenues to follow: manufacturing or quality. To enter operations at a higher level, employees tend to come from discovery research, process development, project management, product development, and technical support departments.
- Consider a position in project management or join a project team. You will learn more about the integral components of product development and gain insight into how companies operate. These skills will make a transition into operations easier.
- Show an interest in the business side of biotechnology. Learn how to manage a budget and oversee purchasing. Learn about product reproducibility and quality. Take finance classes or consider obtaining an M.B.A. degree.
- Learn about the logistics of how products are made and sold and show how you can apply your science background to make processes more robust and efficient.
- Develop your people management skills. Managing employees is a large and important component of operations. If you are in academia, supervise employees and manage labs to obtain initial exposure.
- Take classes in negotiating. This is a highly useful skill in manufacturing and operations.
- Join companies that have distinct operations groups.
- Consider working at a contract research organization (CRO), contract manufacturing organization (CMO), or other type of company that provides a service to biopharma companies. This will provide initial exposure to product development and a broad exposure to many types of products and therapeutic areas.
- Some companies offer internal rotations in operations and manufacturing. Contact human resources for an internship.

## RECOMMENDED TRAINING, PROFESSIONAL SOCIETIES, AND RESOURCES

### *Courses and Certificate Programs*

Classes in business management and strategic planning

Classes on six sigma practices, a system of practices to improve company processes

Stat-Ease (statistics made easy, www.statease.com) is a company that provides classes in Design of Experiments (DOE)

### *Societies and Resources*

Parenteral Drug Association (www.pda.org)

International Society for Pharmaceutical Engineering (www.ispe.org)

American Association of Pharmaceutical Scientists (AAPS, www.aaps.org)

Drug Information Association (www.diahome.org)

American Chemical Society (www.acs.org)

The Association for Operations Management (APICS, www.apics.org)

### *Magazines*

*Nature Biotechnology* (www.nature.com/nbt/index.html)

*Genetic Engineering & Biotechnology News* (www.genengnews.com)

*Science Magazine* (www.sciencemag.org)

*Journal of Pharmaceutical Sciences* (www.interscience.wiley.com)

*Innovations in Pharmaceutical Technology* (www.iptonline.com)

# 15

# Bio/Pharmaceutical Product Development

## The Chemistry Has to Be Good

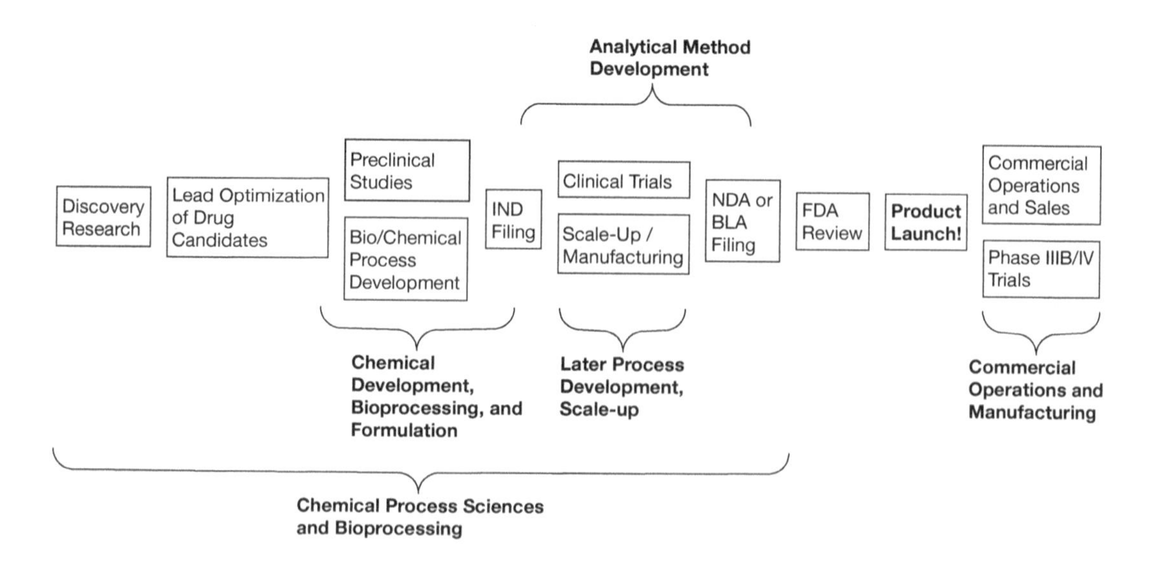


THERE IS A SEPARATE AND PARALLEL TRACK TO DISCOVERY research called bio/pharmaceutical product development, which focuses on the production of a chemical or biological entity. This area is where drug products, biological or pharmaceutical, are created, formulated, and manufactured.

*Product development is for people who want to work on products closer to the final steps of becoming marketed therapeutics.*

Product development tends to attract chemists, biochemists, microbiologists, and molecular biologists who wish to sharpen their practical and theoretical laboratory skills or move up the management track. One of the more attractive aspects of this career is that it allows you to create things and follow them through to the tangible end product. Process scientists work on drug candidates that have already survived a highly selective filtering process in discovery research, so there is a greater likelihood of seeing the results of one's efforts ultimately pay off in the gaining of a product approval.

As a career, product development is scientifically fascinating, because the industry is making vast advancements and the speed of innovation is rapidly increasing. Career opportunities are numerous and job security is excellent.

## THE IMPORTANCE OF BIO/PHARMACEUTICAL PRODUCT DEVELOPMENT IN BIOTECHNOLOGY AND DRUG DEVELOPMENT

*Product development is the bridge that connects discovery research to manufacturing.*

Product development is an essential step in making manufactured goods. The small chemical entities that became the blockbuster drugs of the past and helped create the large pharmaceutical companies that exist today would not have been created without chemical synthesis routes and formulation. The biotechnology industry originated in part with bioprocessing. The ability to produce and develop insulin, the first approved genetically engineered therapeutic, helped launch the biotechnology giant Genentech.

The overall role of product development is to advance the drug candidates that originated in discovery research through clinical trials and, ultimately, to commercialization. In this department, scientists develop formulations and produce clinical supplies to support investigational trials. They also scale up the chemistry or bioprocessing to allow manufacturing for commercial production. Besides making the active ingredient, product development ensures that the drug is safe both for consumers and for laboratory personnel and that the processes are reproducible on a large scale before commercial manufacturing begins.

## CAREER TRACKS IN BIO/PHARMACEUTICAL PRODUCT DEVELOPMENT

There are several distinct career areas in bio/pharmaceutical product development. The sciences of chemical and biological development are vastly different, even if both ultimately result in the creation of therapeutic products. Keep in mind that nontherapeutic products (i.e., reagents, building blocks, catalysts, and industrial enzymes) also require product development.

### *The Core Disciplines in Pharmaceutical Product Development: The Pharmaceutical Sciences*

#### *Process Chemistry, Chemical Development, or Synthetic Chemistry*

*If you react well to chemical reactions, process development is the way to go.*

After a drug candidate has been identified and tested in discovery research, chemical synthesis routes are developed to safely, efficiently, and cost-effectively produce it for preclinical and clinical studies and, ultimately, for commercial-scale production. Commercial production is very different from making compounds on a laboratory scale, so the raw ingredients and reactions must be modified and optimized accordingly.

#### *Formulation*

Formulation scientists develop the most suitable delivery method for the drug candidate's active ingredient. Delivery can be intravenous, oral by pill or liquid, in an ointment, or by inhalation. There are several core disciplines associated with formulation development, such as crystal, chemical, and pharmaceutical engineering.

#### *Analytical and Bioanalytical Chemistry*

Analytical chemists develop and implement methods to ensure that the purity, strength, quality, and stability of the active ingredient satisfies preclinical and clinical research standards before its final release to the market. The bioanalytical group detects ultra-trace amounts of drug substance in the cells.

*The analytical chemists are watchdogs who ensure that the drug is consistently pure and of high quality.*

### *The Major Disciplines for Biological Product Development: Bioprocessing, Process Sciences*

Bioprocessing refers to the development of large-molecule products such as recombinant proteins, DNA, RNA, RNAi, or antibody therapeutics. There are also nontherapeutic biological products, such as reagents and industrial enzymes.

#### *Cell Culture*

Recombinant proteins are made from different types of biological systems, such as microbial hosts, yeasts, and mammalian cell cultures. The gene of interest is inserted into a plasmid which is then used to transfect the appropriate cell line. After an elaborate screening process, the most proficient clones isolated from a transfection pool are selected for culturing. Monoclonal antibodies are often made using this type of "recombinant production system," as well as industrial enzymes and important drugs like insulin. Recombinant technology is also used to produce gene therapy agents in the form of plasmid DNA or engineered viruses. Other types of DNA- and RNA-based drugs are produced through chemical synthesis.

*Small-molecule products are created from chemical building blocks; the challenge is to devise the optimal route of synthesis. Biological products are derived from bioengineered cell lines; the challenge is to purify them.*

#### *Fermentation*

Fermentation development scientists determine the conditions in which to grow cell cultures so that they produce an optimal cell density and high expression of the recombinant biologic product.

### Purification

A host of contaminants are produced in fermented and cell-cultured products. By applying sophisticated filtration and chromatography technologies, purification scientists develop procedures to separate the desired product from the contaminants.

### Bioprocessing or Manufacturing Engineering: Biologics Scale-up

Specialists are needed to handle the numerous challenges encountered when scaling-up the production of biologics. For example, the additional height in cell cultures cultivated in taller and broader tanks creates more pressure on the media and the cells, the dissolved gases in the media are different, and more heat is produced. These factors can significantly affect how well cell lines grow. Additionally, some proteins produced in fermentation are insoluble and form aggregates, which are difficult to purify.

## ROLES AND RESPONSIBILITIES IN BIO/PHARMACEUTICAL PRODUCT DEVELOPMENT

*This is a partial list of the various roles and responsibilities in this field. They will vary depending on the particular division and rank.*

## Technical Roles

### Running Experiments

Product development scientists begin working on new synthetic routes early, typically in preclinical research. Their main responsibilities include running reactions and inventing or modifying synthesis routes. They also work to optimize reaction yields, minimize cost, and develop the final reaction steps into a manufacturing environment.

*Part of the fun of chemical development is coming up with clever ways to make compounds.*

### Reaction and Scale-up Modeling

Before they scale-up reactions from the flask to the liter stage, for example, scientists model chemical reactions with computer simulations to predict how scale-up will affect the kinetics of the reactions.

*Scale-up is taking a medicinal chemistry route and increasing it a thousandfold.*

### Optimizing Processes

Process chemistry strategies change as new chemical entities proceed through development. At first, chemists make small quantities of product for cell culture studies.

When the drug progresses into toxicological evaluation and testing in animals, a process chemist makes sure that all impurities are qualified and that the method is scalable to support Phase I clinical trials. During clinical development, process scientists continue to identify and quantitate all impurities and ensure that methods are reproducible. They also develop synthesis routes for large-scale commercial production. Cost becomes a major issue during manufacturing.

The raw materials, reagents, solvents, and temperatures can be individually optimized for each particular process. Statistical design programs are used to determine which variations are most efficient.

### Analytical Method Development

By the time the company files an Investigational New Drug application (IND), a preliminary analytical method has been developed. During clinical development, the analytical techniques are optimized and validated so that the quality control group can analyze drug batches for strength, purity, and chemical stability.

### Defining Impurity Profiles

*Impurities happen!*

Impurities are commonly found when reactions are scaled up. Toxicology experts predict and identify the impurities and evaluate their potential hazards.

### Stability Studies

Stability studies are conducted to determine how long the drug remains in the proper formulation on the shelf (its shelf life) and to identify degradation products.

### Design and Optimization of Formulation

Drug candidates are formulated for clinical trial supplies and, ultimately, for commercialization. The marketing department might add to the challenge by specifying that the drug substance should be formulated as a pill, liquid, or injectable in order to achieve a competitive advantage.

### Design of Experiment (DOE)

The design of the experiment details exactly how much of each ingredient to add to a reaction and how fast to add it. This information details the range in which reactions can be safely and effectively run.

### Transferring Technology to Manufacturing

After the synthesis route has been optimized, the defined methods are then transferred to manufacturing in a process called "technology transfer." Product development scientists advise the manufacturing team on how to further optimize processes and analyze data. A manufacturing liaison is needed to ensure that both the drug substance (the active ingredient) and drug product (the formulated drug) are made and released on time and in the required manner.

#### Writing Scientific Papers and Attending Meetings

Many companies request that scientists attend meetings, give presentations, and publish scientific papers in peer-reviewed journals. For smaller companies, this is a luxury; for larger companies, it may be a mandate.

#### Analyzing and Testing New Technologies

Particularly true for bioprocessing, as drug development becomes ever more sophisticated, new challenges are encountered that require the testing and application of novel solutions. There is a large industry centered on developing products such as fermentation equipment, chromatography supplies, filters, and much more.

### Project Management Roles

#### Project Team Participation

As team members, product development scientists are alerted to any new compounds in the pipeline so that they have time to prepare for and to explore new technologies and equipment as needed.

#### Writing the Drug Substance Section in IND Applications

Working in concert with the regulatory affairs group, managers in product development may be responsible for writing or reviewing the drug substance and analytical testing sections of IND filings.

#### Serving as the FDA Liaison

Some product development scientists serve as technical development representatives to the U.S. Food and Drug Administration (FDA). They remain in constant written communication with the FDA throughout the entire development process and may be responsible for assembling information for FDA filings. Communications include notices about changes in the production process, impurities that arise, and the potential impact of the impurities.

#### People Management

Supervisors are responsible for motivating employees to be productive. They keep an open line of communication with everyone involved to make sure that they are performing well, enjoying their work, and maintaining the high quality of drug products.

#### Vendor Management

Many elements involved in product development are outsourced to vendors, such as supplying active pharmaceutical ingredients, scaling up processes, and manufacturing, especially for small companies. Extensive personal interactions are needed to manage contract research and manufacturing organizations.

### *Information Management Role*

#### *Information Management and Preparing Technical Documents*

Every time a batch of drug is manufactured or a round of testing is completed, internal technical reports are written. These can be very detailed reports that itemize every test, the results, and any unusual observations or specifications. As many as 20 different assays are run on a batch! Each test is then reviewed by quality control and quality assurance before the product is ultimately released. The amount of paperwork involved can be daunting, but such documentation is absolutely essential to ensure identification or recall of bad batches and to comply with FDA regulations.

## A TYPICAL DAY IN BIO/PHARMACEUTICAL PRODUCT DEVELOPMENT

*Depending on the level and type of position, a product development scientist might be engaged in some of the following activities on a typical day:*

- Working at the bench; designing and running reactions; modifying or reinventing synthesis routes; recording data in laboratory notebooks.
- Recording activities done during the day to fulfill compliance requirements.
- Writing reports based on test results, sometimes for Chemistry, Manufacturing and Controls (CMC) sections of regulatory filings.
- Attending meetings with direct reports or project teams for product development status updates.
- Managing employees, reviewing their data, and discussing results.
- Handling emergencies.
- Managing and keeping track of products supplied by vendors and suppliers.
- Attending local and national chemical development or bioprocessing meetings.
- Training employees in Good Manufacturing Practices (GMP) requirements.
- Attending seminars by visiting professors and job candidates.
- Reading scientific literature.

## SALARY AND COMPENSATION

Over the last five to ten years, chemists and chemical engineers have, on average, earned higher salaries than those in most other disciplines. Chemists tend to be paid more than biologists by as much as 10–20%. This is due to several factors: There is a common perception that chemistry degrees are more difficult to obtain than biology degrees; there are far more biologists than chemists; and, in general, there is a positive correlation between compensation levels and being closer to product sales. Chemists with similar degrees are paid the same whether in discovery research or in chemical development. Synthetic organic chemists, however, tend to be paid slightly more than other chemists, and so do chemists with degrees from academic institutions renowned for chemistry (e.g., UC Berkeley, Stanford, Harvard, MIT, Caltech, UC Irvine, and the University of Illinois at Chicago).

The American Chemical Society publishes yearly salary surveys, which can be found by clicking "Professionals" and then "Careers" at www.acs.org.

***How is success measured?***

Measurements of success depend in part on one's role and responsibility. For chemical development scientists, success is gauged by whether or not development goals such as "weights and dates" targets are reached and by the robustness of the production process. For others, success is when all impurities are accounted for after scaling up to commercial production.

A key sign of success is when you've gained the respect and admiration of coworkers. A person's reputation is built over time as people acknowledge your contributions to projects and your ability to promote cohesive, productive teamwork.

## PROS AND CONS OF THE JOB

### Positive Aspects of a Career in Bio/Pharmaceutical Product Development

- If you enjoy working with your hands and conducting research in the lab, being a bench scientist in this field can be highly gratifying.
- Drug development has tremendous altruistic appeal. Someday you may be able to say that you worked on an approved drug that saves people's lives, or even better, your own relatives' lives.

*The satisfaction of creating products that improve world health makes product development a worthwhile and meaningful career.*

- For process chemists, there is the simple pleasure of designing successful, scaled-up reactions and seeing a tangible product.
- You will be at the fulcrum of drug development, interacting with a broad range of people and exploring new fields across the company. You might be involved in working on products in four or five therapeutic areas during the time a colleague in discovery research works on only one. Every day is full of learning opportunities, and you will rarely be bored.

*There is pure joy and satisfaction when the manufacturing team makes 100 kilograms of 99% pure drug substance by using your scaled-up synthesis route!*

- You will gain an appreciation of the challenges inherent in product development and better understand how development can be a recursive process.

- Product development is scientifically fascinating. There will be many opportunities to thoroughly explore your interest in chemistry, such as learning how to control reactions, maximize yields, and control impurities.

- Unlike some of the other disciplines, this career allows you to apply the practical side of your training and pursue your core competencies and interests.

*If you attended college to become a chemist, this field allows you to remain one.*

- Product development is critical for the success of the company. Your work guiding R&D's most promising products to commercial production adds direct value to the company.

- In this career, you can apply your creativity by manipulating chemical processes in clever and efficient ways. One of the most gratifying accomplishments is realizing that the processes which you designed cannot be improved upon.

*Process development is like art—you want to be known as the artist who created a famous drug synthesis masterpiece.*

- The work environment tends to be less structured and more iconoclastic than in other disciplines. There is much dialog among employees, irrespective of rank and education, and work is data-driven.

- Performance evaluations tend to be fair and objective. Employees in product development are held responsible for meeting timelines and expectations, and it is obvious who is working hard and providing beneficial intellectual input. Advancements can occur slightly faster in product development than in other departments, irrespective of degrees, and you can advance into related areas.

- Solving problems in product development requires a lot of teamwork, which fosters camaraderie and discourages cutthroat competition.

- Compared to work in quality assurance, quality control, manufacturing, and preclinical studies, there are fewer regulatory compliance restrictions.

- There are opportunities to publish results and patent synthesis routes and processes.

- Depending on the company and your position, product development may require less travel than other departments. In general, employees of smaller companies that work with contract manufacturing organizations will travel more than employees of larger companies with their own facilities.

## The Potentially Unpleasant Side of Bio/Pharmaceutical Product Development

- Because projects are timeline driven, the hours can be long and occasionally include weekends and nights.

- Due to the pressure to accomplish tasks quickly and cheaply, work is often rushed. As a consequence, there is the frustration of not having enough time to fully explore new chemical pathways or apply new techniques.

- Product development has less visibility than other disciplines, and the importance of the department is often taken for granted. In general, much of the luster in drug development revolves around clinical trials, and the common assumption is that the major obstacles are fought out in the clinic. Senior management can be unaware of the multitude of potential obstacles and challenges that may be encountered in product development.
- Biologics are complicated and can be technically difficult to work with. Such work may require continual evaluation of new technological approaches and the perseverance to work through difficulties. A molecule can appear to be absolutely wonderful in discovery research, but if the company can't formulate it, or it is too costly to produce, then the project may be terminated.
- For managers, an extensive amount of documentation is required to comply with safety, environmental, and FDA regulations. Some people just loathe this aspect of the job. In big pharmaceutical companies, documentation can consume as much as 25% of your time!

## THE GREATEST CHALLENGES ON THE JOB

### *Working under Pressure and with Restricted Timelines*

*In industry, you learn how to juggle scientific perfectionism with business needs.*

The biopharma industry is under enormous pressure to develop drugs quickly and cheaply. A company could be losing millions each day that a blockbuster drug is not on the market. Each step in chemical and biological development must be managed carefully to prevent delays, and the pressure to produce often falls on product development. To confound the problem, senior management tends to underestimate product development timelines. As a consequence, there may not be enough time to fully optimize processes or develop new routes. It can be frustrating to advance projects that are not scientifically glamorous or efficient enough because of time constraints.

### *The Unpredictable Nature of Drug Products*

*With biologics, you can always expect the unexpected.*

As bigger and more complex molecules are derived from discovery research, the increased complexity of chemical syntheses and the quest to formulate the molecules within solubility limitations are pushing product development science into exotic, unexplored technical areas. This trend is one of the biggest challenges in product development.

### *Safety*

The fast-paced nature of this career and the unstable financial status of young companies that try to save cash by cutting corners can make it easy to inadvertently compromise the quality of the final compound or fail to ensure that production methods are safe for employees and the environment.

### *The Industry Is Changing*

The entire industry is moving from an era of medications with broad applications for chronic diseases to more specific medications aimed at diseases in smaller population groups. As we progress toward more specialized medicine in difficult areas such as oncology and neurology, the manufacturing plants that were designed for high-volume products will be retooled to be smaller, more specialized, and more sophisticated.

## TO EXCEL IN BIO/PHARMACEUTICAL PRODUCT DEVELOPMENT...

### *A Combination of Talents*

Those who are inherently very intelligent, and have an essentially unbridled work ethic and outstanding people skills, can be profoundly successful. Success requires the ability to effectively work in a matrixed environment and the technical expertise to consistently make wise decisions.

### *A Strategic View of Development*

Drug development has become enormously competitive, and there is intense pressure to create medicines for unmet medical needs. Those who excel in this work have the vision to evaluate and integrate a myriad of factors that affect drug development, such as regulatory and intellectual property issues, the cost of goods, deadlines, and competition. For example, a process that may not be the most efficient might make the best business sense given competition and time constraints. This is a skill developed after years of experience.

### *Knowing How Much and When*

It is often not clear in process development how much time and money should be spent on each candidate in clinical stages, because its success or failure hasn't yet been determined. An experienced process scientist develops an almost intuitive ability to know when and how many resources to allocate during clinical trials, and how to tell when things are "good enough" under the circumstances.

### *A Willingness to Try New Approaches*

State-of-the-art product development is highly complicated and sophisticated. Those who remain up to date on all the technology continue to expand their arsenal of tools so that they will have a better selection of products that suit their immediate and future development needs.

## Are You a Good Candidate for Bio/Pharmaceutical Product Development?

*People who flourish in product development careers tend to have...*

*A diligent and dedicated work ethic.* You must be passionate about what you do and driven to finish projects. You should be observant and objective, and you should strive for constant improvement.

*An exquisite attention to detail.* For example, filling out the necessary quality paperwork requires meticulous attention.

*A certain amount of bravado and self-confidence.* In the process sciences, it is important to be willing to take some risk, within safety constraints. For example, you may be asked to scale up a product before you are completely ready.

*The capacity to remain calm and make rational decisions while under pressure or during emergencies.* Deadlines and emergencies are common, and it is easy to feel flustered. It is important to stay calm and think logically during stressful times. If, for example, a batch is found to be contaminated, what do you do? In such cases, risk/benefit decisions need to be calculated quickly to determine whether to continue or abort manufacturing.

*An ability to interact well with people.* Because product development exists in the middle of an interdisciplinary world, you need to be versatile enough to effectively interact with people from different areas, including medicinal chemists, clinical specialists, discovery scientists, manufacturing analysts, as well as quality and regulatory affairs personnel.

*Outstanding communications skills.* Drug development is complex, and a great deal of time is spent sharing information. You should be able to speak and write in a compelling and clear manner.

*If the team doesn't hear or understand what the superstar chemist says, that expertise will be lost in the decision making.*

*Excellent analytical skills.* People in product development tend to be practical and demonstrate an empirical approach to experimental design. They are critical thinkers, not just when interpreting other people's data, but also when reviewing their own data, and are receptive to other people's different interpretations.

*Great persistence.* Projects sometimes fail, but it is important to continue striving for success. Perseverance will carry you through long-term projects.

*An understanding of how product development fits into the big picture.* The biotechnology and drug development industries are for-profit and highly regulated. You need to be a businessperson as well as a scientist and remain aware of the regulatory environment, the higher-level business program, and other issues.

*Mental adaptability and versatility.* Conditions will vary and techniques will change from one situation to another, so you need to be flexible.

*Excellent organizational skills.* This career involves much documentation and interaction with external vendors. All information should be well organized.

***You should probably consider a career outside of product development if you are...***

- An independent worker, not a team player.
- Easily offended and defensive.
- Someone who needs individual recognition or is too competitive.
- Not willing to share your scientific knowledge or resources.
- A perfectionist.
- Someone who detests paperwork.
- Not able to work under pressure or to handle emergencies.
- Not a decisive, logical decision maker.
- Not self-confident.
- Impatient or seeking instant gratification.

## BIO/PHARMACEUTICAL PRODUCT DEVELOPMENT CAREER POTENTIAL

With steady promotions, you may eventually work your way up to senior levels within bio/pharmaceutical product development (see Fig. 15-1). You can also acquire additional company-wide manufacturing and operations responsibilities. It is possible to eventually reach the level of chief operating officer or chief executive officer.

If you want to continue your career growth into other areas, training in product development can allow career transitions to operations, manufacturing, quality, project management, patent law, and regulatory affairs.

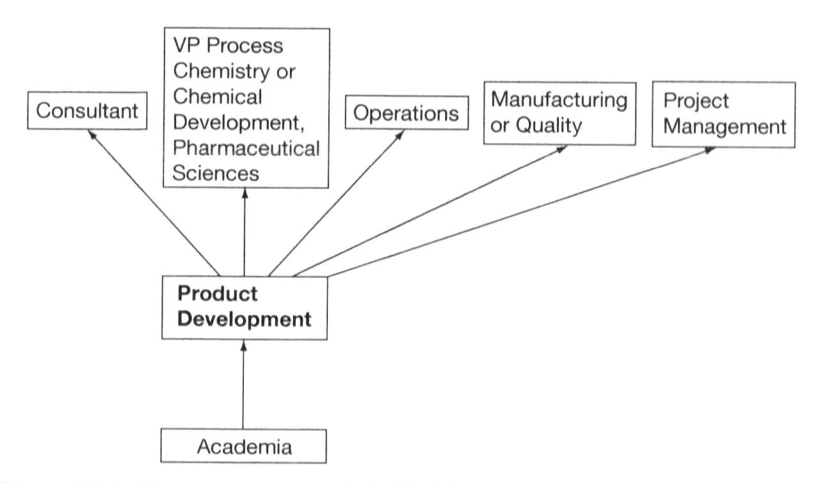


Figure 15-1. Common career paths in bio/pharmaceutical product development.

### Job Security and Future Trends

Job security appears to be high for these careers. During times of economic setbacks, early discovery research programs tend to be terminated, but companies continue to work on products in the clinical stages.

In general, demand for talent tends to be higher than supply, partly because as products enter clinical development, more chemical and biological development supervision is required. Some areas that are currently experiencing an intense demand for talent include mammalian cell expression systems, antibody development, process chemistry, medicinal chemistry, fermentation, and purification. Although the demand is high for biologics product development expertise, there are more small-molecule products being developed than biological products. As a result, there are ten times more jobs in small-molecule product development than in biologics.

Although outsourcing overseas is not currently affecting the growth of product development jobs in the United States, this is expected to change. The economics of cheap labor are irresistible, although intellectual property issues may slow down some outsourcing. Production of biologicals is more capital-intensive and more sophisticated, so outsourcing overseas for biologicals will likely be slower than for pharmaceutical products. Positions for medicinal and process chemists who work closely with discovery and clinical researchers are least likely to be outsourced overseas.

## LANDING A JOB IN BIO/PHARMACEUTICAL PRODUCT DEVELOPMENT

### Experience and Educational Requirements

A B.S., M.S., or Ph.D. degree is required. The vast majority of employees arrive directly from academia, most commonly after completing postdoctoral programs (see Fig. 15-1).

*In general, experience counts for more than education in product development.*

It is beneficial to have had technical training and experience in some of the following areas: chemistry, analytical methods, pharmaceutical chemistry, pharmacology, formulation, and chemical or mechanical engineering. For bioprocessing, it is helpful to have had training in biochemistry, cellular and molecular biology, or microbiology.

Many people will tell you that an advanced degree is not required, but it is highly beneficial to have one. In general, if you seek to reach senior-level positions, it is a tougher path without a Ph.D. degree. If you are considering a master's degree, the extra mile to obtain a Ph.D. degree will pay off.

### Paths to Bio/Pharmaceutical Product Development

Hiring managers seek qualified applicants with training in chemistry or biochemistry basics from the top-ranked schools. For additional practical advice, review Chapters 3 and 7.

- Many of the big pharmaceutical companies offer summer internships for college students. Internships provide an excellent introduction to industry, and if your chemistry is right, a job may be waiting for you after graduation.
- Attend job fairs and campus recruiting events at universities and local and national chemistry conferences where you can network and inquire about companies that are hiring. Meeting company employees may help you refine your career goals and define your cultural fit. The American Chemical Society holds job fairs, and positions are posted on their Web site at www.acs.org under "actions and reactions."
- A few select universities teach process development, including MIT, the University of California, the University of Minnesota, the University of Colorado, and the University of Iowa.
- Work at companies where you will receive the broadest training possible to strengthen your background in analytical, mechanistic, and synthetic chemistry.
- Keep in mind that there are industries that manufacture equipment, industrial enzymes, reagents, and more for bio/pharmaceutical product development. Career opportunities abound for technical specialists or product development research scientists in these organizations. Also consider joining contract manufacturing organizations that service the biopharma sector.

## RECOMMENDED TRAINING, PROFESSIONAL SOCIETIES, AND RESOURCES

### *Courses and Certificate Programs*

Statistics

### *Societies and Resources*

American Chemical Society (www.acs.org or www.chemistry.org)

American Institute of Chemical Engineers (AIChE, www.aiche.org)

American Association of Pharmaceutical Scientists (www.aaps.org): This group has a section focused on pharmaceutical process development, pharmacodynamics, and pharmaceutical development. Within AAPS is an Analysis and Pharmaceutical Quality section (APQ), which combines analytical and bioanalytical sciences.

American Society for Mass Spectrometry (www.asms.org)

U.S. Food and Drug Administration (www.fda.gov): This resource provides a good overview of drug discovery, development, and regulations.

The Williamsburg BioProcessing Foundation (WilBio, www.wilbio.com)

American Society of Gene Therapy (www.asgt.org)

American Society for Cell Biology (www.ascb.org)

### General Biotechnology Societies

Biotechnology Industry Organization (BIO, www.bio.org)

### Magazines and Journals

*Genetic Engineering News* (www.genengnews.com)
*Chemical & Engineering News* (www.acs.org)
*The Journal of Organic Chemistry* (http://pubs.acs.org/journals/joceah/index.html)
*Tetrahedron Letters* (www.elsevier.com)
*BioProcessing Journal* (www.bioprocessingjournal.com)
*Organic Preparations and Procedures International* (OPPI, www.oppint.com)
*Nature Biotechnology* (www.nature.com)
*Science Magazine* (www.sciencemag.org)
*Journal of Chromatography*
*Journal of the American Chemistry Society*

### Books

Friary R. 2006. *Jobs in the drug industry: A career guide for chemists.* Academic Press, San Diego.
Books about pharmacokinetics by Leslie Benet
Books by Malcolm Rowland

### Free Web Sites with Biotechnology News and Meeting Announcements

FierceBiotech (www.fiercebiotech.com)
BioSpace (www.biospace.com)
Biotechnology Industry Organization (BIO, www.bio.org)

# 16

# Life Science Information Management

## The Melding of Computer and Biological Sciences

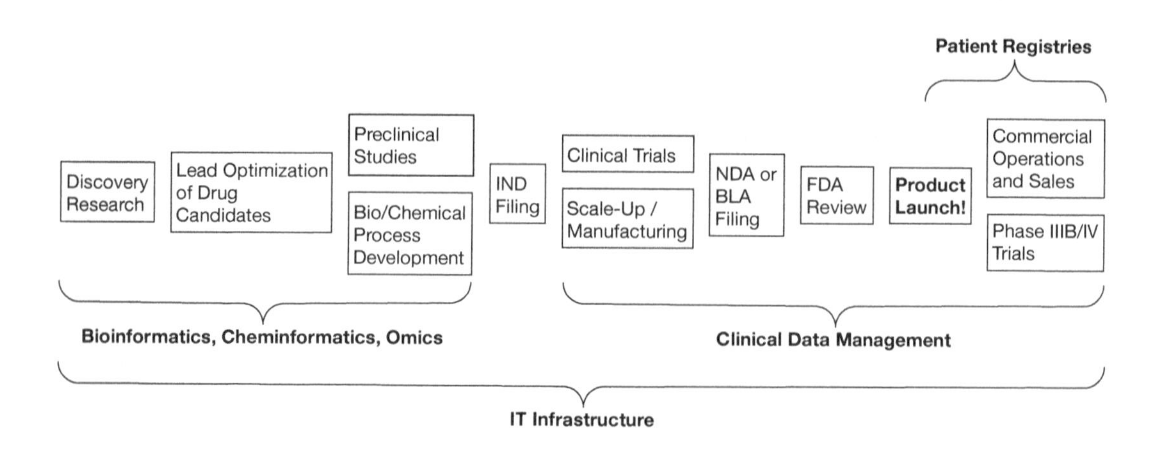


*Trade your test tube in for a keyboard!*

IF YOU ARE A SCIENTIST WITH AN INTEREST in computer science or if you have an information technology (IT) background and want to learn more about careers in the biotechnology and drug development industry, then consider combining your interests by working in life science information management. Career opportunities abound in disciplines requiring specialized expertise in areas such as computer systems validation, data management, algorithm development, software development, quality, and compliance. These jobs pay well and are fairly secure after you have reached a critical threshold of knowledge and expertise. Companies are continuously developing new ways to use computer technology to accelerate drug discovery, and growth in this area is rapid.

*One of the goals of life science information management is to be able to transform a biological question into a task that can be performed by a computer.*

Perhaps the biggest draw to these careers is that they offer the chance to become an integral part of a large industry that ultimately is driven toward developing new drugs to improve world health. It provides a chance to personally "make a difference" while experiencing the excitement and promise of today's biotechnology and drug development industry.

## THE IMPORTANCE OF LIFE SCIENCE INFORMATION MANAGEMENT IN BIOTECHNOLOGY AND DRUG DEVELOPMENT

*Data are king in drug discovery and development.*

The ultimate goal of IT is to develop in silico, or virtual, drug discovery and development. Although this promise has yet to be fulfilled, the proliferation of computer applications has grown exponentially, and IT now plays an essential role in the industry. Biopharma companies are investing in their life science information management systems to increase the speed and efficiency of product development. To cope with the vast quantities of data that are generated, the entire industry is making a massive transition from paper to electronic data capture and management. Computers are being used to better understand chemical and biological processes, to manage business processes such as payroll, scheduling, and report generation, and as tools in the biopharma disciplines. People with experience in both IT and drug development are in high demand, and that demand is expected to grow.

## CAREER TRACKS: LIFE SCIENCE INFORMATION MANAGEMENT

*Depending on whom you speak with, the term "Bio-IT" has many potential definitions.*

Information management is probably one of the fastest-evolving areas in the industry. Because of the speed at which more IT resources are being deployed in the biopharma world, existing job definitions are constantly changing and new careers are emerging. Defining careers in IT is like shooting at a constantly moving target, and the definitions of terms vary considerably. Likewise, titles, roles, and organizational structures of IT departments vary.

Most companies make a distinction between their IT infrastructure organizations and their life science information management (bio-IT) departments. To generalize, with a pure IT background you can find employment in departments dealing with infrastructure and data management, but with combined biology and computer science backgrounds you can work in the bio-IT departments.

When considering career options, keep in mind that there is a big difference between working in regulated versus unregulated life science IT. FDA-regulated areas apply to manufacturing, quality control and process data, clinical data management, document management, change control, software, and hardware validation. In a regulated environment, there is an extra amount of rigorous quality assurance applied to systems and applications. This means that there is more non-programming activity associated with regulated software development, such as documentation, qualification protocol and testing, and making sure that processes and procedures are compliant with regulations, internal policies, and standard operating procedures (SOPs).

*There is a tremendous distinction between working in a regulated versus an unregulated environment.*

***The FDA Regulation for Computer Systems: 21 CFR Part 11***

The FDA regulation for computer systems is called "21 CFR Part 11." It covers many areas including security, protection of data and records, and audit trail features. The goal is to ensure that systems are secure and that the authenticity, integrity, and confidentiality of electronic data are maintained. The regulation allows electronic/digital signatures to be as legally binding as handwritten signatures.

*The following is a hodgepodge of some of the more common vocational areas and departments found in information management in the biopharma industry:*

### *Informatics*

Informatics is a general term that refers to the process of capturing data from instruments and categorizing, annotating, and storing the data. Bioinformatics, cheminformatics, and clinical data management are specific informatics applications.

### *Bioinformatics*

The classical definition of bioinformatics refers to using IT tools to learn more about a particular gene or protein. Bioinformatics uses algorithms to search (or mine) for a particular DNA or amino acid sequence in databases. Bioinformatics includes the development of applications to derive more information about a newly identified gene or protein, such as, what family of proteins it belongs to, where the protein is located in the cell, what the protein's potential function is, what biological cell process it might participate in (e.g., cell cycle, apoptosis), its gene expression profile in diseased versus nondiseased tissue and in various organs in the body, what class of drugs can target it, and more.

*Bioinformatics is the mathematical expression of chemistry and biology.*

### *Cheminformatics*

Cheminformatics, also known as chemoinformatics, is the informatics of chemistry. It involves applying computational science to chemical data to analyze products and to identify promising leads for drug candidates.

*Cheminformatics is the informatics of chemistry.*

Cheminformatics has numerous applications for drug discovery. One such example is the use of computer modeling to make predictions of protein docking based on van der Waals forces. Cheminformatics applications are also used for identifying active components in mixtures of compounds (known as "combichem"). Chemists simultaneously synthesize many molecules, mix them together like a soup, and test the mix for activity. When a mixture works, they use cheminformatics to find the active component.

### The "Omics"

In addition to bioinformatics and cheminformatics, there are other scientific disciplines, such as genomics, proteomics, and systems biology (collectively known as the "omics"). These areas rely heavily on informatics applications and computer systems.

### Medical Informatics

Medical informatics refers to the management and capture of clinical trial data. Those in clinical data management and statistics have the responsibility of interpreting the information for reports and trial results. They are also involved in the validation and testing of the data, as well as designing databases and processes. Clinical data management careers are discussed in more detail in Chapter 10.

### Laboratory Information Management Systems (LIMS)

LIMS is an electronic system developed to capture, manage, and allow company-wide access to laboratory-generated data. Chemists and biologists, for example, can enter their data into a central LIMS database where the information can be tracked and analyzed by team members and by the members of other departments.

### Electronic Data Capture, eNotebooks, FDA Submissions, ePRO

The entire industry is making the transition to electronic data capture. For example, clinical data collection, electronic patient reported outcomes (ePRO), discovery research data entry, and regulatory filings to the FDA are now handled electronically. Electronic notebooks (eNotebooks) are being used to replace the laboratory notebook and provide an electronic view of experiments. As the industry is becoming more automated, machines generate data, and new ways are being developed to capture, retrieve, analyze, and archive the data. Electronic submission of drug safety reports is mandatory with most European regulators and will soon become mandatory with the U.S. Food and Drug Administration (FDA).

### Document Management

A vast amount of information is stored and organized so that it is easily retrievable. With the stringent regulatory guidelines, most departments support a separate function called "Document Management" to serve this purpose. To learn more, see Chapters 11 and 12.

### Patient Registries

Patient registries are informational databases about patients with particular disease states or patients who receive specific treatment regimens. Registries have multiple applications but mainly serve to provide data showing that a therapeutic product was beneficial to patients. It is also used to assist physicians in determining which products work best for particular disease states. This is discussed in more detail in Chapter 11.

### *Business Applications*

There are many business-related computer applications. Computer applications and systems are used in just about every function, all the way from early discovery research to marketing, sales, and finance. For example, customer relationship management (CRM) systems track interactions with key opinion leaders, help develop marketing analytics, and track the frequency of prescriptions written for drugs.

### *Hardware Development and Configuration*

Hardware refers to the computers and chips used by the biotechnology and drug development industry. Most major hardware vendors have a life science division unit. These cadres of pure hardware and life science domain experts work to understand the computational challenges of drug discovery and development and how to deploy their hardware in support of life science applications. There are many vendors and management consulting firms who work in this area.

### *Software Development*

Many software companies and in-house programmers build and design customized software for applications such as bio or cheminformatics, DNA sequencing, gene expression profiles software, interactive voice recognition systems for clinical trials, clinical trial data management software, and more.

### *Corporate IT or IT Infrastructure*

IT infrastructure refers to the way computer resources are configured and managed. These resources include, but are not limited to, networking connections and routers, hubs, switches, data storage, backup, relational database management system (RDMS) engines, E-mail, laptops, and Web sites. In general, these specialists are not biotech-specific, and computer users don't interface directly with them. Many large companies also have an internal IT help desk to resolve computer-related issues.

### *IT Quality and Validation*

Because the drug industry is FDA-regulated, computer systems and processes must undergo rigorous testing and be compliant with the guidelines. IT quality departments are needed to validate, test, and inspect business processes and to ensure that the users, procedural controls, and the computer systems themselves are compliant. The systems that need to be tested and validated are numerous, including laboratory data collection, manufacturing controls, laboratory instruments, clinical data management systems, safety reporting systems, manufacturing systems, and electronic document management systems. Computer validation and IT quality careers are described in more detail in Chapters 10 and 13.

## A TYPICAL DAY IN LIFE SCIENCE INFORMATION MANAGEMENT...

*As this is a chapter describing many different vocational areas, a career in life science information management might be spent working on some of the following activities:*

- Helping users define business or process requirements in ways that can be used to specify a computer-automated solution.
- Assessing appropriate technologies and applications based on defined user requirements and needs.
- Project management or project planning. Developing timelines, making sure deliverables are on track.
- Determining application architecture, coding, and programming.
- Documenting functional specifications, unit tests, and designs. Developing validation plans to define how a system will be tested.
- Designing and testing systems and data collection instruments for electronically based data capture and storage. Testing software and processes. Meeting the user community, recording their requirements and specifications, configuring software, and conducting user testing.
- Training end users to implement new systems or solutions.
- Solving computer-related problems.
- Keeping track of recent computer-related FDA guidelines.
- Auditing and documenting compliance with standards.

## SALARY AND COMPENSATION

In general, salaries in IT are comparable or slightly higher than in quality and discovery research careers, and an advanced degree is not required as it is in discovery research. Compared to professionals in high-tech companies, IT professionals in the pharmaceutical industry might be paid less, but there is more job security. The range of compensation for IT careers depends on the discipline. Because the demand for people with extensive knowledge in both biological and computer science exceeds the supply, they can command large incomes.

***How is success measured?***

Success is measured in part by the degree of increased efficiency of an organization: IT systems succeed by automating or having better control over the data. Passing regulatory inspections and getting authorization to market a product are also indicative of success. Individual performance metrics are useful, including whether or not one has become an expert in a particular area or kept up to date with the latest trends.

## PROS AND CONS OF THE JOB

### Positive Aspects of Life Science Information Management as a Career

*You might be able to earn more money working in a mortgage refinancing company, but the science is much more exciting in Bio-IT!*

- Compared to other IT careers, life science IT can be more interesting and satisfying because of its potential beneficial social impact on world health.
- A career in this area allows you to play a large and visible role in your company by improving efficiency and saving coworkers time.
- There are difficult problems to solve, and the challenges are fascinating and complex. The intellectual challenge of programming can be a lot of fun.
- You will have the chance to learn about (and even improve) a wide range of business practices and principles.
- You can be at the cutting edge of biological, chemical, and computer science. You can explore and test new technologies, different systems, and processes. There is a balanced mix of science and technology.
- You will have the chance to learn a great deal about the pharmaceutical industry by becoming familiar with its computer systems and processes.
- There can be a variety of fast-paced projects and not much routine work.
- An IT career provides a great opportunity to interact with interesting people from diverse backgrounds. You can share the camaraderie of your science coworkers. It is stimulating to be associated with leading scientists who have developed groundbreaking inventions or have published important peer-reviewed papers.
- It is a very results-oriented occupation. Projects have definitive end points and there is frequent completion of various steps along the way. It is exciting to view new data, spot the trends, and learn the final outcome.
- This can be a more stable industry compared to high tech. There is generally better job security and a high demand for qualified and experienced individuals in the pharmaceutical and biotechnology industries, and compared to high-tech companies, they tend to treat their employees better.
- If you enjoy analyzing data without having to do the hard work to generate it, IT is an ideal career. Experiments can be deleted with a button.
- There may be opportunities for some telecommuting, but it depends on your position and the company. In general, for positions that require intensive interactions with biologists, you must be physically present at meetings. Programmers, on the other hand, may be allowed to telecommute.
- There is less travel compared to other positions in the biotechnology industry.

### The Potentially Unpleasant Side of Life Science Information Management...

- Because of their important role in companies, when systems don't work well, users become frustrated and angry. There are frequent emergencies, and the problems need to be solved yesterday.
- The biotechnology, drug development, and medical device industries are strictly regulated; therefore, tedious, detailed documentation is required. This is especially true for the validated fields, where some people spend 75–80% of their time on peripheral activities such as preparing documents and reviewing code. Working under FDA guidelines, you are restricted in ways that make software development more difficult. For example, in a validated environment, problems with a code are more difficult to fix after a study trial has begun.

*In the FDA's mind, if something is not documented, then it did not happen.*

- IT is not a well-understood profession. You may need to explain in detail what you do for people and explain to others why they should pay more attention to you. Because systems are so complicated, companies tend not to adequately plan for contingencies. Problems can cause delays in projects or in meeting corporate milestones, which can be costly to the bottom line.
- The constant deadlines lead to much stress, especially toward the end of big projects when timelines are condensed. Unexpected system problems and lack of adequate planning only increase the pressure to complete projects on time. Computer problems are highly visible, which adds to the pressure.
- Adversarial relationships can easily develop between computer scientists and users. For example, the data management analyst spot-checks the clinical monitor's work, noting points that are missing, errors in data, etc. In addition, scientists frequently have difficulty translating their needs into computer terms or don't have time to explain the reasons behind their specifications.

## THE GREATEST CHALLENGES ON THE JOB

#### *Understanding the Biological Science behind the IT*

Computer science algorithms and chemical synthesis steps can be expressed precisely, but biology is more difficult to understand and explain. There is a deep and vast body of biology-related knowledge that is rapidly expanding, and so much is still unknown that results are often unpredictable, even whimsical. Biotechnology companies tend to work at the cutting edge of biological research. Their technical focus may rely on unpublished, proprietary information that is unavailable even in academia. As a result, it can be daunting to keep up with research demands and to understand the new, sophisticated biological applications.

*Biology is not just a science—it is an art and can be whimsical.*

### Staying Abreast of Recent Information Technology Advances

The IT field moves quickly, and you must take the time to keep up with and evaluate all the new developments. You have to remain alert and be able to adapt to the latest new tools, particularly for security and virus protection.

### Making Good Decisions Based on Limited Data

Bio-IT may sometimes involve making educated guesses with uncertain results.

### Diplomacy

Diplomacy is called for when informing users that they have to rewrite documents because they do not meet FDA standards. This is especially true for people who work in a validated environment. It takes skill to negotiate with people and help them understand and accept what the requirements are and why they are important.

### Staying Updated with the Changing Regulatory Landscape

FDA regulations rarely change, but interpretations and ways to meet them continually evolve. It is important to keep updated and review warning letters sent to other companies by the FDA regarding their computer systems to get insight into the FDA's expectations.

## TO EXCEL IN LIFE SCIENCE INFORMATION MANAGEMENT...

### A Deeper Understanding of How IT Can Be Applied to Drug Discovery and Development

People who do well in life science information management have a thorough understanding of biotechnology and how computer resources can be applied to processes. They see beyond information management's immediate responsibilities and can think about situations from the users' or company's perspective. Because they are in better tune with the users, they can better satisfy their needs and facilitate not only their success, but the success of the company.

### Great Foresight

Outstanding life science information management professionals grasp the significance of their decisions and their potential effects. They understand, for example, that the choice of what data to collect and how to collect them greatly influences the results of analyses. They can envision emerging trends and issues, not just for immediate business applications, but also for developing technology that will help their company outpace its competition.

### The Ability to Be Conversant in Both Languages

Those who excel are fluent in both biological and IT languages. They can discuss details in terms that computer scientists understand and at the same time speak coherently with biologists.

## Are You a Good Candidate for Life Science Information Management?

Because this vocational area is so large and has so many different types of functions and disciplines, there are jobs that fit most personality styles. For example, introverted coders and extroverted analysts can both be successful in this field.

*People who flourish in life science information management careers tend to have...*

*Excellent organizational and multitasking skills.* Organizational skills are needed to put together logical, well-designed solutions for users. There are often multiple projects to work on, requiring an integrated and organized approach.

*Good problem-solving abilities at both the micro and macro levels.* It is important to be able to solve problems at the higher, business process level as well as to deal with the minute details.

*Good "people" skills and a team-player attitude.* Your work will likely involve a lot of collaboration. It is important to be friendly with coworkers and users. This is not true for all IT positions; some programmers, for example, do not necessarily have to interact with people.

*A tolerance for angry customers and the ability to calm them.* Computers are not perfect, and bugs are eventually discovered by users. It is important to know how to manage upset clients without taking their anger personally.

*A service orientation.* Many IT departments provide a service. It is important to be customer oriented, to have the ability to analyze other people's work processes, and to view issues from their perspective. A consultant mentality may be the most effective approach.

*A great deal of patience.* Due to the service orientation, some users take longer than others to learn new skills. It can also be extremely challenging to work within FDA regulations; some programs that take only two minutes to write require two days of work to document. It is beneficial to have a calm personality and to be able to accept that things must be done a certain way.

*Strong analytical, mathematical, and detective skills.* A high level of comfort with technical systems and an aptitude for computer programming are advantageous.

*Knowledge and understanding of the drug discovery and development process.* It is important to be able to use the right terminology and to know which data points are relevant and which are not. You need to understand the data, grasp the overall picture, and know how the underlying regulations will apply to the process. With a broader understanding, it will be easier to coordinate data exchange with people in other departments, such as discovery research and clinical development.

*A flexible attitude.* Changes in priorities and new work assignments are frequent.

*Good communication skills.* Communication skills are needed for interacting with coworkers, critiquing other people's work, and explaining why processes and regulations are important.

*A willingness to continually learn new technology and skills.* Biology and IT are quickly evolving fields. You need to keep up with both of them and continue to sharpen your skill set so that you can use resources effectively.

*An ability to pay strict attention to detail.* Working in a validated environment requires careful attention and documentation in every step of the process. You may need to spend time uncovering hidden information.

*An entrepreneurial, self-starter personality.* Most of these positions require the ability to work independently and take the initiative.

***You should probably consider a career outside of life science information management if you are...***

- Someone who does not think logically; someone who is driven by impulse or instinct.
- A perfectionist or too detail oriented.
- The nonconformist type who wants to take shortcuts, particularly if you are working in one of the validated areas.
- One who cannot work in an environment with frequent deadlines.
- Just in the field to make money.
- Unwilling or unable to understand user needs.

*Working in life sciences IT only for the money is a big mistake—coworkers quickly sense when you are doing this.*

## LIFE SCIENCE INFORMATION MANAGEMENT CAREER POTENTIAL

In the IT field, you can specialize in a specific area as a resident expert and develop highly specialized programming skills without having to manage people. If you are interested in upper management, you can eventually become Chief Information Officer (CIO), Vice President of R&D IT, or Chief Technology Officer (CTO) (see Fig. 16-1). Consultancy is a common career path, as is joining one of the many vendors that specialize in information technology for the life sciences.

If you do not intend to remain in information management, there are other places where you can apply your skills. Life science IT is such a broad discipline that there will be exposure to many different functional areas, such as clinical development, manufacturing, and operations. Another area is patent law, where a computer scientist who is adept at data mining can search for freedom-to-operate clauses.

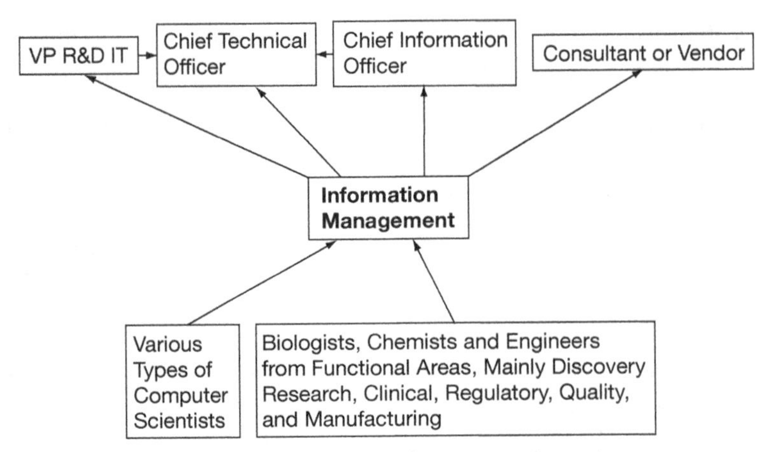


**Figure 16-1.** Common career paths in life science information management.

## Job Security and Future Trends

There is high demand for people with biotechnology-specific IT domain knowledge and experience. It is relatively easy to find employees who have generic computer experience, but those who have computer knowledge combined with experience in drug discovery and development, quality, or science are rare.

*Life sciences IT tends to be trendy and volatile.*

Some fields have experienced boom and bust times. During the genomics era, the industry experienced a tremendous demand for bioinformatics experts. Because of the increased proliferation of bioinformatics programs offered by various universities and the slow and disappointing outcome of most of the genomics companies, the demand decreased. That may change again in the future; the point is to stay ahead of the curve and keep up with the latest technological advances.

There is more job security in life science information management than in discovery research. Bench scientists are specialists in particular therapeutic areas and disciplines, so it is often difficult for them to find positions that match their specific expertise. Life science information management skills, however, are needed in any biopharma company, regardless of its drug strategy or therapeutic specialty.

In biotechnology companies, particularly start-ups, nothing is secure. The industry is fickle and often at the mercy of the regulators and Wall Street. If a drug application is denied or delayed, employee cutbacks are to be expected. IT professionals, however, control the data, and because this is an essential role, people who support infrastructure have more job security.

As for the eventual outsourcing of IT careers overseas, some functions, such as software, LIMS, validation testing, and especially the help desk functions, might go abroad as we witness increased and improved technologies and communications. Companies cannot outsource the interfacing component, because the people who participate in meetings need to interact personally with the end users.

## LANDING A JOB IN LIFE SCIENCES INFORMATION MANAGEMENT

### Experience and Educational Requirements

*Educational qualifications are frequently less important than experience.*

There are many IT-related positions in biotechnology and pharmaceutical companies, but the field is difficult to break into. People in life science IT come from all types of disciplines and backgrounds, and experience tends to be more important than educational qualifications.

For many positions, it is beneficial to have a stronger fundamental understanding of biology than of computer science. Computer science is a very analytical and logical field, whereas biology has many more subtle nuances that take more training to comprehend. The biological field is deep and vast, and the amount of information is growing steadily.

Most employees have science or engineering backgrounds. About half have science training, either in chemistry, biology, or microbiology, and they typically hold Ph.D. or master's degrees, depending on the position. The other half have database, application, and software development experience. It is advantageous to have a computer science or engineering degree. For bioinformatics, a Ph.D. is the norm.

Career paths vary tremendously, and there are many ways to enter the field. With a chemistry, biology, or product development background, you can start from the user side. Areas of relatively easy access are in data mining or laboratory automation, which requires the ability to program robots. Biologists working with large data sets or in data management can easily learn the computer skills necessary to make the transition. For those with computer backgrounds, there are several relatively simple ways to enter life science IT. Infrastructure may be the best place from which to start for those who do not have a science background. Other areas include laboratory information management and quality control development.

Data management entrants are commonly from clinical development. Many have B.S. or computer science degrees. People move into data management from nursing, clinical, IT departments, and data entry work. CRAs with monitoring experience are ideally prepared.

### Paths to Life Science Information Management

#### *Paths for Both Computer and Biological Scientists*

- Consider applying to one of the many tool companies and vendors that provide IT services for the industry: clinical investigators, contract research organizations (CROs), LIMS companies, analytical instrument manufacturers, manufacturing automation engineering firms, clinical trial management systems, ePRO, enterprise resource planning (ERP), interactive voice response services (IVRS), etc. Positions with vendors abound. You can start as an application specialist, an engineer, a business analyst, in sales, in marketing, as a consultant, and more. Working for a vendor also offers a good career path to a chief technical or chief information officer position. In fact, if you are an experienced professional in a biotechnology or pharmaceutical company, you can receive a substantial promotion by joining a vendor. It's generally easier to find employment with a vendor than with a biopharma company.

- Trade shows and industry conferences are terrific places to meet the many vendors and niche service providers. Most of the software and hardware companies and industry giants such as IBM have booths at these meetings.
- Consider joining a management consulting company. They run projects for hospitals, billings, validation, and even clinical IT management for large pharmaceutical and biotechnology companies.
- Most of the industry uses Oracle's database platform and one of several electronic document management systems. Enhance your Oracle and Structure Query Language (SQL) skills. State any Statistical Analysis System (SAS), DB, SQL, Documentum, or Oracle database proficiency in your resume.
- Certificates are available for the various disciplines, such as clinical data management, bioinformatics, document management, and informatics.
- For entry-level applicants, consider a data entry position. Advancement is quick.
- Consider obtaining project management skills or obtain a project management certificate if you want to increase your career advancement chances. See Chapter 9 to learn more.
- The FDA has positions for computer systems experts and inspectors. Visit www.usphs.gov to learn more.

### *Paths for Computer Scientists*

- Apply for positions in which you already have expertise, such as network configuration or Oracle database administrator positions.
- Consider infrastructure, computer validation, or LIMS positions. You don't need a biology background for entry-level positions, and they may be the most accessible jobs.
- Join companies that support the drug development industry, such as Oracle and IBM. Almost every pharmaceutical and biotechnology company uses Oracle databases and IBM hardware servers.
- Become an expert in the types of programs used in drug discovery and development. For example, there are a limited number of viable software systems about drug safety. Learn how to use these programs, earn an internship, or work pro bono for a few months. For clinical analysis positions, learn SAS programming.
- Study and become familiar with medical terminology and concepts, and develop a more comprehensive understanding of the entire process of drug discovery and development.
- If you have an interest in bioinformatics, consider taking science courses, such as general biology, genetics, and molecular biology. You may ultimately need a Ph.D. in the

biological sciences. Bioinformatics master's programs and certificates are available; although these programs prepare you for an academic approach, it is better to have a foundation in basic biology.

- Consider learning more about computer systems. Take computer validation courses, read the FDA guidelines, and become acquainted with the key concepts.
- Gain experience working in a regulated environment. For example, you can work on computer validation in almost any industry. This experience might qualify you for a job in biotechnology.

*Paths for Scientists with an Interest in Information Technology Careers*

- Obtain a job involving the generation, manipulation, or management of data. This could be as simple as using Excel spreadsheets or using bioinformatics applications.
- Consider becoming Microsoft or Java certified so you can qualify as an expert. Some classes are available on-line and can be completed in six weeks.
- Take some computer courses, such as elementary programming, computer algorithms, Java, XML, and C++. Develop your computer programming and database building skills.
- Become fluent in some scripting language and develop tools for answering questions that are detail oriented. You may want to consider an entry-level position in data entry processing. You can be promoted quickly in data quality control or data integration.
- A good way to learn on the job is to take on laboratory automation responsibilities. This refers to programming robots to handle liquids or to read plates in the lab.
- Become an experienced or "power user" of a computer application as a bench scientist and see if that interests you.
- If you are good at sales and working with customers, consider a position as a Field Application Scientist or Applications Trainer for tools and services companies. Some companies develop large software applications. Such positions are described in more detail in Chapter 20.

## RECOMMENDED TRAINING, PROFESSIONAL SOCIETIES, AND RESOURCES

*Courses and Certificate Programs*

Introductory classes on data management, bioinformatics, cheminformatics, and computer validation.

The Society for Clinical Data Management (www.scdm.org) offers a certification program.

Various universities and extension programs. Some colleges offer master's degree programs.

The Center for Professional Advancement (www.cfpa.com). This Web site provides training on the regulations, computer validation, quality systems, and more.

### Societies and Resources

The Drug Information Association (www.diahome.org) is the largest and broadest society for the drug development industry. It has a special interest group for data management and offers special conferences.

The National Center for Biotechnology Information (www.ncbi.nlm.nih.gov) is the NIH institution dedicated to providing library services to scientists and IT tools designers—an indispensable resource.

The Clinical Data Interchange Standards Consortium (www.cdisc.org) is an industry-sponsored organization on data standards for the interoperability of information systems.

The International Society of Pharmaceutical Engineering (www.ispe.org) is an engineering group focused on using machinery and computer technology to automate manufacturing. They offer a series of GMP guides about good automated manufacturing practices.

The Parenteral Drug Association (PDA, www.pda.org) publishes a technical journal and reports and offers classes in quality assurance, quality control, manufacturing, training, regulatory affairs, validation, engineering, and information technology.

The Association for Laboratory Automation (ALA, www.labautomation.org)

The Society for Biomolecular Sciences (www.sbsonline.org)

The Institute of Validation Technology (www.ivthome.com)

The International Quality & Productivity Center (www.iqpc.com) focuses on technology, controls, and validations.

The American Society of Quality (www.asq.org)

### Books and Magazines

Claverie J.-M. and Notredame C. 2007. *Bioinformatics for dummies.* Wiley Publishing, Indianapolis, Indiana.

Alberts B., Johnson A., Lewis J., Raff M., Roberts K., and Walter P. 2002. *Molecular biology of the cell,* 4th edition. Garland Science, New York. The "bible" for molecular and cellular biology.

*IT World*: A magazine about information technology

*Bio-IT World Magazine* (www.bio-itworld.com) offers a free E-mail newsletter and magazine.

*Pharmaceutical Technology* (www.pharmtech.com), a free magazine

# 17

# Business and Corporate Development

## Why Big Deals Really Are a Big Deal

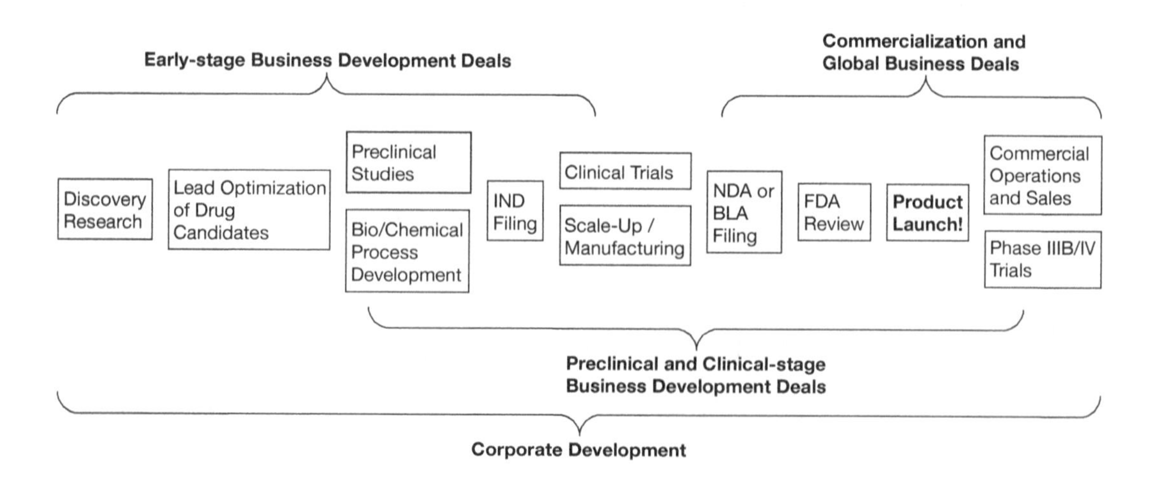


*Business development involves being entrenched in product development, meeting interesting people, and learning about fascinating new scientific discoveries.*

IF YOU WANT TO ESCAPE THE BENCH and work in a field that combines science and business, and if you thrive on interpersonal interactions, consider a career in business and corporate development. It offers high-profile positions that may significantly influence the future of a company. The challenges range from defining strategic plans and identifying deals that will best serve companies' long-term interests to finalizing deals and overseeing their implementation. Skills in analysis, planning, negotiation, contract drafting, and diplomacy are essential, as are decisiveness and the ability to cope with uncertainty, stress, and extensive travel.

## BUSINESS AND CORPORATE DEVELOPMENT IN BIOTECHNOLOGY: A SYMBIOTIC RELATIONSHIP WITH BIG PHARMA

The biotech–pharma partnership can be viewed as a symbiotic relationship. Biotechnology companies tend to be small and versatile, exploring high-risk, cutting-edge technologies, and they typically lack the financial resources to carry out expensive clinical trials. The larger and often much wealthier pharmaceutical companies are more risk-averse and have expertise in manufacturing, clinical trials, and commercialization. However, since their products will eventually go off patent and since potential products can fail during clinical trials, the pharmaceutical companies increasingly supplement their own research efforts by cherry-picking those of the biotechnology companies. Business development professionals identify products with the largest market potential and, through research alliances or other forms of licensing agreements, develop and commercialize these products alongside their own. In doing so, the pharmaceutical companies provide the biotechnology firms with the means and expertise to develop and commercialize their products, while maintaining a constant flow of products to the market. This fulfills their obligation to shareholders, ensuring continued future growth of the company. The establishment of a biotech–pharma partnership immediately increases the value of a biotechnology company and its products, and also generates revenues essential for operational expenses.

## BUSINESS VERSUS CORPORATE DEVELOPMENT

The difference between business and corporate development is nebulous, and the two are sometimes interchangeable, depending on the company. Corporate development is often viewed as the internal team that develops strategy and ensures that there is enough funding and access to other key resources for continued operations. Business development is often seen more as the external face of the company: the deal makers who act on the strategic plan as established by corporate development and the executive management team.

*Business development is about researching and analyzing new business opportunities, transferring that potential value from one organization to another, and ultimately creating real products.*

Note that this may represent an artificial distinction between these two roles. In most emerging biotechnology companies, the business development and corporate development departments are one and the same. As companies mature, business and corporate development tend to become two separate departments, although both continue to provide financial support for the scientists by soliciting financial sponsorship.

It should be pointed out that in more commercially driven companies, particularly service organizations, "business development" is often a colloquialism that really means "sales." So if you are applying for a "business development" position, be sure to check under the hood!

## Corporate Development: The Strategy Makers

The main role of a corporate development department is to work with the executive management team to establish a strategic plan for the company and to raise capital for operational expenses. Corporate development executives may be involved in defining the therapeutic hypotheses or technical platforms that will validate the research programs. They make sure that a progressive, disciplined approach is being taken, and that the company scientists are working on viable programs which are financially sound. In emerging companies, corporate development professionals often work with the Chief Executive Officer (CEO) and Chief Financial Officer (CFO) to raise money, usually from venture capital or corporate partners, to ensure that the company is adequately funded.

## Business Development: The Deal Makers

*Business development is the job of the entire company: A business development professional's role is to orchestrate that process.*

The primary function of business development is to consummate deals that will further the company's strategic development plan as established by the executive management team. It opens doors to potential partnerships and facilitates the flow of information between collaborating parties. It aims to generate revenue by licensing company assets for use by outside companies (out-licensing) or to fill gaps in the product pipeline by acquiring new technologies or products from external sources (in-licensing). In smaller companies, the senior business development executive is a central member of the executive management team.

### *IN- versus OUT-licensing*

The process of selling products to other companies is called "out-licensing," whereas the process of acquiring technology from others is called "in-licensing." Out-licensing typically requires more of a sales mentality: how to find the best customers, how to get the best prices for the technologies. Those on the in-licensing side tend to be more of the scientific ilk, who evaluate and analyze new technologies. They scrutinize the utility and application of the technology, decide whether the concept makes scientific sense, how well it will fit with other programs, whether or not it is in line with corporate objectives, etc. Some people specialize in one area, but it is best to have experience with both.

## CAREER TRACKS IN BUSINESS AND CORPORATE DEVELOPMENT

There are several careers that fall within the category of business and corporate development. These include portfolio management, search and evaluation/sourcing, scientific licensing, technology transfer, and alliance management. Note that in most smaller biotechnology companies, business development professionals might serve in some or all of these roles.

*Many biotechnology companies have a "one-stop business development shop," whereas pharmaceutical companies delineate the various business development roles and responsibilities into several financial and scientific disciplines.*

### *Portfolio Management: Analyzing the Portfolio to Develop Strategy*

When companies have multiple products and limited financial resources, efforts must be focused on those products that have the highest probability of success and the greatest potential for return on investment. Portfolio managers examine the minutiae of each product and present the data to the corporate development and executive management teams, who can then make informed decisions as to which products should be given priority.

Biotechnology companies generally do not recognize portfolio management as a distinct discipline, and it may form part of the business development, commercial operations, research and development, or project management departments, or, more commonly, is outsourced to vendors specializing in portfolio management. In pharmaceutical companies, portfolio management typically has its own separate department or is part of the research and development department, and may involve as many as 30 employees. Portfolio managers must have extensive knowledge of drug development and a strongly mathematical and analytical mind.

### *Search and Evaluation/Sourcing: The Deal Finders*

Large pharmaceutical companies may employ whole groups of experts whose sole task is to "search and evaluate" new technologies to in-license. They travel around the world identifying potential products and conduct technical assessments, which include forecasts of development costs, potential competition, and more. They may also work with the transaction team and participate in business development negotiations.

Careers in search and evaluation can be intellectually rewarding, with exposure to multiple research areas and the latest cutting-edge technologies. The role generally requires a minimum of 10 years' experience in drug discovery and development, because extensive knowledge is required to fully appreciate the related issues. One of the quickest ways to gain that knowledge is to work as a project manager, a job that involves the entire gamut of drug development.

### *Licensing Professionals: The Deal Closers*

Licensing is sometimes carried out by search and evaluation professionals, but it may also be carried out by a separate department. Licensing officers tend to have backgrounds in finance, business development, intellectual property, or law (particularly transactional law). They work on the various factors involved in closing deals, including negotiating and designing payments and arranging the final terms of a deal.

#### *Office of Technology Transfer: Licensing Technology from the Universities*

In universities, licensing matters are dealt with by the Office of Technology Transfer (OTT), which manages the university's intellectual property resources and helps to commercialize new technologies derived from academic research. The OTT manages every phase in the development of inventions. When an invention is disclosed by a faculty member, the OTT assesses its potential usefulness in industry. The technology is marketed to potentially interested parties, and licensing terms are drafted. The technology transfer agent also oversees compliance with the terms and agreements of the license over time.

Many people at the OTT are former graduate students and, more often, former postdocs interested in business and the direct application of basic research to industry. Working at the OTT provides excellent training for future business development or law careers. There is constant turnover in OTT departments from academia to industry, but keep in mind that OTT positions are highly coveted, and some people offer to work pro bono before being granted full-time employment.

#### *Alliance Management: The Deal Implementers*

*Alliance managers are the "glue" that holds partnerships together.*

During negotiations, the department heads and both alliance teams cooperate to draft an integration plan and also explore whether both parties can work together effectively. Each alliance team remains in charge and manages the partnership after the deal is completed.

The alliance management team may be involved in early negotiations, but its main role is to implement the terms of the deal while remaining mindful of the original intent of both parties, and to manage partnerships after the deal has been signed. Its initial job is to provide continuity between negotiations and deal implementation. Alliance managers serve as the main point of contact for internal and external communication for the duration of the partnership.

In smaller biotechnology companies, alliance managers may serve as project managers, overseeing the implementation of the goals and terms of the deal and partnership. In large pharmaceutical companies, alliance managers may be experienced, high-level executives and, as such, will have significant influence within the organization. They usually possess the extensive experience needed to deal with the broad range of issues that may arise from legal, manufacturing, or financial dealings.

Alliance managers must possess exceptionally good conflict-resolution and diplomacy skills. The best alliance managers are skilled at resolving disagreements and finding mutually beneficial solutions. They must understand how to quickly assimilate large amounts of new information, shape assets, and maintain a clear vision that is focused on success.

## BUSINESS AND CORPORATE DEVELOPMENT ROLES AND RESPONSIBILITIES

### *Corporate Development and Strategy*

Before the deal making begins, the executive management team determines the long-term strategic plan for the company. Major decisions are made about whether to go public or to be acquired, for example, and to ensure that the corporate strategy is in line with the company's long-term goals and mission. Other strategic decisions involve the judicious acquisition and out-licensing of assets, the therapeutic focus, and more.

### Competitive Intelligence

Business development professionals devote considerable time to studying industry news. They must keep in touch with the market and its drivers, trends, and catalysts, and with potential competitors, particularly in the company's own therapeutic area and specific technologies.

### *Portfolio Management*

Portfolio managers examine the minutiae of each product and present the data to the corporate development and executive management teams to help them prioritize projects. Calculating the benefits of developing or acquiring certain products in preference to others can be exceedingly complicated. Managers must apply highly structured, logical, and systematic approaches, using sophisticated, and sometimes customized, decision-making computer models. Many factors are taken into account, including costs, regulatory hurdles, and the potential commercial value of each product at each stage of development. The information gathered is used by senior management professionals to make decisions that will ensure maximum returns across the portfolio.

### Sourcing

*Sourcing is about shaking trees and gathering the interesting fruits that drop.*

Search and evaluation professionals, or "technology scouts," spend their time hunting for new technologies and products that might be suitable for in-licensing. In large pharmaceutical companies, a whole team of search and evaluation professionals constantly reviews the range of potential products and brings in the more promising ones for closer examination.

### *Analytics*

*Truly revolutionary deal structures can be made possible with strong financial analyses.*

Financial analysis is a greater part of business development than most people initially presume. Understanding the markets and the financial models allows teams to be able to identify the best "win-win" scenarios for both companies. Without a detailed financial analysis close at hand during negotiations, terms may be set that do not necessarily encompass the deal's true value.

Computer modeling, forecasting, and various other calculations are conducted to produce realistic financial valuations of each asset. Mathematical analyses are used to estimate the return on investments, and, together with forecasting data, this information is used to make better informed strategic decisions. These analyses are also an important means of quantifying and mitigating the risks of product development.

### *Due Diligence*

*It is important in business development to be aware of the warts on an asset!*

Before in- or out-licensing deals are struck, a thorough assessment of the intellectual property (IP) (e.g., patents), technology, management, and more is made in a process called "due diligence." This process minimizes the risk of nasty surprises further down the negotiating line and validates the technology and the value of a deal. IP frequently serves as the foundation for collaborations or strategic acquisitions, and IP lawyers often work with the business development teams to assess the value of the patents and to consider any legal issues that might arise.

### *Negotiating Deals*

The interval between the start of negotiations and the closing of the deal can be long—sometimes nine months to a year—but only a small fraction of this time is spent actually negotiating with the other party. The vast majority of negotiations take place within the company. Business development professionals must gain the support of senior managers by presenting a sound business case for each potential deal. This can be an uphill battle. Managers may have vested interests in supporting certain deals and rejecting others. As a consequence, business development can be a complex political challenge, requiring refined diplomatic skills and thorough preparation.

### *Alliance Management*

*The success of a deal ultimately depends on how well is it implemented.*

The alliance management team ensures that the terms of a deal are properly implemented, with respect not only to the legal requirements, but also to the corporate objectives. In some companies, this may be seen as a project management role, involving the coordination of joint partnership teams, creation of work plans, and organization of meetings, all to make sure that everyone is doing what they are supposed to be doing. The alliance managers are responsible for communicating the partners' needs to their own teams and vice versa, offering solutions to problems and expediting their implementation, and ensuring that the relationship remains positive and productive.

### *Investor Relations and Public Relations*

In some companies, the role of business development can extend to investor and public relations. The staff involved with investor relations (IR) presents the external face of the company to the public or private investment communities. They are frequently called upon to articulate a message and develop a communications platform for investors and future customers. For young companies, which may not yet have a marketable product, public relations (PR) is used to create a buzz and thereby generate interest from the venture capital community. As companies grow, they increasingly rely upon PR agencies to raise their profiles in the media.

### *Legal Interface*

Another role that might be performed by business and corporate development departments involves the company's legal interface. Sometimes highly complex transactions with consultants, clinical research organizations (CROs), suppliers, and partners are handled by business development professionals, although this function is usually carried out in conjunction with lawyers.

### *Networking*

Business development professionals spend a great deal of time monitoring the marketplace and building contacts with other executives in the industry. Networking is useful for learning about other companies' interests and determining how your company's technology might be included in a potential partner's strategic plan.

*Business development is like Brownian motion: The more people you meet and talk to, the greater are your chances of creating future business connections.*

### *Negotiating deals*

When negotiating a deal, it is important to establish that the terms and conditions are in tune with the company's strategic plan. Which aspects of the deal will benefit the company? Do the benefits adequately compensate for the investment being made? Which terms are flexible and which could be deal breakers?

Deals take an average of nine months to complete. They can be derailed by many factors, including competition (one of the biggest factors), an inability to agree on terms, a change in corporate strategy by either party, or new negative data from clinical trials, to name but a few. One of the major challenges is to get all of the necessary functional heads to agree on the exact terms of the deal... a frustrating task that is often referred to as "herding cats" (see page 233, "Greatest Challenges on the Job"). Deals are not signed until the last moment, and relations can turn sour overnight. Because outcomes are generally unpredictable and will often depend on factors outside your control, the whole process can be extremely stressful.

After months of negotiating, the two business development teams will know each other well and will have established good rapport (this is also known as "building relationships"). An understanding of the partners' expectations, priorities, and corporate plans will have been reached, and real trust, built on mutual self-interest, will have developed between the parties. If all goes well and the deal is signed, the alliance management team will then move in to implement the deal.

## A TYPICAL DAY IN BUSINESS OR CORPORATE DEVELOPMENT

*Because of the highly variable nature of the roles in business and corporate development, there is no "typical" day. What follows is a general picture of how a professional might spend the day.*

- Responding to phone calls and E-mails.
- Schmoozing on the phone: renewing contacts, discussing progress, listening to problems, and discussing potential solutions.

*Business development professionals spend a great deal of time on the phone, testing the pulse of the marketplace.*

- Reading industry news to keep up to date with the markets.
- Attending internal meetings with cross-functional core teams at operational meetings, strategic sessions, and board meetings. Discussing the merits of potential deals with senior managers.
- Catching up with analysts, and perhaps financial modelers. Soliciting feedback from functional heads regarding financial forecasts.
- Trawling for new opportunities; scouring the marketplace for in-licensing opportunities and passing them on to appropriate colleagues (usually scientists) for further review and evaluation.
- Preparing for, arranging, and attending business development meetings, both in-house meetings and external conferences and symposia.
- Reviewing draft contracts with lawyers.

## SALARY AND COMPENSATION

Business development positions tend to be more lucrative than comparable scientific positions. For similar levels of experience, a business development professional will earn 10–20% more than a scientist. Director-level business development executives can sometimes earn as much as their VP-Research counterparts! This is partly due to supply and demand, as there are few top-notch business development professionals, but it is also a reflection of their prominent role in the organization: Their transactions have a direct and immediate effect on the value of the company.

***How is success measured?***

In general, a successful business development professional is measured by his or her "deal sheet," and also by the long-term value those partnerships contribute to the company's success or perceived financial impact. It is easy to spend a lot of time on deals that eventually fall apart, and the number of deals signed will not reflect the effort made. In addition, knowing which deals *not* to partner can be just as important as signing the more promising deals.

## PROS AND CONS OF THE JOB

### Positive Aspects of a Career in Business and Corporate Development

- There is rarely a dull moment—a constant supply of new and varied tasks crops up each day, which provides an abundance of mental stimulation. One day you may be working on pricing issues and another day dealing with licensing or financial details.
- It is an excellent way to use your scientific, business, and analytical skills without doing bench work. You are continually exposed to exciting new scientific discoveries and technologies.
- Results come more quickly in business development than in research, and the accomplishments are more immediately tangible. You can be working on several deals in parallel, each of which has a discrete end point and, if successful, will have an immediate and significant impact on the company.
- Negotiating can be exciting. It is like playing poker—you must be able to read the other side's body language without revealing your own. There is a real challenge in being able to understand and accommodate a potential partner's needs, while molding the deal to suit your own company's interests first. Most people have a tremendous sense of achievement when everything falls into place and a deal is finally signed.

*Business development is like playing poker—you need to know when to hold and when to fold.*

- Business development is at the heart of the company: It provides an opportunity to work with many different functional department heads, from science and law to finance, and to learn about the entire product development process.
- There are frequent travel opportunities; a real plus for some people. However, see also The Potentially Unpleasant Side of Business and Corporate Development, below.

### The Potentially Unpleasant Side of Business and Corporate Development

- There can be extensive travel. Business professionals can easily spend more than 50% of their time away from home. Staying in hotels and eating out quickly lose their appeal and can put a tremendous stress on family life, and travel plans are frequently unpredictable.
- Constant networking and schmoozing can be tiring. Too many business lunches and events take their toll, especially if your success rate does not reflect the hours spent.

*A deal can turn on a dime. You don't know the outcome until the very last minute!*

- It is a stressful job. The pressure to get deals done can be extreme—upper management often fails to appreciate how time-consuming and difficult some deals can be to complete. Every day is a crisis; when handling multiple deals, there are multiple deadlines to be met, emergencies to deal with, and "fires" to put out.

- When out-licensing products, you need to develop a thick skin; refusals can seem endless, and it is easy to become discouraged. To succeed, you must have boundless resilience and an unremitting will to generate new leads.
- Others think that they can do business development and that *they* could have gotten a better deal. Be prepared to be second-guessed by people in other departments.
- It is not uncommon for deals to fail, and it is disappointing when they do. You can spend a lot of time and effort on a project, only to see it fall apart, often due to circumstances beyond your control. However, even a failed deal will increase your network of contacts and may lead to something promising later.
- Your deal sheet may be constrained by the company's corporate strategy, budget, or other factors beyond your control. If this happens, it may be time to look for new pastures.
- Contracts can be boring. They are typically 20–100 pages long and must be studied in great detail, a tedious and uninspiring task for many people.
- Business development can be very political; the results of your efforts have a direct effect on other members of your company, and coworkers can feel threatened. For example, researchers who have worked on a product may take it personally, or even lose their jobs, when it is out-licensed. You can easily stumble into a political quagmire if you are not careful.

## THE GREATEST CHALLENGES ON THE JOB

### *"Herding Cats"*

Getting the functional heads of the company to agree with an idea, making them comfortable with the terms of a deal, coercing them into attending meetings, and involving them in due diligence and planning is a near impossible task—it's like trying to "herd cats." Every functional head in the company has a seat at the table, and each may voice an all-too-frequently strong opinion. This process is important, because the success of a deal depends on proper implementation, and the participants need to agree with the deal's terms, because they will have to live with the consequences. Assembling functional heads for meetings for each issue that is raised during negotiations and convincing management that a particular deal is in line with corporate objectives is no easy task, particularly when big egos and personal interests come into play.

### *Timing and Being in Line with Corporate Objectives*

*Make the deals that you can make. What's hot today may be ice-cold tomorrow!*

The business development team is always eager to complete deals, but there are major obstacles to overcome: Management may have over-valued the asset; perhaps the company's strategic plan is limiting; perhaps its budget is too small; it may even be that the technology is out of date and it is simply impossible to generate interest. The most important challenge, however, is to ensure that the outcome of the negotiations is in line with the company's corporate objectives. The outcome will have an impact on the future of the organization, so the deal had better be a good one. If it is not well structured, there could be long-term and large-scale consequences for the company. As mentioned earlier, it is never a good idea to relinquish control of an asset to a partner, but you must make sure that the partner has some incentive to take the product further. If the partner loses enthusiasm, progress will slow, and development might be halted altogether. It is vital that the deal provide both companies with incentives to carry the project through to fruition. Otherwise, the "Not Invented Here" (NIH) syndrome might result.

*The most successful business deals create value for both parties by aligning interests and maximizing value.*

### *The "NIH" Syndrome*

No, not the organization in Washington, D.C., but the "Not Invented Here" syndrome in business development. When a company in-licenses a technology or product, the acquiring research team may view it as a demonstration of their own failure to come up with a product themselves. They may even be forced to abandon their own projects in favor of the new arrival, which can cause a significant degree of resentment and a loss of motivation. In some cases, attempts may be made to sabotage the new product. Fortunately, this problem is now on the wane as large pharmaceutical companies are being much more acceptant of in-licensing.

## TO EXCEL IN BUSINESS AND CORPORATE DEVELOPMENT...

### *It Takes Many Years of Experience*

In general, experience is one of the keys to success in business development. The more relationships you have, the easier it is to make even more, and relationships are what drive deals. Experienced staff are more adept at negotiations. They are better at evaluating people, products, technologies, and company strategies, and they know when to wait and when to act, when to advance and when to retreat. They can engage quickly with potential partners and proceed with negotiations more efficiently.

### *Strategic Negotiation Skills and an Ability to Empathize*

Successful business development professionals are intimately familiar with their company's corporate mission and are quick to recognize the interests of their potential partners. This understanding leads to better relations with partners, and the joint motivation to succeed provides a better chance to anticipate problems. The better the analytical analyses are, the more foresight a negotiator has and the better the "prenuptial" agreement will be; i.e., the better prepared the company will be to deal with problems in the relationship, which may occur years later. If these do occur, and separation really cannot be avoided, then it should be achieved as amicably as possible: no easy task when both companies are vying for the major rights to the same product. If either side relinquishes control to the other, they put themselves at the mercy of their former partner, a potentially fatal position. An experienced negotiator will stand a better chance of coming out on top.

*Business development is like courtship: It's about arranging marriages on solid fundamentals and reducing the likelihood of divorce.*

## Are You a Good Candidate for a Career in Business and Corporate Development?

***People who flourish in business and corporate development careers tend to have...***

***Exceptionally good interpersonal skills.*** (See Chapter 2.) Interpersonal skills are needed for building relationships and facilitating communication internally and externally.

***Excellent diplomatic skills.*** Because of the politics of business development, you should be able to handle yourself well and deal confidently with internal issues. To succeed, you should be able to diffuse personal conflicts, which can often arise during tense negotiations. A really good business development person is able to avoid such conflicts by taking preemptive action.

***The ability to listen well and be highly perceptive.*** To understand the customer's needs and aspirations, you must be able to ask the right questions and listen to the answers. You should also be able to read body language. This will help you to understand how your customers really feel, which is usually more revealing than what they say.

***The capacity to see minute details and the big picture at the same time.*** You have to be able to understand exactly how a deal fits into the corporate strategy and translate that into the actual language of the agreement. This will allow you to concede on items that are unimportant while understanding the "deal-breakers" that cannot be negotiated.

***Excellent communication skills.*** These positions require the ability to communicate the aims and potential of the project, its progress, and achievements; first to internal managers, then to potential partners, and finally to negotiators. You must be able to tell a coherent story and connect with customers in a persuasive manner.

***Strong analytical skills and the ability to think quickly.*** When negotiating deals, it is important to be able to evaluate their financial and business implications on the spot. Portfolio managers need a strongly mathematical and analytical mind to handle such complicated and sophisticated calculations.

***Exceptionally good strategic and problem-solving skills.*** It is an advantage to be able to think quickly and creatively and come up with the most efficient solutions to problems.

*To be good in business development, you need to be extremely strategic, brutally honest, mentally tough, and able to defend your position.*

***Tenacity and patience.*** Business deals can take a long time and can be frustrating and stressful. It is important to remain calm if you want to collect. You must be relentless in your approach and be able to overcome obstacles.

*An outgoing and gregarious personality.* This position requires solid schmoozing and networking skills to develop a stuffed Rolodex of contacts. In addition, excellent conversational skills are a must.

*Visionary leadership skills.* Business development professionals must be persuasive and able to convince upper management, and others, of the virtues of their product. When closing a deal, they must take charge, rally the troops, and lead them through the entire process.

*Efficient multitasking, organizational, and project management skills.* It is not unusual for a business development professional to be working simultaneously on 6–20 deals. Projects must be prioritized and you must be able to cope with an ever-changing schedule.

*A willingness to take calculated risks.* Business development professionals spend a great deal of time recommending proposals to senior management. There is a fine line between an innovative technology that could be a major earner for the company and an idea that will fail at the first hurdle, but the successful professional cannot afford to stick only with the safe options.

*The ability to sell.* This talent is central to the role of the business development professional for out-licensing deals and for internal negotiations.

*Tremendous cultural awareness.* This job involves interacting with people from many diverse backgrounds.

*A broad knowledge of the industry.* This includes not only business, but also science and legal issues. A deep understanding of drug discovery and development is required.

*The ability to be a self-starter who takes the initiative.* You need to be able to pursue deals—they don't just walk in the door.

***You should probably consider a career outside of business and corporate development if you are...***

- Inflexible and easily frazzled by crises, unpredictable schedules, and frequent re-prioritization of tasks.
- Someone who prefers continuity in one's work.
- Too quiet or too talkative.
- Submissive, passive, or indecisive.
- A person who tends to get lost in the details and loses sight of the overall goal.
- Afraid to attend meetings or parties where you do not know anyone.
- Argumentative, obstinate, or stubborn.
- Someone with a volatile or unpredictable personality.
- Arrogant, egotistical, and insensitive to the needs and sensibilities of others.
- Too aggressive or too salesy.

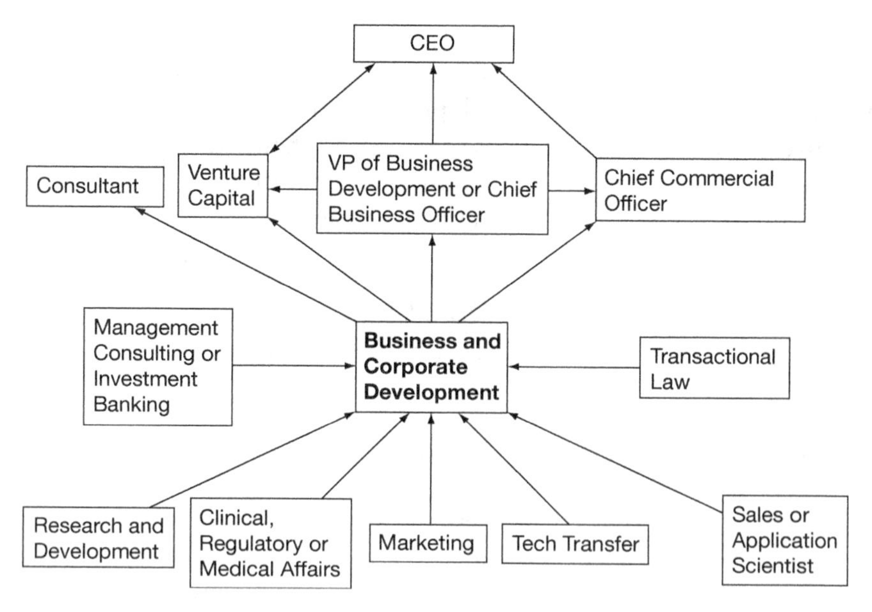


**Figure 17-1.** Common career paths in business and corporate development.

## BUSINESS AND CORPORATE DEVELOPMENT CAREER POTENTIAL

Business development experience provides preparation for a number of executive operational positions, including chief business officer (CBO), chief operating officer, general manager, and CEO (see Fig. 17-1). "CBO" is generally used to define a senior-level business development position on the executive management team. Typically, CBOs are involved in business and corporate development and frequently, but not always, finance, sales, or marketing heads may report to them.

It can also be a stepping-stone to a range of other careers, including management consulting, biotechnology analysis in financial markets, finance and accounting, investment banking and/or fund management, sales, marketing, or executive search positions, to name but a few. Business development experience teaches a deep appreciation for markets, the ability to make well-tuned judgments, and the capacity for making deals...excellent skills for a future in venture capital.

### Job Security and Future Trends

In general, there has been a steady growth in demand for business development professionals. Experienced talent is in short supply, partly because very few companies offer manager or entry-level business development positions, and partly because of the low volume of trade in the biotechnology industry. There are simply not enough deals through which to train people. Once trained, business development professionals become highly coveted by companies.

There is a high turnover rate for business development positions. This is partly because of the limited supply and high demand mentioned above. Business development professionals may also move on rather quickly because they have exhausted their companies' deal-making capacity; i.e., they have put themselves out of work by completing all the available deals!

*Business and corporate development represent a small community—Word seems to get around in about two nanoseconds.*

## LANDING A JOB IN BUSINESS AND CORPORATE DEVELOPMENT

### Experience and Educational Requirements

There is no typical background for business and corporate development professionals, and they can transfer from just about any functional area in biotechnology or pharmaceutical companies. People are often promoted from inside a company, most commonly from research, and also from the sales or marketing departments. People with strong finance backgrounds tend to do well in business development, as do management consultants, investment bankers, and equity research analysts. Another common path is via technology transfer.

There may be classes that can train you in business development, but the best education is through real-life experiences. The best way to learn is to be apprenticed by an experienced professional who can show you the tricks of the trade.

As mentioned earlier, a direct move into business development from academia is not necessarily the best route. It is far better to gain some biotechnology industry experience first, operational and project management experience in particular, before entering business development. Extensive knowledge and the understanding of the complex nuances of biotechnology's commercial nature are highly useful qualities. As a business development professional, you will draw on your industry experiences. A lack of such experience may make it more difficult to move up the business development career ladder.

Once you have some business development exposure, it is advisable to gain experience with both in- and out-licensing, mergers, acquisitions, going public, and venturing. This will increase your marketability.

Most people agree that, for business development, personality attributes are more important than educational qualifications, and there are generally no fixed requirements. That said, hiring managers generally prefer applicants with a science degree and some business qualification. An M.B.A., a law degree, or an M.D. is desirable, and possession of both a Ph.D. and an M.B.A. is generally seen as the optimum. Some positions, however, have more specific educational requirements. For example, the "search and evaluation" teams typically require a Ph.D. or M.D., or a Pharm.D., whereas transactional roles may require an M.B.A. or a law degree.

There is a growing tendency to hire Ph.D.s and M.D.s in business development. Being able to talk credibly, answer questions, and summarize in scientific language are real assets. It is often easier to train a scientist in business sense than to teach a business professional about science. That being said, a Ph.D. scientist working in business development is a business person first. He or she can liaise with research scientists within the company or outside to understand the scientific validity of potential deals.

## Paths to Business and Corporate Development

Business development positions are very competitive, and it can be difficult to get your foot in the door. Entry-level applicants must demonstrate that they are truly interested in business.

- Network, network, and network—it is all about who you know. An extensive network is needed for business development, and contacts can get you in the door. Meet people at business development conferences and ask for referrals. If you network enough, you will find someone who is hiring, or you will be referred to a recruiter with a job opening.
- If you are working in a biotechnology company, ask to take on special projects in the Business Development department. You might evaluate new technologies or conduct due diligence, for instance. Consider moving into alliance, project, or portfolio management positions, which will provide exposure to deals. It is a small and relatively easy step from here into business development. Gaining and demonstrating project leadership capabilities will also prove highly applicable to business development roles.
- You should learn as much as possible about drug discovery and development. This will help you to develop the insight needed to pick winners. One of the fastest ways to learn is to work in project management. As a project manager, you will gain a broader perspective and exposure to the entire drug discovery process. You will also acquire other valuable skills that are directly applicable to business development.
- Consider getting an M.B.A. degree. An M.B.A. coupled with a strong background in science is a sound basis for an entry-level business development position. If an M.B.A. is not an option, consider other ways to gain business exposure. Become familiar with patent law and biotechnology industry basics. Subscribe to industry newsletters and stay informed about what is going on in the field. Learn how the industry works, who are the major players, what are the current terms, and familiarize yourself with typical deal structures, etc.
- Consider interning or working at a university office of technology transfer. Unfortunately, technology transfer positions tend to be highly coveted. You might consider becoming a patent agent and working at a law firm or even interning pro bono to land a position as a technology transfer officer. Entry-level employees generally start as licensing analysts or liaisons and move up to become licensing officers after a few years. With technology transfer experience, you can then join a law firm or go into business or corporate development.
- Consider working at a management consulting company, in investment banking, or in equity research as a way to first gain industry and business exposure. Some of the larger establishments provide mini-M.B.A. programs that could help you bypass the cost of funding your own M.B.A. degree.

- Apply to large, established companies that are financially stable and can tolerate technical and financial risks. Try to apply to companies with a wide array of business development needs so that you can be exposed to a variety of deals. More secure companies can entertain large-scale deals that will change the makeup of the company, whereas smaller companies are more reliant on other companies for their survival and have fewer options and fewer deals to make. This is the best way to achieve total immersion in business development.

## RECOMMENDED TRAINING, PROFESSIONAL SOCIETIES, AND RESOURCES

### *Courses*

The Licensing Executives Society (www.lesi.org) offers a three-day intensive course in business development.

The Association of University Technology Managers (www.autm.net) offers business development courses with classes on statistics and finance.

### *Classes*

Finance, including financial modeling

Marketing and positioning

Patent law classes taught by the Patent Resources Group (www.patentresources.com) or the Practicing Law Institute (www.pli.edu)

IP assessment

Negotiation and facilitation classes

### *Business Development Societies and Resources*

The Licensing Executives Society (www.lesi.org)

The Association of University Technology Managers (www.autm.net)

### *Books*

C. Robbins-Roth. 2001. *From alchemy to IPO: The business of biotechnology.* Perseus Books Group, New York.

R. Fisher, W. Ury, and B. Patton. 1991. *Getting to yes: Negotiating agreement without giving in.* Penguin Books, New York.

W. Ury. 1993. *Getting past no: Negotiating your way from confrontation to cooperation,* Bantam Books, New York.

### *Free On-line News Services*

Signals Magazine (www.signalsmag.com)

BioSpace (www.biospace.com); subscribe to "Genepool" and "Deals & Dollars," free newsletters describing life science industry news and deals

Fierce Biotech (www.fiercebiotech.com), a free daily E-mail service providing industry news and deal flow

The Biotechnology Industry Organization (BIO; www.Bio.org)

### Biopharma News Sources That Cost Money

*Subscription news services cost a considerable amount of money, but provide more in-depth analyses than the free services. Investing in them may not be justified for job seekers, but since most business departments use these sources, you should be aware that they exist.*

BioWorld Online (www.bioworld.com)
BioCentury (www.biocentury.com)

# 18

# Marketing

## Communicating a Message to Customers

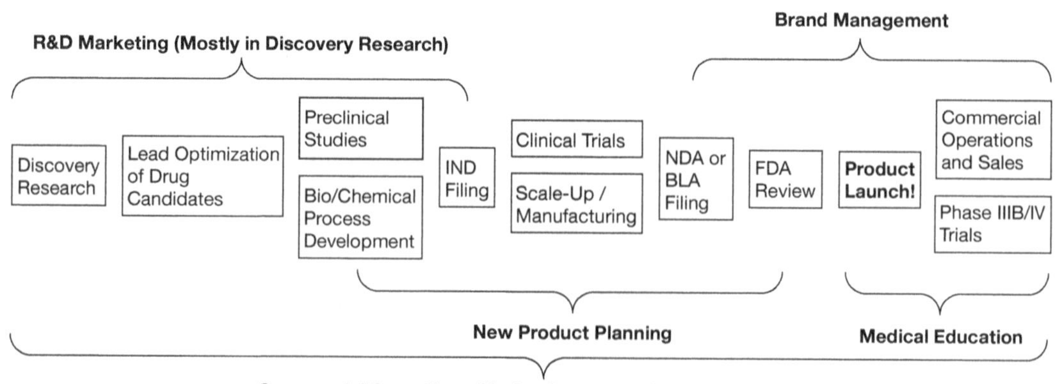


MARKETING IS OFTEN THOUGHT OF AS A SIMPLE MATTER of developing clever slogans or likable advertisements. It is actually a data-driven, rational science. Marketing professionals follow an orderly approach in the development of strategic plans to achieve their objectives. There is a science to and a fascinating psychology behind "positioning" products so that they convey to consumers messages that will result in sales. Marketing professionals may be involved from the earliest stages of a product's development all the way through managing brands after launch. They advise on product development plans and assess the commercial viability of business development deals.

*A good marketing campaign requires tactical and creative brilliance.*

Marketing is a great way to apply your scientific, medical, and business acumen. This field is extensive and offers the potential to move into a succession of very different jobs, all within the marketing department, and all requiring continuous learning and development. Skills acquired in marketing, such as developing strategies and carrying the responsibility for revenues, provide excellent preparation for executive management positions and a high-level business perspective of the biotechnology industry. Biotechnology marketing professionals are exposed to a broader range of job responsibilities and experiences than are sales staff, and their skills are enhanced as they gain an understanding of how value is created and how markets are developed.

Marketers tend to be motivated by the intellectual diversity of their daily jobs. They enjoy building and managing a business and maintaining a long-term commitment to a product. The fast-paced, accomplishment-oriented, collaborative environment makes for an exciting day's work.

## THE IMPORTANCE OF MARKETING IN BIOTECHNOLOGY AND DRUG DEVELOPMENT

*Marketing professionals communicate the value of products to potential customers.*

Customers (patients, doctors, and scientists) are overwhelmed with the volumes of material and ads containing product information, and they struggle to assimilate the most appropriate material for their needs. In the broadest sense, the marketing professional's goal is to accelerate the process by which the medical and scientific communities learn about and eventually adopt novel and superior technologies.

Without marketing, a new technology may go unnoticed, but with a superb marketing campaign and a product that fulfills expectations, it could become a widely applied standard of practice. Of course, if the product falls short, it will not be adopted by anyone, regardless of the quality of the marketing campaign.

*Marketers are the generals who provide direction and support the foot soldiers of the sales force.*

Marketers serve several roles in the product development and launch process. Early in product development, they provide commercial advice and analyze the financial viability of products. They help guide product development efforts in the quest to outperform the competition. Later in the product development process, they provide sales staff with strategic support, sales forecasts, and other resources. In exchange, the sales force works closely with customers and continuously relays customer feedback to the marketing department.

## CAREER TRACKS: BRAND MANAGEMENT AND OTHER MARKETING-RELATED FUNCTIONS

*Marketing is a hodgepodge of disparate disciplines with no inherent structure.*

There is no standard career path in marketing. Most individuals move between sales, marketing, and commercial operations to take on different responsibilities (see Chapter 19).

Marketing is a broad and diverse occupation that includes many subspecialties (see Fig. 18-1). For most people, the word "marketing" conjures up the concept of "brand management," but marketing also includes many other related roles. These roles may be performed in commercial operations departments or in a separate marketing department. For the purposes of this chapter, we will describe them as "marketing," even though in the real world they may be considered as a slightly different function.

### *Brand Managers (also called In-Line Marketing)*

*Each brand is like a mini company.*

Brand managers are commercially responsible for a "brand," i.e., a product. In a company such as Honda, for example, one

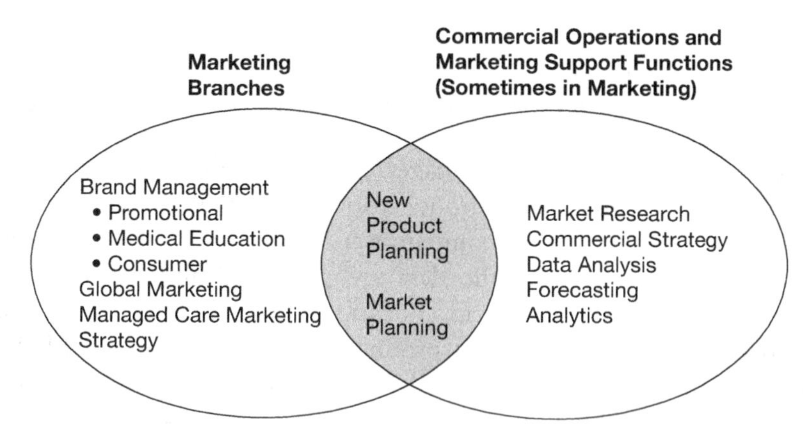


**Figure 18-1.** Representative marketing and commercial development departments.

of the brands is the Accord. The brand manager is in charge of everything associated with the Accord—from defining what new features customers might want, to working with manufacturing in the design of consumer-requested features. The brand manager is also responsible for advertising the brand directly to consumers by highlighting its superior qualities and subtly bashing the competition. The real challenge is to view marketing from the perspective of what a potential consumer might want to hear: the practical benefits of the product, its emotional impact, and the associations that it evokes in people's minds. For example, a Honda Accord is essentially a mode of transport, but its image may be associated with more intangible qualities such as luxury, reliability, and prestige.

*A brand manager is at the hub of marketing activities for a product.*

Brand managers define the overall strategy for marketing a brand and lead multidisciplinary teams to drive projects and sales. These teams can include representatives from the marketing, sales, commercial operations, market research, legal, clinical, regulatory, and manufacturing departments. They can consist of one person or a large, perhaps 30-person, department with additional support groups. The brand manager of larger brand teams may be a product director, business unit director, marketing director, life cycle leader, executive director, or vice president of marketing and sales. These leaders are often responsible for the profit and loss of the entire business unit or brand.

***Within brand management teams there are four types of product managers, who are responsible for...***

***Promotional marketing.*** Product managers in promotional marketing oversee the process of advertising to customers. They create the branding elements and a recognizable "look and feel" of the product. They develop material for sales representatives and are involved in product positioning and message development, which are applied to direct promotions to customers, public relations, and corporate communications. Product managers in promotional marketing support only "on-label" information; i.e., material about drugs officially approved for a particular disease state in the package insert of a drug ("the label"). "Off-label" information, on the other hand, is developed by the medical education department (see below), which publishes newly generated information about approved drugs such as information about drug safety, drug–drug interactions, new disease indications, and more. Once these data become incorporated into the Food and Drug Administration (FDA)-approved labeling, off-label information and messages often evolve into formal product positioning and messages.

*Medical education marketing.* Product managers in medical education communicate newly generated clinical data to physicians through continuing medical education (CME) programs or through meetings with the company's consultants. They invite speakers, organize symposia, and make sure that the data are published and presented at scientific meetings, patient advocacy groups, and educational programs. Because these activities involve data that are not yet within the product labeling (and are not FDA-approved), they often collaborate very closely with the R&D department to ensure that they are representing the data appropriately and are not making false claims. They may also be involved in developing the key messages for commercial publications.

*Patient, consumer, community, or direct-to-consumer marketing.* Direct-to-patient or direct-to-consumer (DTC) marketing plays an increasingly important role in successful marketing campaigns. The increasing patient involvement in treating diseases and the proliferation of patient advocacy organizations have made patients a critical audience for marketers. Although marketing of some of the big brands (e.g., those for cholesterol reduction or erectile dysfunction) still relies on major television and print advertising campaigns, many specialty drugs are advertised on the Internet and in other nontraditional marketing channels, where they can reach their target patient population more specifically. For example, some companies buy ad space from Internet search engines, which then link the advertisements to search terms associated with the drug's targeted disease. Companies also work with patient advocacy organizations to include informational brochures or advertisements in membership mailings.

*Strategy.* Strategy product managers assess the market and determine which factors might affect the brand. They help prioritize strategies to improve the product positioning and minimize negative perceptions of brands. Because the battle for market share has become increasingly competitive and fast-moving, companies have found it useful to have full-time staff dedicated to focusing on near-term strategy.

In some marketing departments, these duties are fulfilled by a director-level brand manager of marketing, while other companies delegate the global strategic and analytical responsibilities to a distinct and separate department. The staff dedicated to the global and long-term strategy will often assist the clinical development team by providing market assessments of possible new indications for a product. They also work with business development to determine factors such as the size of the market and how to drive licensing and acquisition of new products. Strategy brand management may also reside in a department dedicated to new product planning.

### New Product Planning, New Product Development, or Market Planning

In general, the members of the new product planning department have two roles: First, they offer commercial advice (market assessment) on products from the proof-of-concept stage all the way through Phase III clinical trials. Second, during Phase III trials, they are involved in positioning, branding, and raising awareness of specific disease states ("priming the market") before transferring the products to the brand management team for market launch and subsequent management.

*Those in new product planning give each product an identity and a personality before it is handed over to brand management.*

If you are interested in applying your scientific, medical, or business background, new product planning can be a particularly attractive career. New product planning involves assessing the market value of products for business development opportunities. The career provides exposure to a number of therapeutic areas at early stages of a product's clinical development, so a strong science background can be more useful than it is in brand management. It should be noted, however, that this position is typically at or near director level and usually requires prior experience in brand management.

### Product Development Management

There may be a subgroup within new product planning called product development management. Product managers in this department are stationed across the country and typically work from home to identify key opinion leaders (leading experts) who will be aware of and who will discuss and endorse products at symposia, medical meetings, etc.

### Global or International Marketing, Global Strategic Marketing, or International Market Planning Product Management

Global marketing managers build a worldwide marketing strategy. They identify and develop opportunities for sales in new countries and align potential partners to make sure that appropriate resources are in place. Most companies have a brand team and possibly a sales force for each country in which they sell products. By doing so, they ensure that their products and messages are consistent in different languages. International marketing teams are also involved in global pricing and reimbursement issues.

### Managed Care Marketing

Managed care organizations and other payers (health maintenance organizations, preferred provider organizations, government health schemes, and company health plans) represent a powerful force in the market. The adoption of products and procedures by these groups can have a dramatic impact on sales: Fewer products will be sold if care plans will not pay for them, and the price that is paid to the manufacturer can vary dramatically depending on the type of payer. Reimbursement (paying physicians for their procedures and treatments) is becoming an increasingly important issue in drug development and requires a distinct strategy to maximize potential revenues. Attaining reimbursement has become more difficult. To command a premium price, a product must have clear differentiating characteristics and be accompanied by a thorough data package. The managed care marketing department devises and uses health economic reasons to convince managed care organizations to place these products onto approved treatment lists.

### *Commercial Strategy, Analysis, and Planning*

Product managers in commercial strategy, analysis, and planning anticipate the resources that will be needed by marketing and sales teams and use those estimates to determine how best to deploy the sales force.

### *Market Research*

Market research analysts gather data about the market size, the customers, and the customers' level of interest. Market research is generated by surveying the opinions of potential customers in a non-biased way to determine, for example, their interest in the products or the effectiveness of ad campaigns. These data are useful in many marketing functions and are essential for making informed strategic decisions.

### *Health Outcomes Research*

> *If you like talking about science but don't want to do benchwork, you might enjoy a career in market or health outcomes research.*

Health outcomes analysts generate data that are used to evaluate the "reimbursability" of products that are in development or will be acquired. In a role similar to market research, they conduct research to ensure that these products have clearly defined benefits and offer cost-effective solutions (pharmacoeconomics) for reimbursement and health care plans. They survey the "payers" (mostly managed care and insurance agencies) and industry experts to better understand how products are reimbursed and to reveal users' concerns. They help to resolve reimbursement issues for individual brands by developing and publishing pharmacoeconomic models that will appeal to health care plans. They might build a case, for example, to show that even though a new drug may be more expensive than its competition, its superior efficacy will provide significant cost savings to the care plan or insurance company.

Market research and health outcomes analyst positions are common entry-level opportunities for scientists who are interested in careers in marketing. The work is akin to laboratory experiments: A hypothesis is formulated, data are generated, and the results are analyzed. Because the job involves complex scientific concepts, market research firms often prefer to hire people with medical or scientific backgrounds, whose expertise will facilitate surveying doctors and other scientists.

### *Marketing Products to the Scientific Community*

Working in marketing for nontherapeutic biotechnology companies can be a wonderful way to apply your science expertise to business (see Chapter 6). In these positions, the marketed products are used in the research and development world and are sold mostly to research scientists. Such products are not usually regulated by the FDA, so their life cycles are much simpler and shorter. Therefore, many more products can be managed—as many as 100 at a time. Product managers in nontherapeutic companies have overall responsibilities similar to those in therapeutic companies. They assess marketability and competition to help decide which products to launch. They create brochures and marketing campaigns, set strategy on pricing and positioning, set forecasts, train the sales force, and more.

## MARKETING ROLES AND RESPONSIBILITIES

*The roles and responsibilities of marketing and its related functions are many, and the following list includes only the essentials. More information can be found in the countless books devoted to this subject, some of which are listed at the end of this chapter.*

### Generating Research

Understanding the needs of the customers is a significant part of marketing. In general, companies employ the services of market research to determine the number of potential consumers (market size) and to generate revenue forecasts. Data are generated by contacting potential and existing customers and gathering feedback on their current practices, needs, and opinions about new technologies in development.

*Know thy market!*

These data are then used for market assessments and a variety of other purposes. For example, surveys can identify potential products that might interest customers. This information is useful when making "go" or "no" decisions for products at various stages of development. Market research is particularly useful for determining the effectiveness of ad campaigns and identifying the best target audience for a particular product.

Health outcomes analysts research clinical data and study the cost-effectiveness of products in development. These data are used to support bids for health care plans and reimbursement applications. The information is also used by new product planners in the design of Phase II and III trials to incorporate certain end points and by business development managers to analyze the reimbursement potential of particular products.

### Product Development Strategy

Until quite recently, product managers were brought in right before product launches for brand management. Marketing considerations, such as defining the best claims for products and competition, are now taken into account much earlier in the product development process. Marketers may also be involved in making decisions to select product names and design packaging.

Marketers work with the product development team to identify new applications for their products. Clinical departments identify new claims and ways to expand drug indications and lobby for company resources to run additional clinical trials.

### Priming the Market

Years before a product is launched, marketers begin to build a perceived need in the market. They help to produce publications about particular diseases to raise public awareness so that there is a warm welcome for their product when it is finally launched.

Marketers rely in part on the endorsements of key opinion leaders (influential and well-respected experts) to help establish market demand. They build relationships with these experts and encourage them to use the company's products, because the enthusiasm generated will cascade down to other customers.

### Brand Championship and Management

> There is no easy lunch in a market launch!

The heart of marketing is brand management. In a way, brand managers run their own mini-companies: They are ultimately responsible for the performance of their brand(s). Their main goals are to maximize sales and to gain the greatest market share for their product. At the center of the many-peopled activities concerning their product, they create and implement a specific brand plan, much like a business plan, upon which strategic initiatives are developed. They also serve as internal champions and fight to obtain additional resources to fuel the sales engine.

### Positioning

Marketers spend time "positioning" their brands. Positioning involves developing key messages for consumers in order to establish a unique perception about a product's attributes. A good example of a position is Nexium's "The Purple Pill." As the marketplace evolves and new applications for a product are identified or competition intensifies, brands may require repositioning. For more information on positioning and the psychology behind it, read Jack Trout's *Positioning, the Battle for Your Mind.*

### Promotion

To get the word out about products, marketers use various forms of media for promotional messages. These messages are often relayed during scientific meetings, continuing medical education events, on television, or by direct mail. Marketers also develop a "detail folder," which contains literature and marketing materials that sales reps can distribute to their customers.

### Public Relations

Marketers who work in public relations review literature before it is published to ensure that consistent key messages (positionings or claims) are included. Information intended for public disclosure, such as press releases, promotional pieces, and sales aids, undergoes an extensive review process. It is submitted to regulatory affairs personnel and then, in the United States, to the FDA.

### Pricing

Pricing is a complicated business. There is no magic formula, and the final decision will depend on the demand for the product and the cost of development. Because cost is often the deciding factor when products are being chosen by managed care plans, pricing policies are closely scrutinized.

### Competition Assessment

> Know thy competition!

Customers often quiz sales representatives about competitors' products, particularly if the competitors appear to be offering an equivalent product. Marketers help the sales staff respond to these questions by thoroughly studying the competition and providing educational material outlining the superior qualities of their own brand. Competition analyses are also important in new product planning and for business development to determine the potential market and the value of products.

### *Analytics*

Some marketers spend their time conducting financial analyses. These include financial evaluations of various programs, revenue projections, sales forecasts, calculations of the net present value of various brands, and risk assessments. These analyses have become increasingly complicated, and they are often undertaken by specialized teams within the marketing department.

### *Business Development*

Some marketers assist the business development team in the evaluation of in- and out-licensing opportunities by providing their commercial advice on the marketability of products and developing revenue forecasts to estimate a product's value. The marketers consider various aspects of the product, including the ability to differentiate it from, or show superior benefits over, the competition; the risks of possible side effects; and its potential for new indications.

## A TYPICAL DAY IN MARKETING

*A typical day in marketing might include some of the following:*

- Meeting with coworkers in marketing and other departments to discuss strategy and/or create promotional materials.
- Generating and reviewing training documents for field sales teams.
- Managing vendors—mostly advertising or market research agencies.
- Setting up symposia for physicians or patient advocacy groups.
- Conducting or reviewing various market and competition analyses.
- Meeting with doctors or key opinion leaders and consulting with sales representatives.
- Visiting other companies to explore areas of common interest and investigating the potential for collaboration.
- Traveling, sometimes as much as 30–40% of the time.

## SALARY AND COMPENSATION

In general, marketing professionals earn slightly less than their coworkers in sales, business development, and clinical development, but more than research scientists with an equivalent amount of experience.

***How is success measured?***

Marketing professionals contribute to a company's success in many different ways, and it is difficult to assess an individual's impact. If a brand is on the market, it is relatively easy to measure sales figures, but if not, performance can be more difficult to estimate.

Success can generally be gauged by the following:

- Sales, in particular, market share data; how the brand is performing on the market compared to the competition.
- Consumer feedback, which is usually relayed from sales representatives.
- Achievement of personal objectives; execution of tactics and meeting specific goals.

## PROS AND CONS OF THE JOB

### Positive Aspects of a Career in Marketing

- Marketing tends to be fast-paced, and the day-to-day responsibilities are varied and interesting. Because the demand for products and technologies is constantly changing, and a new marketing strategy is needed for each product, there is little job routine.
- Marketing provides a broader range of job responsibilities and experiences than do jobs in sales and many other careers. The potential to move into a succession of very different jobs while remaining in marketing provides the opportunity for continuous learning and career development.

*There is unlimited career potential in marketing.*

- Marketing is a strategic and intellectually stimulating field. The long-term analyses that are conducted to develop a marketing strategy are every bit as challenging as scientific research, and there is a continual need for generating new and creative ways to affect the market.
- Marketing provides an opportunity to be both creative and analytical. Technical skills can be applied to reviewing scientific data, understanding the significance of new technologies, and predicting their commercial implications. It takes ingenuity to begin with a complex scientific product and derive a concise and meaningful message that will appeal to mainstream audiences.
- Brand managers and senior-level marketing executives have a significant impact on the company and therefore carry a heavy burden of responsibility. These positions do receive appropriate recognition when things go well.

- The marketing field provides rapid feedback for judging your effectiveness. An idea for a marketing campaign can be presented and executed in a matter of weeks, quickly demonstrating whether the approach was a good one.
- A job in marketing is an opportunity to interact with a wide variety of people in multidisciplinary functions and provides insight into how the product development process works. Internally, marketers interact closely with sales, clinical, regulatory, research, business development, finance, legal, and manufacturing coworkers. Externally, there are opportunities to meet highly influential key opinion leaders, advisory board members, and more.
- Marketing professionals often travel to attend conferences, medical education meetings, appointments with customers, and meetings with key opinion leaders. Because symposia are generally held at attractive locations to attract doctors, this can be a nice perk of the job.

### The Potentially Unpleasant Side of Marketing

- Too much travel can be stressful. Some people, particularly those in brand management, spend 50% of their time traveling, but the average in marketing is around 30–40%. This is not as bad as sales, but the trips can be equally unpredictable and sporadic, becoming more frequent before and during product launches. The travel can play havoc with normal family life.
- Marketing is not a 9-to-5 type of job, and in some positions, particularly brand management, it can be all-consuming and stressful. There is frequently not enough time to do everything. For perfectionists, compromising standards because of time constraints can be frustrating.

*Your brand is a living and breathing thing. You will think about it and live with it 24/7.*

- The ultimate accountability for revenue falls on marketing. The marketing teams provide sales representatives with the tools for their job, but it is difficult to control their individual performances. If a brand performs badly, marketing takes the blame.
- Many marketing-related roles serve to inform others. This means that the results of one individual's efforts end up on someone else's desk.

## THE GREATEST CHALLENGES ON THE JOB

*Strategic Decision Making in a Changing Environment*

Making strategic decisions in an unpredictable environment is one of the more significant challenges in marketing. To maximize marketing potential, it is not enough to merely react to changes and competition; you must prepare the foundations for the future and develop a long-term marketing strategy.

### *Competition*

External competition can be fierce, and success can depend on displacing other products. There are many other products on the market, each with a team of bright and talented marketing professionals behind them who are trying to entice the same customers. There can be internal competition as well, because sales, marketing, and R&D all vie for the same pool of funds and resources.

### *"Herding Cats"*

As with business development and other careers in biotechnology, marketing also requires influencing people both within and outside the department. The ability to persuade these stakeholders to follow your lead is often likened to "herding cats" (see "Greatest Challenges" in Chapter 17). Because marketing affects the entire organization, the heads of departments involved in strategy often have their own ideas and agendas. Leadership skills are needed to manage their opinions and gain consensus. These problems can generally be handled diplomatically, but they certainly add a challenge.

### *Food and Drug Administration Regulations*

Marketing in the United States has become heavily regulated by the FDA, and the guidelines are continually changing. These guidelines restrict the way in which a marketing message can be presented, forcing an emphasis on literal statements of fact. Because it is known that customers respond best to emotional messages, FDA regulations can be seen as a curb on creativity. Every piece of information that is made available to the public must be approved by the agency, so the speed of publicizing new information is slowed considerably. As a result, marketing teams cannot respond to consumer feedback and to market data as quickly as they would like.

### *Ethics Versus Business Strategies*

Because biotechnology businesses are profit-driven, marketing and sales staff can find themselves under intense pressure to hit revenue goals. Given the levels of competition between companies and the tight legal and regulatory constraints on how companies are allowed to advertise, temptations to bend the rules may arise. It is fairly common for marketing professionals to find themselves in a dilemma between serving the best interests of the company and serving the best interests of society. For example, a company might decide to halt development of promising treatments because the customer pool is either too small or too poor.

## TO EXCEL IN MARKETING...

### *Leadership and Product Championship*

*To work together harmoniously, everyone on the marketing team must be singing from the same score.*

Those who excel in marketing are able to lead initiatives, sell their messages throughout the company, and effectively manage people. Product managers must be able to work with members from other departments in unison to be optimally effective. This includes the ability to manage, motivate, and empower more junior staff to facilitate the implementation of initiatives. Strong leadership skills and the ability to structure rational arguments are needed to gain the trust and team commitment to put plans into action.

### *The Ability to Make Sound, Strategic Decisions*

Because the market is dynamic, and the number of market scenarios is endless, it can take years of accumulated experience to develop a strong marketing instinct. People who excel in marketing have learned to trust their intuition or make educated guesses, because the complete set of data needed for making well-informed judgments is rarely available. Successful marketers are disciplined in their thinking and decision making. They have a clear view of what needs to be achieved and the managerial skills to make it happen. They can develop plans based on five- to ten-year projections and are able to visualize potential consequences when variables are altered (for a more thorough description of strategic skills, see Chapter 2).

*Strategic marketing is like a game of chess: You must have a clear long-term strategy and be ready to fend off attacks from the competition.*

### *Challenging the Status Quo*

Those who excel in marketing continue to strive for improvement. They constantly reassess chosen strategies and make changes, even when times are good.

### *Understanding the Customers*

Good marketers, like good sales staff, understand their customers. They understand their problems and become familiar with their attitudes and the language they use.

## Are You a Good Candidate for a Career in Marketing?

Because of the many roles in marketing, there are positions to suit a wide range of personalities and aptitudes. Companies tend to build teams with mixed skills and personality types that will complement each other. A brand team made up entirely of ambitious, self-confident, outgoing, strong-willed individuals would probably fall apart pretty quickly. In other words, the characteristics described below are generalizations and will not apply to everyone!

### *People who flourish in marketing careers tend to have...*

***Strong interpersonal skills and the ability to work well in a team and matrixed environment.*** (See Chapter 2.) Marketing requires the ability to work in a multidisciplinary team environment. It is essential to be congenial and to enjoy interacting with people. Even in more senior positions, you cannot do all the work yourself, and you must rely on the team's talents and efforts to help achieve your goals.

*Strong leadership skills.* You need to be able to rally the team and others around the brand plan. It is important to be able to engage, inspire, and motivate partners and coworkers, and to communicate your ideas confidently. You must be able to defend your plans in the face of criticism.

*Strong communication skills.* Senior marketers and those in brand management need to be able to effectively communicate vision and strategy. They need to explain complicated ideas and data succinctly and persuasively to different audiences. Good writing skills are required to compose formal written submissions.

*An ability to listen well.* Listening skills are necessary so that you can ask the right questions and listen closely to customers and coworkers.

*Highly organized project and time-management skills.* The ability to multi-task and prioritize is indispensable. Brand management is a hectic, fast-paced job with a lot of time pressure. If you prefer having more time to devote to single projects, consider market research, analytics, outcomes research, and other commercial operations positions.

*The flexibility to adapt.* Marketers should be able to anticipate changes in demand for their products and have the flexibility to adapt their plans in response to new situations.

*An ability to manage stress and be comfortable with responsibility.* Brand management is a highly visible position with tremendous responsibility and stress. It is important to be comfortable reporting both good and bad news to upper management, to take criticism well and not to become defensive, and to be able to justify your actions with sound data and good judgment.

*Stamina and boundless energy.* These are needed to handle the fast-paced, dynamic environment in addition to the travel load.

*Strong motivation and ambition to succeed.* Marketers want to feel that they have accomplished something each day. Goal-oriented "drivers" tend to enjoy a career in marketing.

*Strong analytical skills.* Some marketing positions require the ability to gather and interpret data to formulate strategic plans, make forecasts, and conduct other analyses.

*Creative talent.* The job of developing positionings, ad campaigns, innovative ways to reach the customer, and strategies to grow market share requires a creative mind.

*Brilliant marketers can think "outside the box."*

*Exceptional decision-making skills, particularly when there are inadequate data.* Data can be interpreted in many different ways, and rapid decisions must be made, even with imperfect information; otherwise, the organization could be paralyzed.

*Curiosity and breadth of knowledge.* Most marketers tend to have a strong desire to continue learning. It is important to enjoy keeping up with the constant influx of information.

***You should probably consider a career outside of marketing if you are...***

- An independent worker who prefers to work alone and is unwilling to consult with and delegate work to others.
- Painfully shy or too passive. You may come up with the best ideas, but unless you are brave enough to talk about them, no one will ever know.
- Too aggressive, or lacking tact and diplomacy. As a marketer, you depend on the cooperation of your team, so you don't want to alienate its members.
- A perfectionist. Those who have difficulty making decisions based on imperfect data or don't like working within tight time constraints might be frustrated.
- In need of constant positive reinforcement.
- Inflexible. Situations can change rapidly, and if you are not adept at adjusting priorities or reconfiguring your schedule on a moment's notice, it may be difficult to work as a brand manager.

*A competitive nature.* Competition can be fierce, and your success may be measured against other companies' products for market share (see page 253, "Greatest Challenges").

*An interest in people.* Marketing involves human psychology and understanding how people think and behave. Marketers spend a great deal of time studying customers' beliefs, value systems, and objectives, as well as what makes them react. If you are empathetic and able to think in terms of the customers' perspective, it is more likely that you will be able to persuade them to buy your product.

*An optimistic attitude.* Not all of the products in the pipeline or ad campaigns will succeed, but you must prepare for success regardless.

*The ability to work in an innovative environment.* Particularly true for new product planners, marketers are often at the cutting edge of science. There is often no single right way to approach a problem. Innovation is not a smooth road and may require refining as projects progress.

*The ability to pay attention to detail and see the big picture at the same time.* It's important to stay focused on the strategic objectives yet remain tactical in order to flawlessly execute the plan.

## MARKETING CAREER POTENTIAL

There is no single common career path in marketing. Because there are so many different roles, and because sales and marketing are so closely intertwined, there is an endless array of career path possibilities.

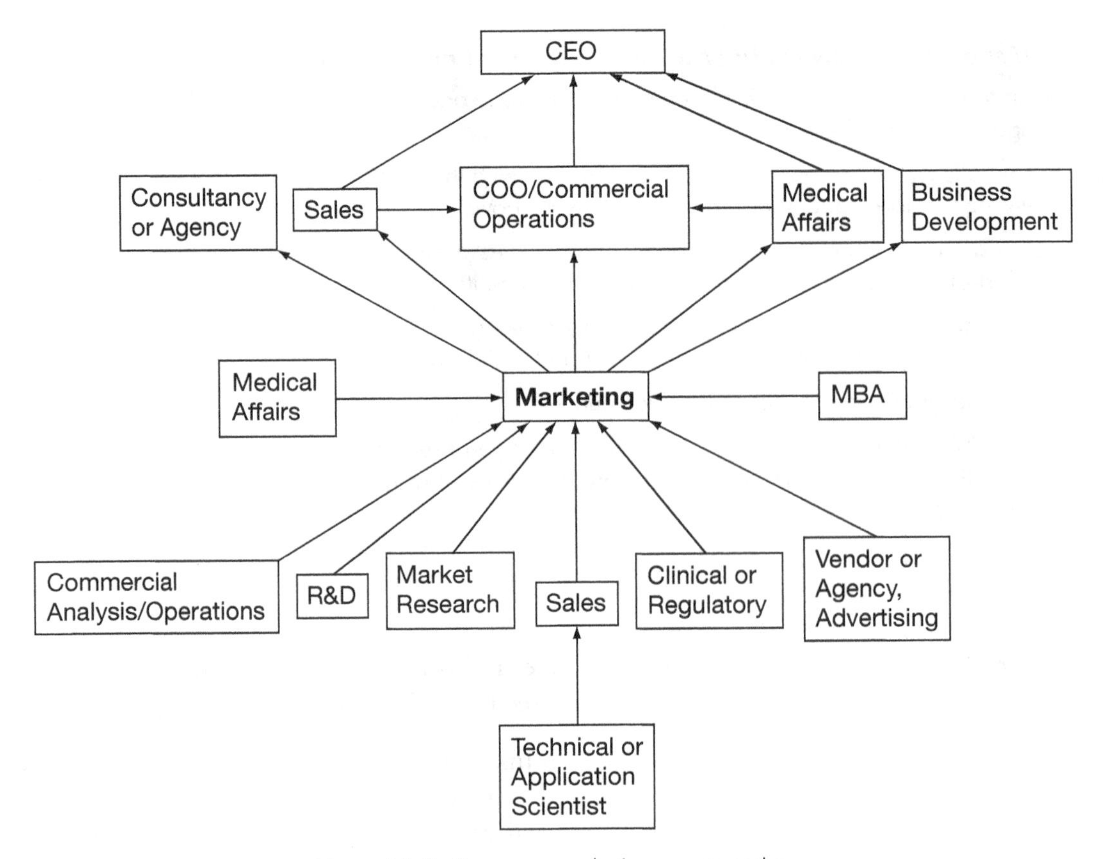


**Figure 18-2.** Common marketing career paths.

The commercial training garnered in marketing provides excellent preparation for executive-level positions, such as VP of commercial operations, chief operating officer, and CEO. In fact, most of the CEOs in large pharmaceutical companies have spent significant time in commercial operations or marketing.

If you want to remain in marketing, numerous possibilities are available within the career, and because marketing and so many other functions are interrelated, it is easy to make lateral moves. As diagramed in Figure 18-2, areas in which marketing experience can be directly applied include careers in sales, medical affairs, and business development. Also possible are portfolio planning, project management, alliance management, training, analytics or finance, corporate communications, and investor or public relations. You could also join an advertising, medical education, or other agency, or consult on your own or with an established firm.

## Job Security and Future Trends

Job security depends in part on the company and the life cycle of brands. A large company is more likely to have a number of promising products in development, so that if one fails, the brand team will be called upon to manage the next.

## LANDING A JOB IN MARKETING

### Experience and Educational Requirements

People who advance in marketing have generally gained experience in sales at some point in their careers, usually early on. This is because understanding the customer's needs and behaviors and the sales process is essential. Many are hired straight out of M.B.A. programs, from market research firms or departments, management consulting, or strategic planning. Regardless of one's background, it is a good idea to spend some time in sales. It is critical to develop a sympathetic appreciation of both the customers and the sales force.

There are usually no fixed educational requirements for marketing positions. The most common background is an M.B.A. degree. In companies, a mix of educational backgrounds is seen, including psychology, history, and nursing. Most have at least an undergraduate degree in the life sciences, and many possess an advanced degree, such as a master's, Ph.D., M.D., Pharm.D., or Master of Public Health degree.

Although an advanced degree is extremely helpful, it is not required. By improving your ability to understand and explain the science behind the brands, you will increase your effectiveness. A science background will provide more credibility with customers, key opinion leaders, and clinical partners, and it can add tremendous value to the marketing team.

For marketing positions in nontherapeutic companies that develop and sell products to basic and discovery researchers, a science degree and laboratory bench experience are especially advantageous. Bench experience is helpful for seeing the customer's point of view and being able to understand how your company's products can directly benefit research. As a result, there are many more people with Ph.D.s or advanced science degrees in marketing for biotechnology tools and services companies.

For commercial analysis positions, particularly market and outcomes research, advanced degrees in science, economics, public policy, health services and policies, statistics, and epidemiology are common. The educational backgrounds for professionals in new product planning commonly include Pharm.D., master's, or Ph.D. degrees. Again, these qualifications are advantageous but are not usually required.

### Paths to Marketing

*Getting into marketing is half the battle!*

Positions in marketing are highly coveted. The biggest challenge is overcoming the initial hurdle of securing a job and obtaining experience. There are few entry-level marketing positions, and there are vast numbers of sales professionals competing for them. Consequently, there is a glut of capable job seekers. For those with science backgrounds, getting a foot in the door can be difficult. Hiring managers prefer to hire people with M.B.A. degrees, sales representatives, and market research analysts.

*You may have to bite the bullet and go into sales first.*

Additionally, companies tend to impose conservative hiring practices for positions that are held accountable for revenues.

Therefore, they are less likely to take risks on people with no prior experience. Once you have gained some marketing experience, however, the opportunities are limitless.

*Here are some tips on how to obtain a marketing position:*

- If you are determined to get into marketing, if you want to live and breathe the occupation, you will be more likely to succeed if you start in the trenches as a sales representative. It is a great way to learn about marketing, to observe how customers make purchasing decisions, and to be exposed to the issues that sales forces face. In fact, many marketing hiring managers won't hire people unless they have sales experience. Many people, especially scientists, may be intimidated by the prospect of working as a sales representative. Remember, however, that sales reps work to educate doctors or scientists about products. Some people discover that they love sales and never leave (see Chapter 19).
- Obtain an M.B.A. degree from a top school or work in a management consulting firm. Try to conduct an M.B.A. summer internship in a biotechnology or pharmaceutical company. Summer internships often translate into full-time jobs, depending on your performance and fit in the corporate culture.
- Apply for market research or outcomes research analyst positions. These are great entry-level positions and provide an overview of biotechnology and marketing. Such organizations typically prefer to hire people with advanced science or medical degrees.
- If you just can't fathom sales, consider a role in medical affairs. Your scientific or medical qualifications may be more valued in a position such as a medical science liaison. These positions tend to be easier to obtain than jobs in marketing (see Chapter 11).
- Join the marketing department of a biotechnology tools, services, or medical devices company, then move laterally to a drug discovery company, or vice versa (see Chapter 6).
- Apply for a position such as technical service representative, field applications specialist, or project manager, or join a product development team. These positions allow you to work closely with marketing teams and, again, the transition to marketing will be easier.
- Consider analyst or associate positions at consultancies or agencies that provide services for marketing departments. Join vendors that specialize in advertising, market or outcomes research, publications planning, communications, public relations, or competitor analyses, or organizations that develop valuations of new markets. These functions are typically outsourced to vendors who prefer to hire people with technical backgrounds or industry knowledge.
- Think about other ways of working more closely with customers, such as working in a contract research organization (CRO), perhaps as a clinical monitor, where there is direct contact with doctors and patients.
- If you are already employed in a biotechnology company, request an internship or rotation in the marketing department. Many companies will allow you to return to your

original job if you decide marketing is not for you. Sometimes marketing support functions are available. For example, members of clinical groups might be asked to support a clinical brand.

## RECOMMENDED TRAINING, PROFESSIONAL SOCIETIES, AND RESOURCES

### *Societies and Resources*

Medical Marketing Association (www.mmanet.org)

International Society for Pharmacoeconomics & Outcomes Research (www.ispor.org)

American Marketing Association (www.marketingpower.com) (not biotech specific)

### *Books and Magazines*

Bazell R. 1998. *Her-2: The making of herceptin, a revolutionary treatment for breast cancer.* Random House, New York.

Gladwell M. 2000. *The tipping point: How little things can make a big difference.* Little, Brown and Company, New York.

Lehmann D. and Winer R. 2004. *Product management,* 4th edition. McGraw-Hill/Irwin, New York.

Ries A. and Trout J. 1981. *Positioning: The battle for your mind.* McGraw-Hill, New York.

———. 1994. *The 22 immutable laws of marketing.* HarperBusiness, New York.

———. 2005. *Marketing warfare,* 20th edition. McGraw-Hill, New York.

Trout J. 2000. *Differentiate or die: Survival in our era of killer competition.* John Wiley & Sons, New York.

### *Classes*

Economics

Biostatistics

Clinical Drug Development and Research

# 19

# Sales

## Generating Revenue and Educating Customers

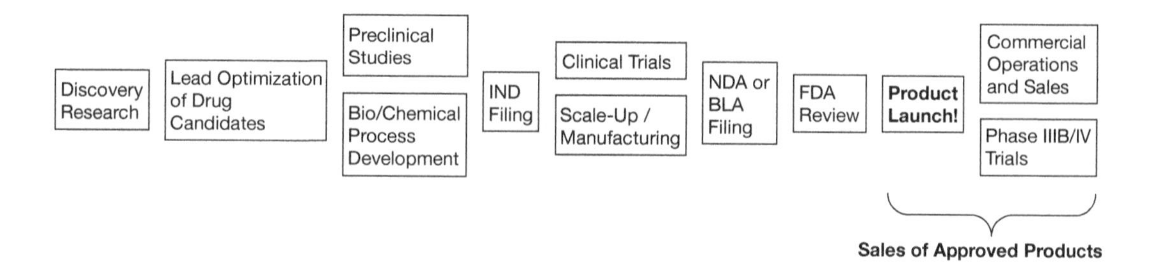


IF YOU PREFER INTERACTING WITH PEOPLE rather than doing bench or office work, are outgoing, enjoy explaining scientific details, and have an entrepreneurial attitude, then sales might be the career for you. Working in sales allows you the freedom to operate independently, to be individually accountable for your success, and to be financially rewarded for your results. Training and experience in sales provide an excellent foundation for many other careers, particularly in marketing and business development.

*Traveling to new places, linking science to business...sales is an adventure!*

### THE IMPORTANCE OF SALES IN BIOTECHNOLOGY AND DRUG DEVELOPMENT

No matter the product, one important function of a sales professional's job is to inform. It is his or her job to educate consumers about the special features and benefits of new products that can solve problems and facilitate consumers' work. The best salespeople know their products inside and out and can link features to the customer's needs.

From any company's perspective, sales are needed to generate revenue for continued operations and the development of new products. In public companies, increasing revenue is a primary driver for stock valuation and shareholder satisfaction.

*Sales personnel are like mitochondria that generate ATP for the company.*

### *Scientific, Drug, and Medical Device Sales*

There are three main types of sales in biotechnology and pharmaceutical companies: scientific, drug, and medical device sales. For scientific sales, the end consumers are research and development scientists. The processes of drug discovery and development have become ever more sophisticated, thanks to quantum leaps in technology. With recent technological advances, there is a vast amount of information to assimilate. Scientific sales professionals are responsible for relaying this complicated information to customers and describing the potential uses and benefits of the new technologies. With all of this information, it is critical that clients and potential customers understand the real-world value that a product brings to their research efforts. Within scientific sales, there are opportunities in the commodity (i.e., reagents and glassware) and capital (i.e., expensive instruments and software) industries.

For drug and medical device sales, the immediate customers are physicians, and the end consumers are patients. Doctors are extremely busy and overloaded with information about drugs. Sales representatives (sales reps) promote products and keep doctors up to date, so that they can make the best-informed decisions about which drugs to offer their patients.

## CAREER TRACKS IN SALES

### *Account Management and Field-based Sales*

The titles and roles vary among companies, but in general, account managers, also known as sales reps, territory managers, or business development managers, are field-based and serve in the trenches of the sales force. They are the key point of contact with customers and generate revenue for the company. They are responsible for maintaining existing customer accounts and building new ones in assigned geographic territories. In some companies, account managers are not directly responsible for sales, but instead serve as reimbursement experts or people who provide business proficiency to accounts.

Sales groups typically operate as hierarchies. Junior sales reps and inside sales personnel move up to territory or account managers, who move up to senior or key account managers. At the sales manager level, higher-ranking sales reps are responsible for providing resources and strategy to more junior sales reps. They may oversee a particular product line or territory and manage the corresponding sales force. In pharmaceutical sales, a regional director often has 5 or 6 district managers, each of whom might oversee 8–12 junior sales reps (see Fig. 19-1).

### *Senior Sales Management*

At the senior sales level, business unit directors, vice presidents of sales, or vice presidents of franchise manage a cost center or business unit and are responsible for successful sales

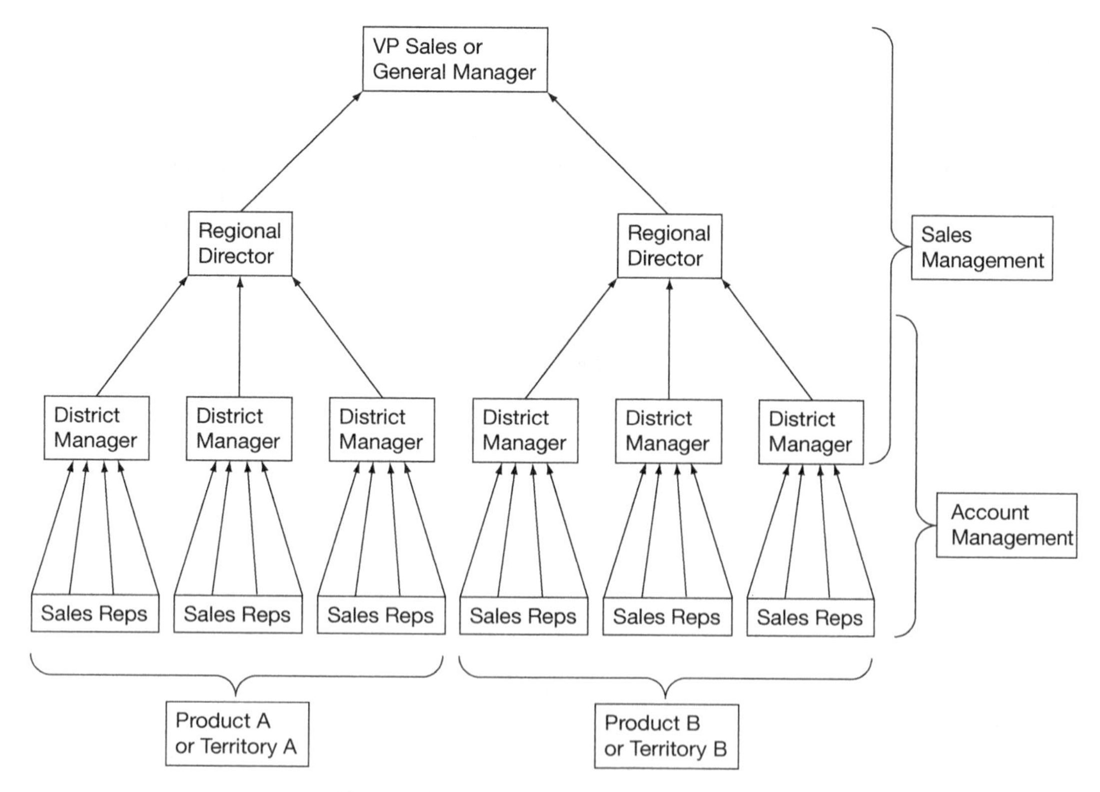


**Figure 19-1.** Representative sales structure.

and certain product launches. They may be responsible for the profit and loss of an entire business unit or brand (see Fig. 19-1) and are typically based at the corporate office.

#### *Drug Sales Representatives*

In biotechnology and pharmaceutical companies, sales reps (also known as therapeutic or clinical specialists) answer questions, supply information, and sometimes offer free samples of their products to physicians. Each rep may be in charge of a territory covering 50–100 doctors. They may also organize symposia or other educational programs or meet health care professionals casually over lunch or dinner to discuss and promote the drugs that they represent.

#### *Sales Operations*

Within some sales groups are people who provide the sales force with training, administrative support, and other resources. They help ensure that the sales processes run smoothly and efficiently.

#### *Inside Sales*

Personnel in this group answer and make sales calls over the phone from company locations and do not travel for account management.

**_Primary versus specialty care sales_**

Drug sales reps work in two basic areas of sales: primary care and specialty care. Primary care sales reps provide doctors with mainstream products, such as painkillers and allergy relief products, for well-understood diseases. There are many thousands of primary care sales reps, and it is a highly competitive occupation. These reps need a minimal understanding of diseases and products, and they are trained to administer information in "sound bites" to doctors. As such, it is easier to recruit and train entry-level, college graduates for these positions.

Specialty care sales reps sell niche drugs in specific therapeutic areas, such as oncology, ophthalmology, or infectious diseases. These products don't deliver a mass-market message, so the sales approach is different. Because of their specific and extensive knowledge, these reps tend to play a more consultative role and build stronger personal relationships with the doctors who practice in their therapeutic areas. They may also provide extra services such as expediting reimbursement processes, explaining to nurses how to reconstitute medicines, and monitoring patients' responses to therapy.

#### *Technical Sales Specialists*

In a team-based approach, technical sales specialists help account managers with complex deals. They serve as experts on particular product lines and work with customers on a consultative basis to demonstrate technologies and discuss the benefits of product features. Technical specialists differ from field application scientists (see below) in that they provide more pre-sales support and have revenue objectives and commissions. They typically hold a Ph.D. or have had significant laboratory experience.

#### *Field Application Specialists or Field Application Scientists*

Like technical sales specialists, field application specialists (FAS) work with account managers in a team-based approach and serve as technological experts who provide mostly post-sales support. They have acquired a thorough understanding of product lines and provide consultative advice for clients. They run pre-sales demonstrations, give presentations, and help customers assemble and use instruments and solve technical problems. The responsibilities of field application scientists are described in more detail in Chapter 20.

## SALES ROLES AND RESPONSIBILITIES

*The overall role for sales reps is to generate revenue and increase brand equity. This is managed by reaching your sales quota or "hitting your numbers."*

#### *Prospecting*

Sales reps prospect for new customers and follow leads. They spend a lot of time on the telephone making new contacts and trying to identify the purchasing "decision maker."

### *Territory and Account Management*

This is the heart of sales. Sales reps manage existing clients, or accounts, and identify new ones in an assigned territory. These reps identify leads, give presentations, deliver proposals, negotiate contracts, and close deals. As they handle the business relationships, they invite trainers, technical sales and/or field application specialists, or medical science liaisons to help customers. One important component of account management is to clearly and convincingly demonstrate the value proposition; i.e., what is the value of the product to the buyer.

### *Sales Management*

Internal efforts are made to ensure that the sales process runs smoothly. The managers monitor staff performance, develop strategic initiatives, and conduct analyses to determine how to improve sales force effectiveness. Managers are also involved in staff motivation and salaries, implementing short-term sales and motivational contests. Additional responsibilities may include training staff and coordinating sales meetings, reviewing customer relationship management (CRM) software entries, and more.

### *Community Outreach Programs*

Community outreach efforts help build the company's reputation and expand awareness of the company and its products. Many sales reps work with local societies and other organizations to support the community and popularize a particular therapy, product, or company. They might provide educational support for a program or meeting and sponsor invited speakers.

## A TYPICAL DAY IN SALES

*Depending on the type of sales position and the company, a typical day might be spent doing some of the following:*

- Devoting 1–2 hours each day to phoning prospective clients. This is known affectionately as "cold calling."
- Arranging and attending meetings with potential clients, following up on previous meetings, and working to close deals.
- Traveling to meetings. This can occupy 25–70% of the working day.

*You gotta love to travel!*

- Attending networking events, trade shows, and conferences.
- Reviewing accounts with managers, discussing pending business, and building sales forecasts.
- Visiting doctor's offices, responding to questions, distributing educational information, and offering free samples.
- Dining with customers, nurses, or primary care physicians to educate them about the virtues of a product or drug.

- Organizing and running promotional dinners, medical board meetings, or symposia.
- Managing patient outreach programs, such as free screenings for particular diseases in specialized therapeutic areas.
- Coaching and mentoring individuals to help them develop better strategies and increase their revenue-generating skills.

*A typical day for sales management and more senior sales reps might include:*

- Training or motivating the sales team. Spending time on account strategy by providing specific objectives. Allocating resources and disseminating appropriate information to the sales force.
- Interacting on an executive and board level. Working with peers and the board of directors to put strategic plans and programs in place.

## SALARY AND COMPENSATION

*You can expect to double or triple your pay as compared to a research position!*

Sales can be a lucrative career. Because success in sales is crucial to the company's survival, the head of sales may be one of the highest paid individuals in a company, next to the chief executive officer (CEO). Sales professionals typically earn more than those in marketing and business development. Salaries can be double or even triple those of research scientists with an equivalent number of years' experience.

Drug sales may earn more commission than scientific sales, but those selling products with the largest price tag have the potential to earn the heftiest commission. Therefore, sales of capital equipment and instruments are the most lucrative, whereas the research commodities industry (reagents) is at the lower end of the salary spectrum.

If you are just starting out, do not expect to earn a substantial income immediately. Depending on the position, it may take several years to develop the reputation that will provide you with well-deserved rewards.

Sales compensation plans tend to be complicated and vary from company to company. In general, sales reps receive a low base salary supplemented by a performance-based commission. Additional bonuses and incentives are to be expected if one surpasses sales quotas.

**How is success measured?**

Success is typically measured by whether or not sales quotas are reached. Other factors taken into consideration include the generation of new business accounts, the number of meetings set up with potential clients, customer feedback, and improvements in the market share.

In sales management, success can be measured by progress made toward financial objectives as well as market share analyses.

## PROS AND CONS OF THE JOB

### Positive Aspects of a Career in Sales

- The income potential is remarkable. Depending on many factors such as the type of company and market conditions, a job in sales can be one of the most lucrative positions in biotechnology and drug development companies with products on the market. When things are going well and you are making your numbers, you will experience not only financial abundance, but a great sense of accomplishment as well.

- There is freedom to operate independently. Because they operate on their own most of the time, sales reps generally control their own work schedules. For example, you can choose which accounts to focus on and which meetings to attend. This freedom adds an entrepreneurial dimension to the job: Typically, the harder you work, the more money you can earn.

- Additionally, most sales professionals work from home and enjoy all the benefits of a "five-second commute to the home office." This affords a broader freedom in the choice of where to reside, which can be particularly useful when accommodating the needs of family members. Another perk is that coffee breaks and lunch hours can be used to complete home chores such as laundry, or to take a break and work in the garden. An additional benefit of this autonomy is that the complicated politics often experienced in an office setting are not a distraction.

- Excellent opportunities exist for meeting interesting people and learning about new technology. You may meet Nobel laureates, innovators of the top-selling drugs in the world, industry thought leaders, and others. In addition, these people may rely on you to provide specific ideas about how to solve their problems or how to better serve their patients. It is also an outstanding way to learn about the latest and most effective technologies as well as the inner workings of research and development.

- Your technical background can be applied in countless situations. It can be quite rewarding to help those who prescribe drugs or conduct drug discovery research and to know that you can personally improve the way people work by providing new products to enhance their productivity.

- There is an abundance of job variety. For those who do not like daily routine, sales will suit your work style.

- Gratification for one's efforts is more immediately realized than in most other biotechnology jobs. The sales cycles are much shorter than in business development. A deal can be initiated and closed in 3–12 months, depending on the cost of products. There are discrete deliverables with measurable end points, as compared to research projects, which tend to languish.

- Sales has a large impact and high visibility within the company. The success of a company may be contingent upon the sales team meeting the revenue forecasts.
- The sales training in pharmaceutical companies is tremendous. In addition to sales effectiveness training, there are workshops on situational leadership, coaching, training and writing skills, and more.
- Internal sales meetings can be fun and motivational. Sales people tend to develop strong camaraderie. It is entertaining to meet other sales personnel, share war stories, and make new friends. Sales coworkers can be stimulating, bright, highly energetic, and motivated people.
- If you enjoy traveling, sales can be a wonderful career. Traveling can be particularly rewarding if you have a territory that suits your recreational interests. You can schedule a weekend vacation around a business trip and take the opportunity to explore new areas of the country that you would not otherwise have visited. You can often use frequent flyer bonuses to pay for vacation travel.

*Try to find a job where your sales territory is in Hawaii or near the major golf or ski resorts!*

- A job in sales is a chance to learn about other people's cultures. You will be introduced to people from many different backgrounds.
- Added benefits often include a company-paid car and an expense account to dine customers. If your sales quotas are exceeded, many companies offer large rewards, such as all-expenses-paid vacations to exotic places.

## The Potentially Unpleasant Side of Sales

- There is constant pressure to reach your quarterly sales quota. Internally, market research is used to forecast the sales expectations for your product. If you are selling less than your quota, you are under intense daily pressure, particularly if you are on probation; you can even lose your job. Even if you have had a tough quarter and just manage to reach your quota, the sales cycle starts over for the next quarter, so it never ends. It also takes time to develop a reputation, and the first couple of years on the job can be extremely challenging.

*The need to meet your quarterly sales quotas is like the Energizer battery: it goes on and on and on...*

- For scientific sales, too much airline traveling can be exhausting and put a real strain on family life. Despite these problems, you need to appear bright and enthusiastic with clients, even if you have had no sleep. The average amount of travel is about 50–70%, but this can go up to 90% in some positions. It is possible to limit air travel if there is enough business in your local territory. Drug sales reps tend to have smaller geographic territories and, as such, travel less.
- Working from home can be a blessing and a curse: You cannot escape work. Expect to work 10–12 hours per day and sometimes on weekends.

- As in most biotechnology careers, there can be a considerable amount of corporate paperwork in sales. This may include submitting expense reports, making monthly reports, maintaining a customer database, or documenting the number of client visits per day.
- There is a stigma associated with the sales profession. People generally tend to distrust sales reps and often do not understand the level of education required to do the job successfully. Once rapport has been established with a customer, and a level of respect has been achieved, the job becomes easier, but the first couple of years can be difficult.
- The work can be repetitive, particularly in drug sales. Sometimes the same sales pitch must be made many times before it yields results, and it can be a challenge to remain positive and motivated.
- Performance measurements can be frustrating. There is no perfect measure, and it is difficult to predict sales accurately. Each sales rep might be working equally hard, yet their performances might be perceived differently because of variables that skew the measurements.

## THE GREATEST CHALLENGES ON THE JOB

### *Reaching Quarterly Quotas*

The greatest challenge in sales is reaching the quarterly quotas. Even when the economy is slow and cuts are being made in industry and academia, you will still be held accountable for meeting your sales goals.

### *Too Many Sales Reps*

*There has been explosive growth in the size of sales forces.*

It has been shown that there is a direct correlation between the size of a company's sales force and the number of prescriptions written for that company's drugs. This has led to an explosive growth in sales forces as each company strives to keep up with the competition. Unfortunately, the many sales reps vying for appointments have become irritating to clients, and some doctors have responded by imposing "no-see" policies. In response to this and other factors, the industry is now swinging in the other direction, and large pharmaceutical companies are cutting back their sales forces.

## TO EXCEL IN SALES...

Sales is essentially about building enduring relationships, trust, and rapport with customers. To do this, you must build credibility and understand your clients and their objectives.

*Building Credibility*

Although "relationship selling" may be less important than it once was, sales reps must be able to demonstrate expert knowledge and personal integrity and present themselves as reliable and credible. If customers are uncomfortable and doubt whether their needs have been understood, they will be less likely to make a purchase. This task is made easier by working for an established company with recognized brands and a reputation for quality.

*Know thy customer! Telling is not selling.*

*Understanding the Clients' Needs*

Great sales professionals induce sales by demonstrating value to the client. This may include identifying specific solutions to customers' problems or presenting a solution that the customer had not previously considered. Examples of value propositions might include helping the client make products faster, better, or more cheaply, or providing them with a competitive edge.

*Each customer has a problem. It is up to you to find a satisfactory solution.*

## Are You a Good Candidate for Sales?

*People who flourish in sales tend to have...*

*Drive, integrity, and a strong ethical character.* You need to be highly motivated (the financial incentives help with this) and honest in this job.

*Discipline and persistence.* Lasting relationships with clients are keys to success in sales. This takes time and a great deal of persistence, diligence, and discipline. Successful sales reps are always looking for new markets and leave no stone unturned in the pursuit of potential clients.

*Persistence, discipline, and diligence are essential.*

*Resilience.* You will hear "no" many times, but you must remain upbeat and enthusiastic and not take rejection personally. Physicians make decisions based on the best practice for their patients' health, not necessarily because they like the sales rep. It is important to be perceptive and respect your customers' time constraints.

*Excellent communication and presentation skills.* Because much time is spent talking to customers, communication skills are essential. They are important for listening to customers' needs, giving presentations, and writing to clients. Due to the high level of customer interaction, those who are outgoing and gregarious tend to perform well.

*Outgoing, motivated, and highly energetic people apply here!*

*Excellent time management and organizational skills.* There will be multiple accounts to manage; therefore, being detail oriented and responsive to clients' needs is required.

*Strong interpersonal skills (see Chapter 2).* You must present yourself as trustworthy, dependable, and credible. An easy and likable manner helps to earn clients' trust so that they will enjoy working with you and will return for future business. A sense of humor is also a plus!

*The ability to make customers feel comfortable is one of a sales rep's biggest assets.*

**Enthusiasm for the products.** Sales reps who are excited about, and who believe in, the technologies they are selling are generally the most successful. This is something to bear in mind when looking for a job.

**Specific knowledge about the product and the marketplace.** Having a thorough understanding of the technology you are selling, and being familiar enough with its applications to explain its most complex features, are invaluable. It is also useful to have a good grasp of other products that might best accommodate your client's changing needs.

**The ability to think quickly.** Very often, you have only one chance for a presentation. If you are ill prepared and fail to answer questions on the spot, you may lose a sale.

**An entrepreneurial attitude.** As you move up the sales ladder and take responsibility for junior staff, it is like running your own company. Successful sales people learn to be enterprising and entrepreneurial.

**Boundless energy.** In sales, you are constantly on the go, with traveling, arranging appointments, and meeting customers in order to reach your sales goals.

**Flexible thinking.** In this business, an open mind and a willingness to consider novel solutions are assets.

**A strong ability to solve technical problems.** For scientific sales, being able to sympathetically solve customers' technical problems helps you develop rapport with clients (and the potential for a career move later on!).

**The ability to quickly identify clients' personality styles and "read" human behavior (see box: "Personality Styles").** Many highly successful sales professionals learn how to quickly understand what motivates clients and to recognize a customer's personality style. To expedite the sales process, it is important to put the customer at ease by responding to his or her behavior appropriately.

*You must be able to evaluate your client's personality style in a nanosecond.*

**A polished personal presentation.** It is important in sales to dress well and be presentable.

*For sales operations in particular, you need...*

**Strong quantitative and analytical skills.** This position requires the ability to analyze data and draw conclusions, as well as the ability to understand the nuances of the data.

**Good management and motivational skills and the ability to enjoy facilitating the success of others.** It is gratifying to see deserving coworkers accept promotions when they have built a track record of success.

***You should probably consider a career outside sales if you are...***

- Unable to handle uncertainty. There is a lot of unpredictability in the job; for example, it's hard to know when a client will purchase a product.
- Unable to handle rejection. Many people will say "NO" before one may say "YES."
- Painfully shy, or the reverse: overly friendly, insincere, or hyper-extroverted. Being too bubbly can be seen as disguising a lack of self-confidence, and being too withdrawn can lose a sale.
- Unable to understand or convey the value of the products to the customer or to match its benefits to the customers' needs.
- Too focused on technical details or lack the assertiveness to close a sale.
- Focused on yourself. You need to be sympathetic and empathetic with your customers.
- Egotistical or arrogant. There is little glory in sales, just financial and career incentives.
- Too aggressive or out to make a quick buck, manipulative and self-serving.

***Personality styles***

People have different personality styles (refer to the book *People Styles at Work* listed at the end of this chapter). It is important to be able to quickly gauge a client's style so that you can adapt your approach and spend your time focused on the problem at hand, rather than on easing personal friction. If clients feel uncomfortable with your style, they will be looking for an escape route instead of the purchasing form.

Some clients, for example, take a long time to review data and make decisions, whereas others are in a tremendous rush and want to act quickly. Many need to feel comfortable with a sales rep before they get down to discussing business. In sales, as in other vocational areas, there is an art to conducting a business interaction while making it a pleasant experience for all parties.

## SALES CAREER POTENTIAL

There are three main career tracks in sales: account management, sales operations, and sales and marketing. Most senior sales professionals have held marketing positions at some point in their careers, and most marketing professionals have had experience in sales. The fields are intertwined, and lateral moves between the two are common.

Those who wish to remain in sales may want to move up the ranks and go into sales management. However, it is also perfectly respectable, and lucrative, to remain as an account manager.

Being accountable for revenue and "carrying the bag" is excellent preparation for executive-level responsibilities, such as chief of commercial operations (CCO), chief operating officer (COO), CEO, or entrepreneurship (see Fig. 19-2). In fact, many CEOs, particularly in large pharmaceutical companies, began their careers in sales.

For those interested in gaining sales experience as a venture into the biopharma industry, the career possibilities are endless. Sales can be a launching pad into marketing, com-

**Figure 19-2.** Common sales career paths.

mercial operations, business development, product development, or entrepreneurship. Some sales staff go on to become training agents, purchasing agents, or field application scientists. In addition, with sales training, you are not limited to biopharma—you can expand your horizons and sell in just about any industry.

## Job Security and Future Trends

*A job in sales is high risk, but highly rewarding.*

For both scientific and drug sales, experienced professionals can be highly valued and marketable, especially if they are recognized for particular areas of expertise.

For scientific sales, as long as research budgets in industry and academia flourish, products will continue to be bought—and commissions earned! Drug sales may be less stable than scientific sales. Drugs can be unexpectedly pulled off the market or cut from managed care plans, or new generics can be introduced, which frequently results in the downsizing or reorganization of sales forces.

Jobs in specialty care sales might provide more security than those in primary care. There is an overabundance of primary care sales reps because large pharmaceutical companies can recruit attractive college graduates and train them in three months. However, there is a limited pool of truly knowledgeable specialty care sales reps in the marketplace.

The one area of sales that is in threat of being outsourced overseas is the administrative function of inside sales (taking purchasing orders).

## LANDING A JOB IN SALES

### Experience and Educational Requirements

#### *Scientific Sales*

The best preparation for a career in scientific sales is basic research experience. Most scientific sales staff are former researchers from either private or academic laboratories who were frustrated with mundane bench work and sought more human interface.

It is often more important to have hands-on experience and a good understanding of the technology than a handful of qualifications. In some instances, however, hiring managers prefer more specialized sales employees who hold advanced degrees. Technical specialist positions typically require a Ph.D. or a master's degree. M.B.A. degrees are also advantageous but are generally not mandatory.

#### *Drug Sales*

Most reps start out in primary care sales. Here, a college degree in any area will suffice, although science graduates have an advantage.

Work in specialty sales requires a deeper understanding of the science, the competition, and the disease state. A science, medical, or business background is preferable. Although qualifications vary, most specialty sales staff have at least a college degree, and the most commonly held advanced degrees are M.B.A.s. There are many specialty sales reps with master's degrees (for example, in public health or health industry management) or nursing degrees.

For sales management, a business qualification is required, which includes finance, economics, statistics, and business analytics. An M.B.A. is preferred and a science degree is advantageous, although neither is mandatory.

### Paths to Sales

- Establish relationships with sales reps or talk to sales personnel at large industry meetings. You may find open positions. Call sales reps and ask for informational interviews. Most sales reps are friendly and enjoy talking about their jobs.

- Try to spend a day with a sales rep or account manager. There is no better way to determine whether this career is for you than by experiencing a typical day in sales.

- Consider applying for a field application or technical specialist position. You will work closely with account managers and be exposed to sales processes without being directly responsible for revenue generation.

- Apply for sales positions in companies that you respect and that sell products you trust and have used. The job will be easier if you believe in your products, and companies prefer to hire end-users (customers). Also, keep in mind that it is often easier to sell products for companies with an established and respected reputation.
- Consider applying for positions with contract manufacturers or vendors who sell services to sales operations. These vendors are typically smaller companies that cater to pharmaceutical and biotechnology companies' needs by providing data, services, and management such as performance measurements, territory alignment, and incentive compensation. Joining one of these smaller companies can be a good first foray into sales.
- Contact your doctor or other physicians you know and ask them for referrals to their favorite sales reps. This will provide an excellent introduction to successful sales reps, and they may even be flattered by the referral.
- Remember that biotechnology companies tend to hire experienced sales personnel, whereas pharmaceutical companies train entry-level applicants. Therefore, unless you have experience, it might be easier to start with a pharmaceutical company (see next point).
- Apply to companies willing to invest in your sales training. Most large pharmaceutical companies have excellent training departments and will spend time and money developing their employees into professional sales executives. Expect at least the first three months to be mostly training. Small biotech companies tend to be cash starved and have less inclination to train their staff.
- If you have a background in science, apply to companies that sell biologics or scientifically complicated, cutting-edge drugs. Your scientific expertise and education will be most valued where consultative selling is required.

## RECOMMENDED TRAINING, PROFESSIONAL SOCIETIES, AND RESOURCES

### *Courses and Certificate Programs*

Sales and marketing certificates are offered at local universities.

Siebel Sales Methodology (www.siebel.com) provides sales training.

Miller Heiman (www.milllerheiman.com) offers world-renowned sales system training.

MBA programs, classes in business, finance, economics, statistics, and marketing

Courses for learning how to use Microsoft's Excel and PowerPoint programs

### *Societies and Resources*

Toastmasters International (www.toastmasters.org) is highly recommended for training in public speaking.

Society of Pharmaceutical and Biotech Trainers (www.spbt.org)

**Books and Magazines**

Bolton R. and Bolton D.G. 1996. *People styles at work, making bad relationships good and good relationships better.* Amacom Books, New York.

Fisher R. and Ury W. 1991.*Getting to yes: Negotiating agreement without giving in.* Penguin Books, New York.

Rackham N. 1988. *SPIN selling.* McGraw-Hill, New York.

Zoltners A., Sinha P., and Zoltners G. 2001. *The complete guide to accelerating sales force performance: How to get more sales from your sales force.* Amacom Books, New York.

# 20

# Technical Applications and Support

## Getting Paid to Be the Expert

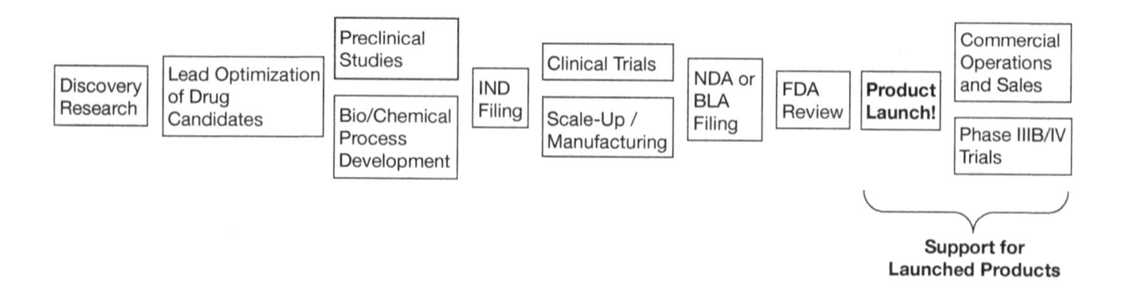


ARE YOU THE GO-TO PERSON WHEN COWORKERS SEEK technical advice in the lab? Do you enjoy teaching and explaining complicated concepts? If you have excellent interpersonal skills and like working with people, you might enjoy a career in technical applications and support. You will have the opportunity to exercise your communication skills while maintaining a connection to scientists at the bench. There is daily job diversity, little or no bench work, and you will not be responsible for making direct sales. In addition, these jobs tend to be less stressful than many other careers, so you can lead a more balanced personal life and even enjoy weekends off.

*Technical applications and support are about enabling customers to be successful with their products or technologies.*

### THE IMPORTANCE OF TECHNICAL APPLICATIONS AND SUPPORT IN BIOTECHNOLOGY AND DRUG DEVELOPMENT

It is important to develop long-term and productive relationships with clients so that they will refer others and return for repeat business. This is accomplished in part by effective pre- and post-sales support. High-quality support

*Productive customers are happy customers, and happy customers return for more business.*

will help clients realize the full potential of a company's products and will increase customer retention. This ultimately helps drive the sales engine and enhances the reputation of the company.

## CAREER TRACKS IN TECHNICAL APPLICATIONS AND SUPPORT

This chapter describes three main types of careers: field application specialist positions, technical support, and technical training. These positions are similar because they each involve the company/customer interface, mostly through post-sales support. They require technical competence but do not involve direct sales (closing deals). Technical support serves clients via the phone or by E-mail, whereas trainers and field application specialists typically interact with customers in person. Other career tracks include field technical engineers and business analysts.

### *Field Application Specialists or Field Application Scientists*

*Field application specialists are technically savvy solution providers.*

As biotechnology advances, customers find it difficult to keep up with the flood of new technologies and methodologies. They frequently challenge sales reps to demonstrate the customer-specific benefits and particular nuances of sophisticated products. These customers, who are typically Ph.D.s, often feel that fellow Ph.D.s will have a better appreciation of their needs and a better understanding of how to apply a company's products. In response, companies have created the field application specialist (or field application scientist) (FAS) positions. FASs are technical experts who maintain a thorough understanding of the products with which they work.

The role of FASs varies among companies. Typically they provide mostly post-sales support, but they also work closely with sales representatives to help potential customers evaluate products before purchase and to provide consultative advice regarding customers' experiments.

*Being a field application specialist could be the most interesting and productive postdoctoral training that you'll ever experience.*

### *Technical Support*

*Technical support representatives are the "answer people."*

Externally, technical support reps serve as the front line, answering technical, product-related questions and solving customers' problems by telephone or E-mail. Within the company, they help direct product development efforts by providing customer feedback. In addition, these positions serve as a rich training ground from which to internally recruit employees who are knowledgeable about the company's products and inner workings.

### *Technical Trainers*

Trainers teach clients how to use products, and they enable people to do their work better by teaching them new techniques and skills.

#### *Field Technical Engineers or Service Engineers*

These are the fix-it technicians, experts who explain, install, and repair complicated hardware and/or software and help address problems that arise. These positions are most common in companies that sell instruments or capital equipment.

#### *Business Analysts or Subject Matter Experts*

Subject matter experts (SMEs) typically work in consulting companies as expert scientists. They provide scientific advice about business processes or projects. The resident expert may assist commercial or technical groups and interact closely with clients.

## ROLES AND RESPONSIBILITIES IN TECHNICAL APPLICATIONS AND SUPPORT

*In larger companies, the roles and responsibilities of FASs, technical support representatives, and technical trainers may be clearly delineated, but in smaller companies, they may be combined. The following is a generalized list.*

### Field Application Specialists

#### *Pre-Sales Support*

FASs represent the company from a technical standpoint to help promote sales. They discuss the features and benefits of the products based on the customers' needs. Typically, they conduct demonstrations, give technical presentations and training sessions, answer questions, and help troubleshoot. They might support clients while they are evaluating a product, review customers' data, or help them design experiments.

#### *Product Adoption*

FASs develop consultative relationships with customers and may even help them conduct experiments using the specialist's products. Their published results serve as third-party endorsements that can be used as promotional material for future sales presentations. FASs also develop relationships with "industry thought leaders" (leading academicians and industry experts). Product endorsement by these leaders cascades down to other researchers in the field, eventually driving sales and product adoption.

#### *Post-Sales Support*

After products have been purchased, FASs help set up equipment, answer questions, troubleshoot, maintain instruments (the role of a field technician or service engineer in larger companies), and manage customer relationships.

#### *Product Development and Marketing*

Inside the company, FASs are a rich source of information for the product development, marketing, and quality assurance departments. Because they work closely with customers, they have an intimate understanding of customers' needs and problems. FASs often test early versions of products. They assess competition, trends, and new technologies that might affect product sales.

#### *Keeping Technically Updated*

FASs need to stay current with the latest scientific trends in their assigned technical areas. They must remain up to date on their customers' and competitors' publications so that they can explain why their products are superior to or different from those of the competition. Occasionally, opportunities arise to conduct original research and publish papers.

### Technical Support

#### *Telephone Assistance Cases*

Technical support reps spend most of their time taking care of problems over the phone or by E-mail with customers. Some calls can be as simple as finding a catalog number, whereas others can be very involved and take weeks or even months to resolve. Technical support reps take complete responsibility for each case until the customer is satisfied. For difficult cases, they may consult with specialists or the product development team.

#### *Reporting Malfunction Problems to Manufacturing*

Technical support reps report problems to the manufacturing group and are sometimes responsible for accurately determining the root causes for malfunctions.

#### *Product Development Experts*

Most technical support reps are "product specialists" and provide expertise in a particular group of related products. They serve as key point people when their coworkers are confronted with difficult problems. Specialists are assigned to new products before market launch. They interact with manufacturing and R&D on new product development, bringing fresh insights and the customer's perspective to design and development. They might review instruction manuals to ensure that directions are clear and concise. After the product goes to market, the specialists continue to interact with manufacturing and R&D to refine the product.

### Technical Training

#### *Course Design*

Trainers develop courses. Although they don't usually design courses from scratch, they do update and supplement previous course materials based on new product entrants and industry's current methods and uses.

### *Training Delivery*

When customers request training, trainers initially conduct a needs assessment so that the content is pertinent to the audience. Training may be conducted at the customer's site or at the company's headquarters. Afterwards, trainers use customer evaluations to gauge the course's effectiveness.

### *Internal Training*

Trainers attend R&D meetings and work with the product development teams to gain a full understanding of their products. They frequently teach other employees within the company, including FASs, technical support reps, technicians, service engineers, sales reps, and others.

### *Sales Support*

Trainers are often called in to help sales reps present technical material to potential clients. They can also indirectly promote more sales by interacting with clients and determining whether they need additional products.

## A TYPICAL DAY IN TECHNICAL APPLICATIONS AND SUPPORT

*A technical applications and support professional can expect some of the following activities on a typical day:*

### *Field Application Specialist*

- Spending time with customers for presentations or demonstrations, pre- and post-sales support (~25–60% of the time).
- Solving customers' technical problems and communicating solutions by E-mail or phone.
- Developing presentations for new products.
- Attending conferences.
- Keeping up with the scientific literature.

### *Technical Support*

- Resolving customers' problems on the phone or by E-mail (~90% of the time).
- Attending product development meetings.

### *Training*

- Designing courses and updating materials (~30% of the time).
- Teaching courses, preparing the training venue, and traveling (~40% of the time).

## SALARY AND COMPENSATION

Field application specialists earn base salaries that are comparable to those of bench scientists in industry with an equivalent number of years' experience, but they also earn bonuses. Bonuses are not guaranteed, but when sales go well, they may increase salaries by 10–20% a year. Seasoned application scientists can earn six-figure incomes, but the base salary of a FAS with a Ph.D. may be about the same as the starting salary for a sales rep with a bachelor's degree.

Technical support reps generally earn 10–20% less than researchers in industry with the same amount of experience. Technical support is viewed as an overhead cost by upper management, so salaries in this area are frequently a focus for cost-cutting.

Technical trainers generally earn 25–30% less than researchers in industry, but their salaries are still far higher than those in comparable positions as lecturers at junior colleges or universities.

***How is success measured?***

*Repeat business is the best measure of success in these careers. Customers are less likely to return for more products if they don't receive proper support.*

It can be difficult to measure field application specialists' contributions to sales, because their role is just part of the sales equation. Metrics often applied include performance evaluations, customer surveys, and the number of sales that result from presentations, as well as the number of customer visits, seminars, or training sessions.

Evaluation of a technical support rep's performance is somewhat more subjective. Supervisors may consider the number of cases completed and review case histories and customer surveys.

For technical trainers, the metrics used to assess performance include evaluation forms, the number of people trained, sales after training, requests for additional training, and referrals from clients. Training might result in reducing the number of technical support calls as well.

## PROS AND CONS OF THE JOB

### Positive Aspects of a FAS Career

- There is tremendous job variety. Work can be unpredictable, dynamic, and variable. Opportunities to tackle a variety of topics, questions, tasks, and goals occur daily.
- There is constant exposure to science, across a wide range of disciplines. It is a great way to develop expertise in particular areas into which you can eventually transfer your skills.
- FASs have the flexibility and freedom to operate as they see fit. Most of them work from home, plan their own schedules, and set their own priorities.

- You will have the opportunity to meet incredibly bright and talented scientists in industry and academia. Most customers are enjoyable to work with, and there are opportunities to build relationships and a large network of contacts.

*There is so much to learn as a FAS that at first, it is like drinking water from a fire hose!*

- You will be exposed to and trained in business fundamentals ranging from sales and marketing to business development and contract negotiations.
- Your work could have a significant positive impact on scientists' research. Ultimately, you may be contributing to the advancement of science for the betterment of human health.
- You can enjoy extra perks, such as traveling, wining and dining customers, a car allowance, a company-paid cell phone, a laptop computer, and more.

*In technical applications and support, you can have one foot in the lab and the other foot in business.*

## The Potentially Unpleasant Side of a FAS Career

- There can be a great deal of travel, which can put a strain on your personal life. Depending on your territory, you may spend half of your time traveling, mostly to visit clients and to serve on booth duty at conferences.
- It is often difficult to measure the financial impact of your efforts. As a consequence, you constantly need to emphasize the importance of your department, particularly during difficult financial times.
- Most FASs are at the bottom of the sales department's organizational chart. This reporting structure can be frustrating, because although FASs maintain a deep understanding of the customer and product, their voices and insights might not be heard by management.
- Occasionally there are disagreeable customers with bad attitudes or arcane questions to problems that are difficult to solve. In these cases, diplomacy, tact, and tolerance are essential.

## Positive Aspects of Technical Support as a Career

- Technical support can be a wonderful job for people who don't want to sacrifice their outside interests or family life for their career.
- In some companies, the atmosphere at work is friendlier and more collegial than it is for other industry positions. Technical support reps work together as a team in a relaxed, family-like environment.
- You can contribute to the forward progression of science by using your extensive technical knowledge and people skills to help others.
- There is a lot of daily variety. The continuous development of new technologies means that there will always be new product information to learn and new technical problems to solve.

*In technical support, you can take the weekends off, you don't need to work nights, and when you go home, the work remains at work.*

- You may learn about the cutting-edge technologies being developed in your company and contribute to product development.
- It can be rewarding to satisfy demanding customers in a way that makes you feel you did the right thing for both the company and the customer.
- There is little or no travel required.

## The Potentially Unpleasant Side of Technical Support

- The job can be repetitive. Although you will occasionally field questions that call upon every bit of your training, many of the questions will be banal and uninteresting.
- As with FASs, it is often difficult to measure the financial impact of your efforts. As a consequence, you constantly need to emphasize the importance of your department.
- Increased government regulations have resulted in more bureaucratic paperwork and stricter adherence to government-imposed restrictions.
- Because of the nature of technical support, some customers are frustrated or angry when they call. Although difficult customers do not call often, the day grows longer when you receive more than one such incident.

*The customer is always right, even when he is wrong.*

- Because someone always needs to manage the phone lines during operational hours, it can take a while to adjust to the need for an attentive adherence to the work schedule. You may not be able to leave your post whenever you feel like it. You can take breaks, but coworkers must be informed when you do.

## Positive Aspects of Technical Training as a Career

- This career is especially rewarding for those who enjoy teaching, and the pay is better than for the equivalent academic lecturer posts.
- Clients *want* your help, and their appreciation for your efforts is gratifying. It can be rewarding to provide them new knowledge and skills that might transform their work.

*The thrill of training is when you see those light bulbs in people's heads go off, when people finally get it. It's that "aha" moment.*

- Training provides a great opportunity to meet interesting people and establish a network of contacts. The mutual respect that forms between you and your clients might eventually lead to long-standing collaborative relationships.
- Your schedule can be flexible, especially when you are not teaching. You can schedule classes far enough in advance to accommodate vacation plans.
- If you enjoy travel, you may have opportunities to visit places where you otherwise would not go.

### The Potentially Unpleasant Side of Technical Training

- The extensive travel that is sometimes required can become tedious after a while.
- As with field application specialists and technical support, it is often difficult to measure the financial impact of your efforts and your importance in the company.
- Training is often perceived by upper management as a Band-Aid for other more deeply seated problems. Sometimes clients expect to accomplish more than is reasonable in a given period of time.
- Trainers are typically at or near the bottom of a company's organizational chart and tend to be undervalued. As a consequence, they are usually paid less, comparatively speaking.
- When teaching, there can be complete inflexibility of scheduling. You *must* attend and lecture, as there are usually no backup trainers if you are sick or if there is an emergency.

## THE GREATEST CHALLENGES ON THE JOB

#### *Understanding What the Customer Wants*

The biggest challenge in technical applications and support is to understand what the customer wants or needs and to be able to constructively respond by providing pertinent content. Often, even the customers do not know what they need.

*Know thy audience! Information is only as good as its usefulness. Efficiently gauge your audience so you can effectively engage them.*

## TO EXCEL IN TECHNICAL APPLICATIONS AND SUPPORT

#### *Field Application Specialists: Business and Science in One Package*

The FAS's ultimate goal is to expedite the sales process by providing technical assistance to clients. People who excel in this career can move a conversation fluidly between science and business. By doing so, they can provide technical expertise and at the same time answer sales questions and prepare the client for considering the purchase, thereby making it easier for the sales rep to close the deal. In addition, great FASs know how to anticipate their clients' needs and then demonstrate how their products will fill those needs.

*People who excel in technical applications and support have exceptional interpersonal skills and the ability to teach difficult concepts in a clear, easy-to-digest form.*

#### *Technical Support: Empathy for Customers*

The ultimate goal of technical support is to solve customers' problems. People who excel in technical support tend to be empathetic to the customer's situation. If customers believe that a representative is really sharing their frustration and sincerely wants to help, they will be more satisfied with their customer support experience and more likely to say positive things about the company.

*Technical Training: Enabling the Client*

The trainer's goal is to ensure that customers who have purchased products will be successful when using them. People who excel in training don't just answer questions; they also teach clients to resolve problems on their own. They tailor the training programs to fit their clients' specific interests and learning styles, and they can clearly explain difficult concepts in an easy-to-comprehend fashion.

*A good trainer can inform. A great trainer can show the client how to find answers on his own.*

## Are You a Good Candidate for Technical Applications and Support?

*People who flourish in technical applications and support careers tend to have...*

*The heart of a servant.* These careers are for people whose passion is to help others and who receive satisfaction from assisting people.

*Strong interpersonal skills.* You will need the ability to quickly develop rapport, credibility, and trust with clients and coworkers (see Interpersonal Skills in Chapter 2). Once a client is comfortable and relaxed, you can more readily focus on solving his technical problems. In the context of this chapter, some valuable people skills are:

- A tremendous amount of patience. Every time you teach a course or answer a question, you need to be able to present it as if the client is learning it for the very first time. Not every client learns the same way or at the same speed, so creative training approaches sometimes work best.
- Exceptionally good listening skills. This includes paying attention to what clients do *not* say as much as to what they do say.
- Sensitivity and compassion. Empathy can go a long way toward connecting with clients or comforting upset customers.
- Superb diplomacy skills. Clients can be easily offended when they think they are not being taken seriously. You also need to be able to tactfully keep difficult people from dominating training sessions and to be able to skillfully defuse disagreements.
- Objectivity. Clients will consult with you again if you provide informed and unbiased advice.
- The ability to quickly gauge and adapt to customers' personalities and learning styles (see Personality Styles in Chapter 19).

*Technical proficiency and strong, broad science backgrounds.* These are technically difficult positions. You need to be able to quickly acquire and process large amounts of information about new products and technologies. If you are up to date with the latest technologies and can think on your feet, you will be able to respond adroitly to your clients' queries.

***Excellent communication and presentation skills.*** You should be able to clearly explain complicated ideas. FASs and trainers need to be comfortable speaking in front of audiences. Excellent writing skills are important as well, because these positions often require the writing and editing of training manuals.

***Exceptionally good problem-solving skills.*** These careers are a good match for creative people who enjoy the challenge of finding innovative solutions to problems.

***The ability to provide balanced responses to customers who disagree.*** You shouldn't be argumentative, but if you are too agreeable or obsequious, you may lose credibility. You need to be able to respond to customers' criticisms objectively and truthfully (this will be easier if you are current with the scientific literature).

***Superb multitasking and time management skills.*** There may be multiple demands on your time and many types of tasks to perform.

***The ability to be flexible and adaptable to change.*** It is important to be able to learn new technologies and processes, and to be able to adapt to adjustments in your schedule, reorganizations in companies, and changes in leadership. Even the more regular schedule of a telephone support rep can be thrown off by a case that takes days or even months to resolve.

***A "team-player" attitude.*** These positions often require a collaborative approach to solving problems, so you need to maintain a positive, supportive attitude toward your coworkers. This is especially true in the collegial work environment of technical support.

***A responsive, diligent attitude toward work.*** You need to respond quickly with solutions for clients.

***Thick skin.*** Customers can be brutal at times. They may be frustrated by your company's product and may want to take it out on you personally. You need to be resilient and self-confident enough to not be upset by their behavior, while at the same time remaining sensitive to the causes of their frustrations.

***Credibility with customers and groups within your company.*** You may serve as a bridge between the clients and the product development team, so you need to be able to converse meaningfully with both sides.

***Drive, self-motivation, and the ability to work independently.*** You need to be self-disciplined enough to work in an unstructured environment, particularly if you work from home. You should also be diligent enough to report back to headquarters consistently.

***Honesty and integrity.*** Employees should be responsible for their actions, particularly telephone support personnel and FASs. You may lose credibility and a sale might be jeopardized if you are proven incorrect or if you promise a customer something that can't be delivered.

***Enthusiasm.*** If you are excited about your products' benefits and the information that you are conveying, you will help sell more products.

***You should probably consider a career outside technical applications and support if you are...***

- Territorial or possessive about your data, expertise, or clients.
- Someone who takes negative interactions with upset customers personally (this is particularly true for those in technical support).
- Arrogant, impatient, or one who easily becomes defensive.
- Unwilling to care about or help with the sales team's efforts (this is particularly true for FASs).
- Unwilling to take the time to understand the client's needs.
- Unable to adhere to a tightly structured phone schedule (if you are in technical support).
- Too negative or a complainer.

## TECHNICAL APPLICATIONS AND SUPPORT CAREER POTENTIAL

Technical applications and support positions provide great entry-level career opportunities for scientists, and many people in this field enjoy their jobs so much that they never leave. Unfortunately, these positions tend to offer limited upward career potential beyond management of other technical services personnel. For those who are ready for new challenges, jobs in technical applications and support can serve as a springboard to other vocations. People in these positions have the opportunity to explore the industry by interacting not only with customers, but also with personnel in manufacturing, R&D, marketing, and sales.

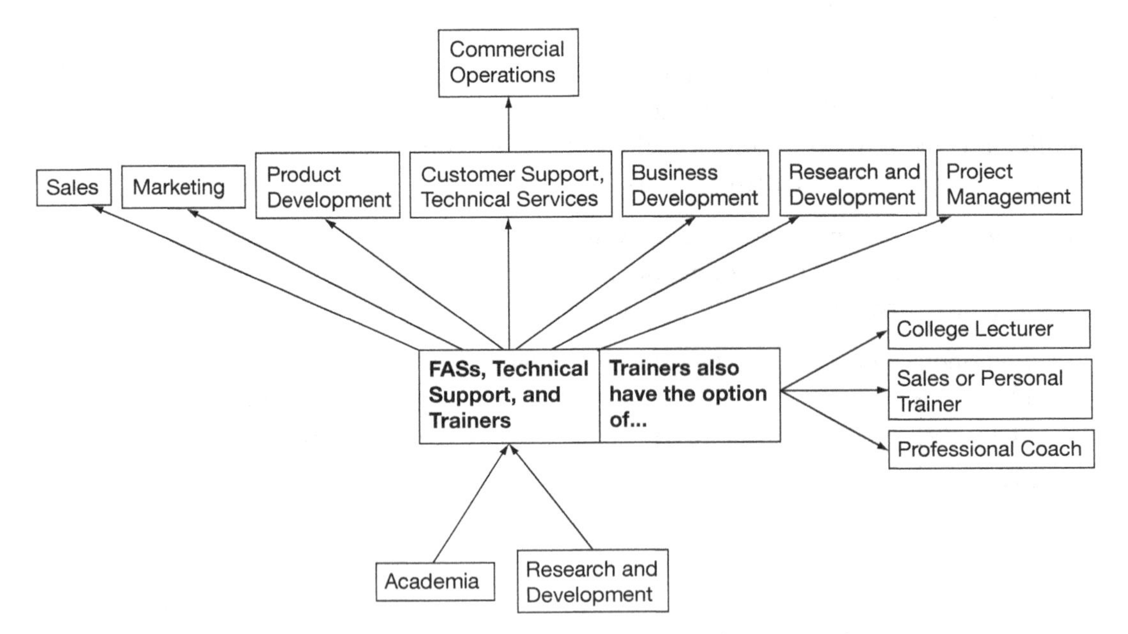


**Figure 20-1.** Common career paths in technical applications and support.

For FASs, moving into sales is a natural progression. Their experience can also lead to careers in product development, marketing, business development, customer service, and training (see Fig. 20-1). FASs are often recruited by their customers because of their technical expertise.

Product development and marketing are the most common career progressions for technical support reps, and they are frequently recruited into these departments from within the company they already represent. Some people also transfer to sales, R&D, training, manufacturing, business development, operations, or project management.

Experience as a trainer provides useful skills for management roles, such as communicating messages, motivating employees, delivering excellent presentations, providing feedback, and coaching people to lead teams. Career options for trainers are typically in FAS positions, sales management, marketing, customer support, project management, product development, and user design. Other possibilities include careers as performance consultants, professional developers, and executive coaches.

### Job Security and Future Trends

As products become more complicated, people will be needed to clearly explain how they work and how to use them. Support functions, however, tend to be viewed by upper management as a luxury with only indirect effects on revenues, so they are quickly affected by economic downturns.

Because FASs and technical support reps are often recruited to other positions, there are high turnover rates in these areas. Although the overall number of positions remains fairly stable, there tends to be a constant demand for qualified people to fill them.

For those who are concerned about outsourcing of these careers overseas, keep in mind that FASs and trainers interact directly with customers and within companies. Because the biggest market for discovery research is in the United States, these positions will not likely be outsourced. For similar reasons, it is likely that technical support will remain at company headquarters. Support representatives work closely with product development teams and sometimes work in the lab, especially when they are trying to resolve customer problems. They also serve as beta testers and provide customer feedback, two functions that are best facilitated by close interactions with their coworkers.

## LANDING A JOB IN TECHNICAL APPLICATIONS AND SUPPORT

### Experience and Educational Requirements

For these positions, companies tend to hire candidates with significant laboratory research experience, such as industry researchers, postdoctoral fellows, and, sometimes, graduate students. In general, having prior industry experience is an advantage, and it can be a great advantage to be a customer who has purchased and extensively used the products.

Educational requirements depend largely on the company's needs, position, and the type of product. For most positions, a B.S., master's, or Ph.D. degree will suffice, as long as one has had substantial laboratory experience. Those with a B.S. degree may require up to ten years of laboratory work, whereas Ph.D.s may already have had enough laboratory experience. Candidates with Ph.D. degrees are generally preferred, because they often possess a broader knowledge of science and have encountered a wider variety of situations where creative, problem-solving skills were required. Additionally, those with Ph.D. degrees tend to have greater credibility with customers.

### Paths to Technical Applications and Support

- Talk to FASs or account executives who are manning booths at large industry meetings. Inquire about potential job openings. This is a great place to drop your resume and make a good first impression. Dress as if you are going to an interview.
- Apply directly to the companies in which you are interested, particularly the ones whose products you have used.
- Consider working in discovery research to initially gain industrial experience, and then transfer. Become a "power user" of the product that you want to represent.
- If you are interested in a technical support career, call a company's technical support department and ask whether they are hiring. The person who answers your call will be sensitive to their hiring needs.
- If you are interested in technical training, contact your own training department and see where you can help. Volunteer to help teach and design classes. Observe training events or serve as a "subject matter expert." Provide suggestions and recommendations for designing courses. Consider teaching at junior colleges and universities to show your sincere interest in training and to hone your skills.

## RECOMMENDED TRAINING, PROFESSIONAL SOCIETIES, AND RESOURCES

*Societies and Resources*

Toastmasters International (www.toastmasters.org), to improve public speaking skills

American Society for Training & Development (www.astd.org)

Books for trainers are available at The Bob Pike Group Web site (www.bobpikegroup.com).

VNULearning (www.vnulearning.com) offers electronic newsletters and a training magazine.

*Classes for Trainers*

Langevin Learning Services (www.langevin.com)

Friesen, Kaye and Associates (www.fka.com)

Project Management Institute (www.pmi.org)

Classes and books on instructional design

# 21

# Corporate Communications

## Communication between External and Internal Worlds

*Corporate communications provides newsworthy information for a variety of audiences and returns feedback to senior management.*

CORPORATE COMMUNICATIONS OFFERS opportunities for people who enjoy writing and who can eloquently distill and interpret technical information for varied audiences. These careers provide an opportunity to participate in the shaping and building of a company. They encourage individuals to establish relationships with the outside world, such as the media, government, and the investment community, and internally with company leaders and employees.

## THE IMPORTANCE OF CORPORATE COMMUNICATIONS IN BIOTECHNOLOGY AND DRUG DEVELOPMENT

*Corporate communications is the guardian of the company's reputation and credibility.*

Corporate communications generates interest in the brand and faith in the company's ethos. Communications are needed to transmit news and, ultimately, to make the company become widely known and respected. A company's reputation and credibility must be protected to maintain investor confidence, and information must be disclosed in ways that are compliant with regulatory authorities. Corporate communications transmits the CEO's message to the media and the employees in an effective and responsible manner and plays a central role during crisis management.

Additionally, corporate communications not only helps determine the corporate messages that define the company's strategy, but also relays external feedback to senior management (see Fig. 21-1).

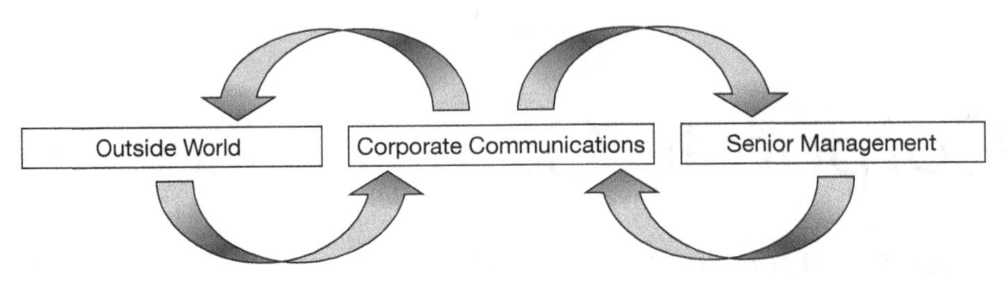


**Figure 21-1.** Communication flow in corporate communications.

## CAREER TRACKS IN CORPORATE COMMUNICATIONS

Corporate communications is an umbrella term covering several interrelated departments. These departments are substantially larger in public companies. In fact, most small start-ups do not have a corporate communications department. They typically hire consultants to write and distribute press releases, and they parcel out the other functions internally until the company is large enough to afford communications expertise of its own. As a company grows, each of the following functions becomes more specialized:

### *Investor Relations (IR): Marketing the Company to the Investment Community*

The investor relations personnel serve as the point persons in the company for investors. They also create and develop corporate brochures and annual reports to generate interest from prospective investors. IR strategists tweak corporate messages to ensure that public communications are of current interest and are appropriate.

*Investor relations officers serve as liaisons between the company and the investment community.*

In the early stages of a public company's life, when sales have not yet been generated, the value of the company is largely determined by the potential of its product development programs and the size of the markets they address. How well these are perceived by the investment community drives the company's stock price, which directly affects the company's ability to raise capital.

### *Public Relations (PR) and Media Relations: Marketing the Company to General Audiences*

Public relations officers write and disperse newsworthy information to press release services, trade press, newspapers, business media, television ads, and radio. It is their role to carefully cultivate relationships with a diversity of journalists and media professionals.

### *Public and Corporate Affairs*

Generally considered part of PR, public affairs interacts with government institutions and consumer and patient advocacy groups. Its function may also include philanthropic and corporate social responsibilities to enhance the company's reputation.

### Community Relations

People in this department work to enhance the company's reputation within the community. They set up grants and volunteer programs for local nonprofit societies, schools, hospitals, social service agencies, and other organizations. Employees are encouraged to volunteer for social causes.

### Government Relations or Affairs

Large biotechnology and pharmaceutical companies often establish a governmental relations branch. This involves building relationships with legislators and working with people who influence policy. People in this group explain how the industry's opportunities benefit the public and the economy, and they also address any controversial technical issues.

### Employee Relations or Internal Communications: Communicating Internally

Employees are considered "investors" in the company, so it is important to help them feel motivated and empowered. In partnership with the human resources department, employee relations may be responsible for providing internal publications to employees in addition to explaining the overall corporate goals and strategies.

### Marketing Communications (Marcom): Communicating Product Information

*Corporate communications specialists impart technological and scientific details in simple ways so that nonscientists understand why the company's technology is important.*

Marcom is the name for communications sent to the public media that are directly relevant to the sales and marketing of products. Marcom's job is to support sales by communicating the features and benefits of the product. They plan events, trade shows and exhibitions, seminars, briefings, and user group and advisory board meetings. Information is also communicated to the medical or scientific communities. Marcom is generally under the purview of commercial operations (see Chapter 18).

### Technical Communications

A technical communications or technical writing team produces user manuals, on-line product tutorials, and the text for brochures.

### Graphics

Some companies hire graphic artists to handle Web site design, corporate brochures, and trade booth development. This work is often outsourced to niche vendors.

## CORPORATE COMMUNICATIONS ROLES AND RESPONSIBILITIES

*Below is a list of general responsibilities that vary depending on the type of company and one's position.*

### *Developing Corporate Communications Strategies*

The executive management team and the head of corporate communications work together to determine the overall communications approach. This may involve anticipating future newsworthy events, transmitting financial performance, making sure that disclosures are made in accordance with regulations, and planning ways to enhance corporate culture.

### *Integrating Communications and Developing Messages*

The corporate communications group is responsible for making sure that information and messages delivered to many varied audiences and media vehicles remain consistent.

*Communication is not done in a vacuum—messages must be consistent and well integrated.*

In addition, most companies have an executive disclosure committee that decides which information should be publicly disclosed. According to the U.S. Securities and Exchange Commission (SEC), companies are required to report "material events" about anything that is informative to investors.

### *Shaping the Company's Business Strategy*

Corporate communications executives may assist in developing the company's business strategy by collecting feedback from outside the company about potential business affairs. For example, if the company is considering an acquisition, an experienced corporate communications officer will explain the potential fit with investors and gauge their responses.

### *Developing Messaging*

Before any company news is released, the core messages are strategically determined. The messages may be about presenting the vision of the company and value proposition of the products to consumers.

### *External Communications to General Audiences*

*Those who excel can transform a collection of diverse information into a coherent story that fits within the company's overall strategy.*

The company disperses its news to the rest of the world through press releases and other media vehicles in order to increase visibility. When there is news about events such as product approvals, the release of clinical trial results, or executive leadership changes, press releases are written and reviewed by the legal team before the information is released. They also work to have the company or its products mentioned in trade journals, in financial and various general business news media, and at conferences.

### *Developing Relationships with the Investment Community*

Investor relations officers build relationships with investors and equity analysts and serve as their main point of contact. They explain the company's strategy, disease focus, development status, and financial performance in quarterly earnings announcements, annual reports, and other information vehicles.

### *Fund-raising*

The investor relations head often joins the CEO and chief financial officer on "road shows," where they meet investors and analysts in person to increase awareness of the company and raise capital. The IR team helps to set up meetings, guide and attend the road shows, create presentations, and coach the CEO on what information should be disclosed.

### *Creating Brochures and Corporate Presentations*

Corporate brochures and annual reports are created by the investor relations team. This team is often responsible for corporate presentations given by the CEO at investor meetings and for writing scripts for the CEO for quarterly investor conference calls or Webcasts.

### *Creating a Corporate Identity*

The corporate communications group develops a company's identity. They write the company's mission statement, design the Web site and company logo, and devise anything externally visible, such as stationery or signs.

### *Internal Communications*

People in internal communications ensure that employees are up to date on company news and feel personally involved. They relay information about recent business development deals, clinical trial results, and other pertinent company activity. They help employees cope with changes that result from events such as layoffs. They coordinate employee meetings with the senior executive team, manage an intranet site as a communications portal, and create and distribute newsletters.

### *Event Planning, Corporate and Industry Events, and Trade Shows*

The corporate communications team plans holiday parties and company picnics. It is also involved in sponsoring or hosting conferences, trade show booths, seminars, and technology workshops.

### *Media Training*

One of the roles of corporate communications is to train designated spokespeople, such as senior executives, so that they clearly know what information is appropriate to disclose. Because the SEC carefully monitors a company's communications to ensure compliance with regulations, it is imperative that spokespeople do not reveal nonpublic material.

### Executive Coaching

The head of corporate communications may coach senior management before investor presentations and other public speaking engagements in order to refine presentations, maintain the consistency of messaging, and relate what information can be divulged. He or she alerts the executive team of investor expectations, identifies key investors, and explains their importance to the company's fund-raising activities.

### Crisis Management

When negative events of critical importance to the company occur, such as when a clinical trial fails or a product is recalled, this information is communicated to employees and investors. During these emotionally charged and highly stressful times, the company's credibility and the management team's reputation must be preserved. Sometimes coaching is required to reassure employees who are reluctant to accept the news. In addition, basic crisis management plans for specific types of setbacks and foreseeable problems are prepared in advance.

### Public and Community Relations Activities

Some companies support academic research and offer fellowships to provide positive public relations and attract high-caliber employees. They provide grants as incentives for excellence to local high school students interested in biotechnology research, for example, and senior members volunteer for nonprofit societies in their field.

### Establishing Disclosure and Company Ethics Policies

Every publicly traded company has to have a disclosure policy and code of ethical conduct including an anonymous, protected process for employees to report wrongdoing. Senior management and the board are involved in establishing these policies, but they are generally guided by the compliance officer or general counsel.

### Managing Vendors

Corporate communications personnel work closely with vendors who provide services in advertising, graphics, and more.

## A TYPICAL DAY IN CORPORATE COMMUNICATIONS

*Depending on your role, a typical day might include some of the following activities:*

- Surveying the market and the biotechnology industry news. Viewing biotechnology stock indexes and preparing the management team for anticipated key developments.

- Traveling, perhaps as much as 50% of the time.
- Writing press releases, coordinating with media, and coaching executives on how to express the news.
- Meeting with potential customers, advocacy groups, or other organizations.
- Designing, writing, and reviewing material for annual reports and corporate brochures.
- Conducting conference calls and attending internal meetings to discuss strategy and the implementation of programs. Attending company status meetings. Holding brainstorming sessions and coordinating the results.
- Writing scripts for quarterly conference calls and Webcasts. Helping the CEO prepare for these calls and organizing the content.
- Preparing corporate presentations for investor conferences.
- Managing vendors for product launches and other events.
- Adding new information to the company Web site or overseeing changes in the company logo or identity.
- Coordinating investor and analyst breakfasts at industry meetings.
- Arranging corporate events such as holiday parties and company picnics.
- Meeting and talking to investors, analysts, and investment bankers.
- Updating relevant company information for employees.
- Attending industry conferences and investor meetings.

## SALARY AND COMPENSATION

Compensation in corporate communications is similar or perhaps a little lower than in marketing and business development, but higher than in discovery research. Positions in biotechnology companies offer higher salaries than comparable positions in high-tech companies. Compensation surveys can be found at the National Investor Relations Institute Web site (NIRI, www.niri.org).

In general, private companies pay less than public ones, but private companies offer potentially lucrative pre-IPO stock options. This is because once a company goes public, the communication demands increase exponentially. Private companies are concerned with publicizing key events such as the release of clinical trials results, whereas pre-IPO and public companies are concerned about investment bank selections, relationships with analysts and investors, and the scrutiny of the SEC.

***How is success measured?***

The easiest and least reliable way to measure the success of corporate communications for public companies is on the basis of the company's stock price and performance. You can have a successful communications program, but it may not be reflected in the stock price. This is in part because the stock price is influenced not merely by company events, but also by external, uncontrollable, and unpredictable micro and macro economic factors.

There are other ways to measure success, depending on the department and what is being communicated:

- The image of the company and its success in reaching various key audiences.
- The company's ability to retain financial sell-side equity analyst coverage (see Chapter 24).
- The mix of the company investors.
- Getting articles written by major newspapers and trade journals.
- How often the company is mentioned in the news.
- Customer perspectives of the company, including responses to advertising campaigns.
- Employee attitudes, employee retention, and job satisfaction.
- The maintenance of a credible reputation for management.
- Adequacy of preparation for key events, and how well they were handled.

## PROS AND CONS OF THE JOB

### Positive Aspects of a Career in Corporate Communications

- Work is interesting, constantly changing, and fast-paced.
- The biotechnology industry is fascinating, and people are passionate about their work. The management team and employees tend to be vibrant and collaborative. Motivation is high, because the primary goal is to advance science and human health on a global scale.
- It is gratifying to relate an exciting story. Providing interesting news to the rest of the world and to employees is rewarding. Your efforts will contribute to the company's well-being and make it a better place in which to work.

*It is more rewarding to release biotechnology news with a broad social impact than to relay information about widgets.*

- This job is important; you will be interacting with many prominent people in the company as well as the media and investment community. The number and quality of people you will meet will provide you with an extensive network of connections and an increasingly competitive edge.
- People in corporate communications guide corporate strategy and messaging; they have impact on the growth of the company.

- Establishing positive relationships with the outside world is rewarding. Regardless of the news, when investors and analysts are promptly well-informed and treated respectfully, there is a sense of accomplishment.
- Writing can be a great pleasure. Composing the company's story in prose that is clear and informative while providing interesting news to employees and the rest of the world is gratifying work.
- Jobs in corporate communications can be highly creative and strategic. You might be involved in establishing new concepts, selecting new logos, or developing ways to position products.
- Corporate communications professionals have their fingers on the pulse of the company.
- Working in corporate communications is a productive way to leverage your scientific or business background. You can foster advances in science at large by informing scientific audiences and the general public of the company's developments.
- You might be able to telecommute as much as one or two days a week from home.

*If you believe that the mission of your company is worthwhile, then helping it prosper is highly inspirational.*

## The Potentially Unpleasant Side of Corporate Communications

- Senior-level officials travel as much as 50% of the time. Investor and public relations professionals travel the most, primarily for medical, investor, and industry meetings.
- This is a "hot seat" job. Being important means being under pressure, particularly during times when there is negative company news.
- Work can be intense and nonstop. Expect to continue to work at home during busy times. The demand for news never ends.
- As in most biotechnology jobs, deadline pressure and moving at a fast pace can be stressful.
- Executives can be ego-driven and competitive. CEOs, in particular, are sometimes unmanageable, and they often harbor unrealistic expectations. It can be difficult to offer constructive feedback.
- In general, the value and significance of corporate communications is not easy to measure and not well understood by executive management. Despite its importance to the company, corporate communications can be viewed as a cost center and not a core function.
- Most small biotechnology companies perform in a resource-constrained, cash-starved environment until they can sell products. A big part of your job might involve helping to raise capital.

## THE GREATEST CHALLENGES ON THE JOB

### *Managing News Flow in the Biotechnology and Drug Development Industry*

Investors and analysts are short-sighted and think in terms of months, not years. At the same time, the health care biotechnology industry suffers from long development timelines. When there is little newsworthy information to report and your audience has a short attention span, it becomes a challenge to preserve their confidence and interest. It can be difficult to convince them to view your company from a long-term perspective and to envision the landscape many years from now.

### *Getting the Message Out*

*You have to recognize what is truly newsworthy—what will make the front page and stand out against the rest of the background noise.*

It is difficult to obtain wide exposure, particularly for private companies that lack cash-rich communications budgets. An overwhelming amount of electronic communication already overloads investors, and rising above the noise is challenging.

### *Keeping the Message Consistent for Varied Audiences*

There are a multitude of audiences to reach, and the same message needs to be sent with a different spin for each audience. For example, news about an upcoming acquisition might focus on growth strategy and development targets when it is presented to investors, whereas the same news, when presented to employees, might focus on the benefits of sharing resources and the synergy of the two companies.

### *Providing Constructive Feedback to the Executive Team without Being Fired*

Artistic talent is required to adeptly relate negative feedback or constructive criticism to management without causing them to respond defensively. It takes diplomatic skill to convince management that you are really on their side and are merely serving as an objective translator. Management may believe that you have a negative attitude, or worse yet, accuse you of being disloyal! You may need thick skin to be able to do this job well.

## TO EXCEL IN CORPORATE COMMUNICATIONS...

### *Being a Creative and Strategic Communicator*

*Developing a core message is an art that requires elasticity of mind.*

In younger biotechnology companies, it is important to understand the company's technology, the image and values of the company, and what makes it unique. Communications professionals who excel work with management to create a company story that is sensible, yet flexible enough that the company can grow with it. They understand the core story and reconcile all points of view so that it continues to remain fresh and relevant.

#### *Being an Objective and Perceptive Translator*

It takes years of experience to develop relationships with the media and Wall Street that lead to honest feedback. Highly successful communications professionals can monitor the external world's perceptions of the company's performance and translate that into meaningful information which can be used to re-craft the company's message.

#### *Fostering Credibility*

Enhancing and protecting the credibility of the company and its management team is critical. Part of establishing credibility involves serving as an advocate for communications not just for the company, but also for the external world. It is better to consider all the constituencies and present a fair and objective story as opposed to just serving as a mouthpiece for the company.

## Are You a Good Candidate for Corporate Communications?

*People who flourish in corporate communications careers tend to have...*

*Outstanding communication skills.* Written, oral, and nonverbal skills are extremely important. You need maturity, intelligence, eloquence, and the ability to articulate. You don't necessarily need to be a good public speaker, but you should recognize how to reach people in an effective manner.

*It is important to be able to communicate persuasively, clearly, and succinctly.*

*Superb writing skills.* It is important to be a flexible writer, so that you know when to be serious and when to be light or slightly irreverent, depending on the desired mood you want to create.

*Exceptionally good people skills.* A likeable personality and excellent people skills are needed to build enduring relationships both within the company and externally. You need the ability to communicate effectively with institutional investors, the media, and people from various technical disciplines. To get things done effectively, you may need to rely on others, not just your staff, and this requires a collaborative team orientation.

*A strong ethical sense and the ability to withstand pressure from authority.* As a corporate communications professional, you are responsible for ensuring that the company divulges information appropriately. If someone does or says something that is inconsistent with your recommendations or against regulations, you need to have the self-confidence and ethical sense to stand up to him, even if that someone is your CEO.

*A great sense of humor.* You can't take yourself too seriously in this field. After all, your audience is human, and you need to connect with them. A sense of irony and wry humor is sometimes needed when dealing with investors, who will often articulate unrealistic expectations for your company. You will need to maintain a generosity of spirit and a sense of humor to override this and maintain productive relationships.

***The ability to multitask and prioritize.*** During a typical day, you could be working on three to five different projects simultaneously, so you need to be able to handle rapid changes and prioritize accordingly. The ability to track and manage projects and remain diligent is invaluable.

*Corporate communications is like working in "managed chaos."*

***Self-motivation.*** Initiative is often needed to drive programs.

***An ability to manage conflict.*** It helps to be an active and objective facilitator for resolving debates and conflicts.

***An ability to handle criticism.*** If you offer advice to the senior management team, they might have advice for you in return. You should be able to handle their criticism with equanimity and humility.

***Ability to think and plan strategically.*** Many people believe that corporate communications carries out a cheerleading and promotional function, but in reality, it is an important part of the company's strategy. To successfully sway opinions, you need to be interpretive, analytical, and critical. You have to understand how your audiences think and what motivates them. To lead the department, you need to think strategically about the company's direction and remain tactful when dealing with people.

***The ability to handle crises well.*** Crises, although rare, are stressful, and during these times you must remain calm and in control.

***Empathy.*** It is useful to be able to understand your audiences and know what they want to hear. Particularly true for employee communications, you need to be sensitive when you relate information, and treat your audiences as you would want to be treated.

***Perceptive listening abilities.*** It is essential to be a perceptive listener so that you can sense the nuances and subtext of what others are saying. You need to listen carefully to understand how to simplify highly technical descriptions into clear messages for general audiences. When attending investor meetings, it is important to pay attention to subtle feedback from audiences so that you better understand how the company is perceived.

*You need discerning listening skills to understand what is important and to creatively construct a story with plenty of punch to it.*

***Having a service-oriented attitude.*** In particular, being responsive to investor and equity analyst needs is crucial.

***A basic understanding of science and finance.*** If you want your audiences to appreciate the science, you first need to understand it. Professionals in investment relations need an understanding of finance so that they can converse appropriately with the investment community.

***Gregarious personalities.*** In general, socially outgoing people are called for in this career because it is a public position. You might have to cold-call investors and introduce yourself. You may be invited to social events and should feel comfortable spending your time at parties promoting your company.

*General knowledge.* In this career, a little knowledge about many things and a continuous interest in learning are more advantageous than deep expertise on a limited number of topics.

*Mental flexibility.* A degree of mental openness and the ability to compromise are helpful. As a communications specialist, you don't necessarily own the words. Sometimes you may even need to deliver a message that you do not agree with.

*A scrupulous eye for detail.* You need to learn and apply standards consistently. It is essential to create high-quality and unflawed work.

*Creativity.* Sometimes the company's science can be too cut-and-dried. Creativity is required to invent a new brand positioning or an interesting angle for your story.

*Creativity is needed to figure out how to best tell a company story in an interesting way to different audiences.*

*The ability to be a visual thinker.* You may need to convert technical slides into investor-friendly presentations using a variety of audiovisual tools.

*Strong aesthetics.* An aesthetic sense is needed for selection of the most attractive design for the company logo, Web site, and corporate brochures. You need to be able to develop and deliver crisp and compelling messages that are attractive and striking.

***You should probably consider a career outside of corporate communications if you are...***

- A slow and methodical perfectionist.
- Someone who doesn't communicate well.
- An overly shy person.
- Someone who prefers a predictable or prescribed work environment with consistency and structure in their calendar.
- Someone who can work on only one project at a time.
- Inflexible or unable to compromise.
- Overly process-oriented and rigidly bureaucratic.
- Unable to handle conflicts or debates.

## CORPORATE COMMUNICATIONS CAREER POTENTIAL

You can eventually work your way up to a lucrative position as head or vice president of corporate communications, public affairs, or investor relations. There are additional opportunities to join agencies or become a corporate communications consultant (see Fig. 21-2.). Senior heads eventually serve on executive teams and report directly to the chief financial officer, chief administrative officer, or CEO.

Corporate communications experience can help you prepare for transfer into other careers, most commonly into marketing, advertising, event management, sales, and

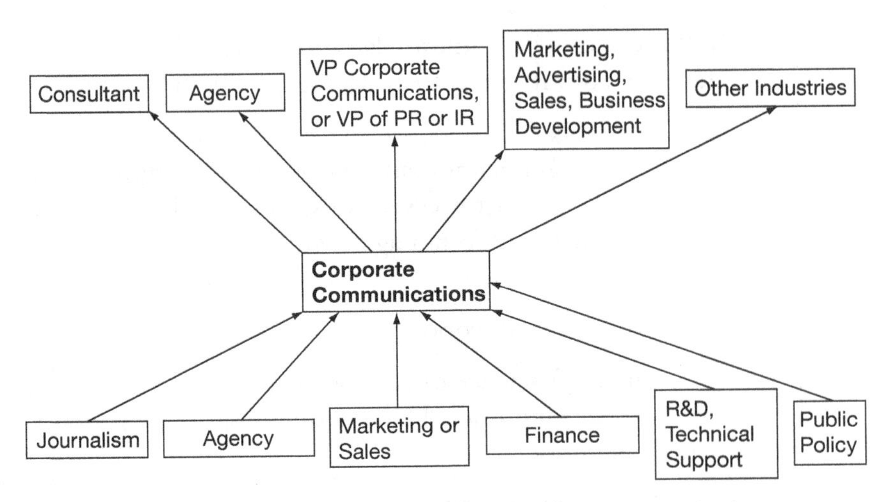


**Figure 21-2.** Common career paths in corporate communications.

business development. Other less common transitions include regulatory affairs, medical writing, investment banking, or philanthropic foundation work. There are opportunities to work in policy development at nonprofits such as the Organisation for Economic Corporate Development or in lobbying. It is also possible to transfer into other industries, because corporate communications principles are basically the same, regardless of the particular business.

## Job Security and Future Trends

Job security in the biotechnology industry as a whole is unstable. There are micro and macro economic factors, and drug development is never a sure thing. Even the most promising science can have disappointing clinical results that might spell the end of a small company. When cutbacks are made, corporate communications is one of the areas hit. As with other departments that are not part of research and development (R&D), it can often be viewed as expendable.

Small companies can exist without even having a communications department. The chief financial officer can serve as the investor relations contact, the vice president of human resources can oversee employee communications, marketing can oversee the marcom functions, and consultants or business development personnel can write press releases. A corporate communications function is essential, however, for public companies and is needed when products are launched.

There is a surging demand for experienced biotechnology communication professionals, and at the same time, there is an overabundance of junior-level people. As biotechnology companies continue to mature, the demand for experienced personnel will increase.

## LANDING A JOB IN CORPORATE COMMUNICATIONS

### Experience and Educational Requirements

*People tend to find their way into corporate communications by circuitous routes.*

Experience and educational requirements for corporate communications vary widely. People come from a variety of backgrounds, and no particular degrees are required. Many communications employees have backgrounds in communications, journalism or scientific writing, marketing, finance, corporate and business development, scientific affairs, or science. Experience requirements depend in part on the department and type of company. If the company has a strong technology base, for example, it is advantageous to have marketing or science backgrounds for crafting technical press releases.

Educational backgrounds vary tremendously in companies, but it is generally accepted that at least a B.S. degree is required. A law or M.B.A. degree is also particularly advantageous. People can obtain training in corporate communications and investor relations programs. Depending on the company, around 15–50% of those in corporate communications have scientific backgrounds, and only half of those have Ph.D. degrees. Those in government and corporate affairs often have Ph.D. degrees. Science degrees are not necessary for success, but they are very useful in technology-oriented companies. Those with science backgrounds tend to initially migrate to marcom positions, where they can use their knowledge to explain the technology and product benefits.

*People with excellent communication skills and a good understanding of science are highly sought after.*

Corporate communications is one of the few occupations in biotechnology that allows easy transfer from other industries. For senior-level executives, however, an understanding of the nuances of the U.S. Food and Drug Administration (FDA) regulatory environment and developed relationships in biotechnology-specific communities are essential.

### Paths to Corporate Communications

- Entry-level applicants should consider taking positions that involve writing press releases and running trade show booths, such as PR communication, technical communication, or marcom positions.
- Consider obtaining an internship in appropriate companies or volunteer to work on a biotechnology project to gain experience. Opportunities are available to contribute to company newsletters, Web sites, and videos.
- Consider working in a graphic design studio to learn how to produce marketing collateral for companies, or specialize in design, media, or advertising.
- Many large research institutes and universities offer corporate communications, scientific affairs, and public affairs positions. These can provide excellent entry-level experi-

ence in an environment full of scientific and technological news to publicize and will allow you to grow your portfolio before seeking a job in industry.

- If you are interested in government affairs, apply for an American Association for the Advancement of Science (AAAS, www.aaas.org) fellowship. This organization offers a policy fellows program in Washington, D.C. You will have the opportunity to work in Congress and help elected officials construct science policies.
- Consider positions at PR Newswire or Business Wire, the two main news agencies. Such experience will give you a good general introduction to corporate communications and the intricacies of press releases.
- Work in an investor or public relations agency with a biotechnology or pharmaceutical practice, where you will be exposed to a broad array of useful activities.
- Develop a portfolio of your work. Write editorial articles for industry trade magazines. Construct a Web site to advertise your portfolio.
- Consider taking classes in writing, English philosophy, business, and language. Learn how to create stories and develop your fundamental communication skills.
- Likewise, if you lack a science background, take biotechnology and science classes or executive development biotechnology courses. You will learn to "talk the talk," be introduced to current biotechnology issues, become familiar with the challenges of science, and meet some of the people involved.
- Consider obtaining an M.B.A. degree, or take business classes to understand financial analyses and balance sheets and to be conversant with investors and analysts. A fundamental understanding of business will help make you credible.
- Consider obtaining project management experience. This will help you develop credibility and garner skills valuable in corporate communications.
- Attend investor meetings to observe people's attire so that you know how to dress appropriately for interviews.
- An excellent way to develop your career is to join local and national industry organizations and attend their events so as to develop your network and learn about the industry.
- Try to obtain work in a large biotechnology or pharmaceutical company with a large corporate communications department. You will have many more different roles and responsibilities to explore, which can lead to lateral job transitions.
- Look for companies that respect the corporate communications function and consider it an integral part of other management functions. Avoid companies which think that corporate communications is about spin, packaging, or even worse, promoting the CEO.
- Read company press releases and corporate information. Analyze why this information was compiled and what critical sets of points were presented. It will provide an insight into the way that you may be required to think and write in order to be effective in this

job. For scientists, the scientific method is presented backward in press releases: The conclusion is presented first, followed by how the conclusion was reached.

- Try to obtain both public and private company experience. It is especially good to experience taking a company through a public offering.

## RECOMMENDED PROFESSIONAL SOCIETIES AND RESOURCES

The best organization to join if you want to learn about this career is the National Investor Relations Institute (NIRI, www.niri.org). The society has local chapters and conferences, and some events are biotechnology-specific. It offers book suggestions, certificate programs, career educational courses and material, and job opportunities. The alumni and network are very supportive.

### *Other Societies and Resources Include*

Public Relations Society of America (www.prsa.org)

International Association of Business Communicators (IABC, www.iabc.com) is a good professional organization to be affiliated with. The society also offers a certification program and awards.

Local chapters of Healthcare Communicators. Conduct a Google search to find one near you.

American Association for the Advancement of Science (www.aaas.org)

Organisation for Economic Co-operation and Development (www.oecd.org), to obtain international experience for scientists interested in government affairs.

Medical Marketing Association (MMA, www.mmanet.org)

American Marketing Association (AMA, www.marketingpower.com)

# 22

# Executive Leadership and Entrepreneurship

## The Business Builders

*Industry converts abstract knowledge into valuable products that can help people.*

EVERY MANAGER WILL, AT SOME TIME during his or her career, have to choose between becoming deeper and more expert in his or her discipline, or broader and more versatile in the business side of things. Those who choose to become more involved in business can follow the executive career track and eventually become a chief executive officer (CEO) or an entrepreneur. CEOs serve the important role of steering the managerial, financial, and operational functions of the company. Entrepreneurs conceive a vision and originate great technology concepts. They both share the thrills of building a legacy, setting the vision and culture of the company, and seeing their ideas become tangible, valuable products.

### THE IMPORTANCE OF EXECUTIVE LEADERSHIP AND ENTREPRENEURSHIP IN BIOTECHNOLOGY AND DRUG DEVELOPMENT

*It is tremendously exciting to take your own initial concepts each step from discovery research through development and finally to the marketplace.*

Executive leadership is critical. Because the product development cycle is so long, and the capital requirements in biotechnology companies are so enormous, having the right executive management team often is the difference between a company's success and failure.

Start-ups are the fundamental building blocks of the biotechnology industry. They are where the truly innovative and experimental projects occur. These new companies bridge the gap between discovery science and the creation of commercial value.

# CAREER TRACKS IN EXECUTIVE LEADERSHIP AND ENTREPRENEURSHIP

*This chapter generalizes about executive leadership positions and entrepreneurship. These are not official "career tracks" but rather career directions.*

### *Chief Executive Officers*

Chief executive officers lead people, set the strategy, and serve as the company spokesperson with investors. They guide the company's strategic direction and establish the corporate goals and culture. From a legal perspective, they carry the ultimate responsibility for the company.

The role of a CEO in large, established public companies is vastly different from the role played in a young start-up. In the early phase, the CEO and executive team write the business plan and develop a strategy, hire the most promising people to implement that plan, and put operations and processes into place. They interact with the financial community and potential corporate partners. As the company grows, the work of the CEO dramatically changes, and if truly successful, the CEO will manage the company as a public entity under the scrutiny of the U.S. Securities and Exchange Commission (SEC). During this process, the management team changes profoundly.

### *Executive Leadership, the Executive Management Team*

The CEO surrounds himself or herself with the heads of various disciplines who serve as experts in their fields. Together, they are involved in the key decision making for the company. Depending on the size and type of company, there may be a president, chief operating officer (COO), chief scientific officer (CSO), chief commercial officer (CCO), chief medical officer (CMO), chief financial officer (CFO), and a vice president of business development. Other heads include vice president of sales and/or marketing, vice president of pharmaceutical development, vice president of human resources, and legal counsel. Small companies typically can afford only one or two heads who serve multiple roles, whereas large pharmaceutical companies employ numerous executives.

### *Entrepreneurs and Founders*

*Entrepreneurship: to do things that no one has ever done before.*

Entrepreneurs are those who establish a company; they could be the inventors of the technology and/or business executives. They often operate as CEOs during the early stages of a start-up company and then remain as members of the executive team during later stages. Their most important roles are to explain the business plan and strategy in order to attract investors and a talented staff.

Most small companies have limited resources, which means that they operate with a minimal staff and little if any market research data. Entrepreneurs must do everything, even things such as making sure that the door keys work, that the electricity is on, and that the phones are operational. Virtually no one will know about the company, and for operating in stealth mode, that's usually a good thing.

## ROLES AND RESPONSIBILITIES

*Roles and responsibilities vary depending on the type of position and company. The following is a generalized list of some of the responsibilities assumed by CEOs, entrepreneurs, and the executive management team.*

### *Writing the Business Plan and Raising Capital*

*A business plan is like the company's graduate thesis before oral exams.*

Most founders and CEOs spend a great deal of time writing a business plan. The business plan describes, in great detail, the entire process of how the company plans to go from idea to profitability. It describes the company's goals, technology platform, operating plan, management team, and more. Armed with a business plan, entrepreneurs then raise capital by "pitching" their concept to potential investors.

### *Cheerleading*

*You could call entrepreneurs the chief cheerleaders...or the chief worry officers!*

One primary role of executives is to be a cheerleader. They encourage their employees during both good and bad times. When projects simply don't work, they acknowledge that the team gave its best effort, evaluate what has been learned, and explain how to benefit from the experience. With an optimistic spirit, they continuously challenge themselves and those around them to deliver something extraordinary.

### *Setting Expectations and Establishing the Vision*

*The executive team is the communicator of the vision.*

The executive team articulates the "vision" of the company to employees and potential investors. The vision should be exciting, enticing, and practical, such as creating value or solving an important technical or medical problem. The executive team establishes a sense of urgency for that vision and proposes strategies for developing products as quickly as possible. At the same time, they establish realistic expectations and prepare for likely outcomes.

### *Planning Corporate Milestones and Budgets, Developing an Operating Plan*

*The CEO is the financial custodian.*

Executive management ensures that the overall program makes sense and that the company is progressing toward marketable products. They make sure that things are being done in a timely fashion, that the quality of the work is at its highest, and that critical questions have been addressed. They compose and implement plans for recruiting talent and dealing with facilities, finance, and legal issues. This is rarely a solo act, as the executive management team surrounds itself with technical and business experts who advise and assist in making decisions.

### *Building and Leading Teams*

*As in football, the team with an outstanding coach is more likely to win.*

One important role of executive leadership lies in building the most productive team with the requisite talent, experience, and ability to fit into the company's corporate culture. Particularly in start-ups, hiring qualified employees who can assume multiple roles and are flexible is critical. In addition, companies try to hire complementary skill sets. If the CEO has a strong science background, for example, it is important to hire someone with strong business or finance experience. If, on the other hand, the CEO has some business background, a more junior-level person would be more qualified for the job, as start-ups rarely can afford a lot of chiefs.

### *Managing Employees*

The executive team spends a great deal of time managing employees and making sure that the group is working as a productive, smoothly run unit. Employees need to be talked to and mentored, and sometimes their behavior requires modification. Managing employees also includes recognizing and rewarding high performers with bonuses, promotions, and public recognition.

### *Establishing Corporate Culture*

Corporate culture represents the company's style, the way employees interact with one another, and how employees are viewed within the company. Every company's corporate culture is unique, and it usually starts at the top with the CEO. For more about corporate culture, see Chapter 3.

### *Market Assessment, Strategy*

The executive management team is ultimately responsible for the company's financial outcome. Using sound business fundamentals, they assess the market to determine what unmet needs the technology might address and whether there is a large enough market to make a profit.

### *Building and Managing Boards*

An emerging company (a funded start-up) is made possible in part by investors who have entrusted the executive management team with money. They typically serve as board members to oversee their investment. For the CEO, this means that major directives and resource-intensive programs are initiated with the board's approval. It is the CEO's responsibility to build, manage, and appraise the board and listen to its insightful advice. The CEO typically also builds and manages a scientific advisory board.

### *Human Resources-related Activities*

Executives set up a multitude of human resources activities, such as health care coverage and 401(k) plans, if they exist in those companies. Large companies typically employ vice presidents of human resources to manage this function.

## A TYPICAL DAY IN EXECUTIVE LEADERSHIP AND ENTREPRENEURSHIP

*Depending on the type and stage of the company, the executive team may spend time on a typical day...*

- Talking to customers and analyzing the market.
- Networking, attending investor meetings, and identifying leads in the venture and corporate capital communities.
- Writing the business plan and fund-raising; describing the plan's business opportunities to prospective investors.
- Attending and running meetings. These include staff meetings, planning meetings for development projects, status updates, or one-on-ones with direct reports.
- Strategizing and establishing corporate goals with the board of directors and team members.
- Conducting business development activities.
- Traveling and attending conferences.
- Doing administrative work.
- Hiring and talking to consultants, attorneys, and other service providers.
- Recruiting and hiring employees.
- Organizing and delegating activities to direct reports.
- Talking with and assisting employees; mentoring them or modifying their behavior; making sure the team is working in a collegial and productive environment.
- Managing crises; working with staff and providing investors with the appropriate information.
- Making sure that the company is still at the cutting edge; benchmarking the company against others.
- Reading business and scientific literature and industry-related news.

## SALARY AND COMPENSATION

Compensation for CEOs and entrepreneurs varies widely depending on one's experience and the size and type of the company. Public companies tend to offer higher salaries, but private companies offer potentially lucrative pre-IPO (before the initial public offering) stock options.

### CEOs

> *The aim of the game is not only to develop a successful company, but also to retain a high percentage of the stock and ownership!*

CEOs in large biotechnology and pharmaceutical companies can earn from six-figure incomes to millions of dollars each year, depending on whether the company is private or public. CEOs are typically, but not always, the highest-paid employees in a company. Vice presidents of sales, marketing, and clinical development can sometimes earn more.

Salaries for CEOs in start-ups tend to be modest. Many initially work without a salary; compensation may arrive years later and only if the company is successful. Biotechnology CEOs in pre-funded start-ups may initially work pro bono, with the assumption that they will be rewarded through their stock options. Those who work in emerging companies (funded companies) are paid substantially more.

### Entrepreneurs

> *Stock options are the most promising incentives in the early days of a start-up company.*

Entrepreneurs in start-up companies need to show commitment by being willing to work unsalaried and to accept stock options in lieu of compensation. Experienced serial entrepreneurs (see page 323, "Career Potential") can expect more shares of stock and a higher salary, whereas unproven entrepreneurs are typically paid low salaries. If unproven entrepreneurs request high salaries, the venture capitalists may question whether they understand what a start-up is all about.

**How is success measured?**

Ultimately, company success is measured by the ability of investors to convert their stock into cash (a liquidity event) by going public or by being acquired. For a vast majority of entrepreneurs, success is measured by accumulated wealth, by the amount of stock they own, and by being associated with a successful start-up.

Venture capitalists invest in very few, carefully selected companies. Getting that first investment, especially from a top-tier venture capital firm or a corporate partnership, is an early measure of success. It signifies that someone believes in your concept and the company. As the company progresses, subsequent measures of success include the assembly of a productive team, demonstration of key proof of concepts, reaching milestones, obtaining larger rounds of financing at higher values, and building revenue and profit. More importantly, a successful company fills a genuine, unmet medical need.

## PROS AND CONS OF THE JOB

### Positive Aspects of Executive Leadership and Entrepreneurship

- Building something of real significance and impact can be an incredible thrill. It is an extraordinary feeling to be part of history. It can be extremely satisfying to work on a drug or medical device that might save peoples' lives or benefit the world in a meaningful way.

- It is immensely rewarding to create something out of nothing. Being able to see a dream come to fruition, for instance, or watching new product opportunities develop into promising medicines is fundamentally gratifying.
- Working at a start-up is adventurous! You might be at the cutting edge of science; you could be in a perpetual state of thrill and excitement during the process of successful financings, important negotiations, and more.
- It's fun to be in command! It is also deeply rewarding to know that you were the person who made the company successful by remaining at the helm throughout the hardships while keeping the company stakeholders satisfied.

*A new business "venture" is really an "adventure!"*

- You will have the chance to work with terrific people. The biotechnology industry is populated with people who are highly educated and deeply motivated. Many have strong entrepreneurial spirits and are enjoyable to work with.
- There can be a strong sense of camaraderie. This is particularly true in small, young companies, where more people are working toward the goal of building a successful company and, in contrast with the situation in many larger companies, fewer colleagues are pursuing their personal agendas.

*Entrepreneurship is about working with exceptional employees and constructing something extraordinary.*

- The potential to get rich via stock options in a start-up is tremendous. There are many ways to become wealthy, and the general public views entrepreneurship as a deserving way.
- As an entrepreneur, you have an opportunity to "wear many hats." You can take on many more types of responsibilities than would be allowed in large companies where you are more likely to be "pigeonholed" into functions befitting your educational background and experience. You have the chance to become involved in business development, operations, finance, human resources, fund-raising, and more.
- Decision-making paths are shorter in small companies. Companies are forced to make critical decisions far more quickly than in big pharmaceutical companies.

## The Potentially Unpleasant Side of Executive Leadership and Entrepreneurship

- Entrepreneurship is risky, and many more companies fail than succeed. Failure can be demoralizing.
- Executive leadership and entrepreneurship are stressful occupations. The CEO is the one on the hot seat and has to accept the ultimate responsibility for the company's final outcome. Many people's livelihoods and careers will be riding on decisions for which you are accountable, and even something as simple as financing the monthly payroll can be a source of recurrent anxiety.
- Expect to work hard, for long hours. Expect 100-hour workweeks, which can significantly interfere with one's personal life.

- The executive team tends to spend a lot of time on human resources–related issues. Much time is consumed building teams and managing difficult people. When giving personal reviews, it can be a painful experience to inform employees that they are not performing well. Most scientists are not trained for this—they tend to prefer devoting their time to scientific pursuits and don't want to be distracted with personality-related issues.
- Because most biotechnology companies are financially starved, CEOs are constantly in a fund-raising mode, leaving little time for the other important aspects of running a company. Ironically, they are often so busy that they don't have time to recruit and hire suitable employees.
- The multiple roles required of the entrepreneur can limit his or her ability to become an expert in any one discipline.
- The CEO can be a lonely position. Some things just can't be shared with the team or even the board.

*In the beginning of a start-up, you can forget about taking limos and staying at fancy hotels!*

## THE GREATEST CHALLENGES ON THE JOB

### *Fund-raising, Fund-raising, and Fund-raising*

Most start-up CEOs and founders spend a great deal of time just raising money. Even if you have previously carried major profit and loss (P&L) responsibility in a large company, learning how to raise venture capital is quite different. Biotechnology is a capital intensive business with long-term horizons for payoff, so continuously raising money is imperative for survival. This is often more difficult than it sounds—factors beyond your control, such as the biotechnology market, could affect fund-raising efforts. It is important to raise enough money at each stage, but not so much that you give away more stock and ownership than necessary.

*Raise money when the market is ripe—timing is everything!*

### *Managing the Many Facets*

A myriad of activities must be planned and implemented to run a successful company, and frequently there are not enough resources, time, or people to do so. Establishing timelines and milestones, deciding how much money to raise and when—CEOs and entrepreneurs are responsible for all of these processes and more. In addition, they must create an enjoyable company atmosphere where employees feel appreciated, where they feel that they are doing something important, and where they understand their roles. In total, it is an enormous but worthwhile challenge.

*Entrepreneurship is about spending as little money as humanly possible while still reaching your milestones on time.*

### *Hiring the Right People*

The executive team spends a great deal of time searching for people with the most appropriate talents and personalities,

*It is very important to know who to let "on the bus" and who to let off!*

because the project's or company's success may depend on it. It is important to identify those who share similar corporate cultural styles and can work with passion. When the going gets rough, you want to be able to rely on employees who will help solve problems instead of being disruptive.

### *Making Difficult Decisions*

*Investments are like marriages—they start off with the best intentions.*

Sometimes painful decisions need to be made. You might have to terminate a promising program if it doesn't fit corporate goals, or you may have to let go of an entire division. These are extremely difficult choices, and making such changes can feel like an assault on your personal accomplishments and visionary decision-making abilities.

### *Managing the Board of Directors*

Even the CEO needs consensus and buy-in from the board (which is composed of investors, advisors, and stakeholders). Managing the board includes establishing trust and open communication and being clear about objectives. Most importantly, you need to gain the board's buy-in, because if your goals are not aligned with the board's, a battle for agreement on a common direction will likely ensue. Convincing a disparate group to make decisions as a team can be very labor intensive, and is referred to as "herding cats" (see Chapter 17).

### *Passing the Baton*

Once a founder has established an exciting technology and raised some money, venture capitalists will frequently bring in an experienced CEO to manage the company, and it can be difficult for founders and CEOs to pass the baton. Common situations requiring new leadership include times when the company is ready to launch a product or is preparing to go public. It can be emotionally upsetting to be replaced after having taken such great risks and given so much time and devotion to a company.

## TO EXCEL IN EXECUTIVE LEADERSHIP AND ENTREPRENEURSHIP...

### *Being Highly "Bankable"*

*Obtaining financing is the lifeblood of start-up companies.*

There is almost an ethereal quality, a certain intangible presence or aura, that investors can sense in a CEO or entrepreneur who will make them a lot of money. These leaders emit enthusiasm and excitement. They express a passion and a sense of ultra-urgency. They have an extraordinarily good understanding of what they are trying to achieve and how to make it happen. They have the ability to convince others to join them in the effort to make their dream a reality, and just as importantly, they have the winning ability to sell their vision to investors.

### *Motivating and Empowering Employees*

> *The driving force for entrepreneurs is not merely to make money—it is to actualize their vision.*

Some executives have an exceptionally appealing leadership style: Employees will go to "the ends of the earth" for them because they have created a company and jobs that are personally meaningful. Successful leaders manage this in part by communicating honestly and openly, mitigating risks, and providing strategies. They credit teams with success, give recognition to exceptional employees, listen to their input, and make them feel empowered in the company.

### *Honesty*

> *You can't be a great CEO if you are not honest—you can be successful, but not great.*

Credibility is everything for the CEO and executive team. Investors bet their capital and their reputations on trust, so it is important for the CEO to be honest with the board members.

The head of an organization sets its tone and determines the type of people who should work there. He or she should provide a role model for success that colleagues will admire and want to simulate. In addition, great CEOs are very perceptive about distinguishing who to promote and encourage and who to let go.

## Are You a Good Candidate for Executive Leadership and Entrepreneurship?

***People who flourish in executive leadership and entrepreneurship tend to have...***

> *Personal ambition is one of the most powerful forms of human motivation—it ranks high with revenge and love.*

***Great ambition.*** Personal ambition is an extremely powerful motivating force, and it is probably the number one attribute of successful leaders. CEOs and entrepreneurs enjoy the challenges of their jobs. They tend to be very competitive and are often obsessed with success; some are driven by the fear of failure.

***Charismatic and evangelistic leadership.*** Successful leaders are almost religiously zealous about their vision and pursuits. This type of enthusiasm is infectious and often makes others want to follow.

***Passionate, yet realistic, about their company.*** Entrepreneurs believe completely in their company, but at the same time, they are willing to listen to critical feedback from those around them.

> *Those who excel have the ability to convert vision into reality.*

***Strategic vision.*** Entrepreneurs must be able to visualize a core idea and foresee how it can proceed from concept to the marketplace.

> *Entrepreneurs are those who after a failure can state, "Well, we learned a lot—now let's start the next enterprise."*

***Great resilience and the ability to bounce back.*** If one project fails, it is important to be prepared and willing to offer alternatives and recover quickly. Things can and often do go wrong and you should be ready to adjust your course when necessary. If your company fails, you should view the failure as a learning experience whose lessons can be applied to your next job.

*The ability to be tough.* Being a leader is not for the faint of heart. Leaders must be comfortable making strategic decisions that have real effects on every stakeholder in the company. They need to be strong enough to kill projects that are not meeting corporate goals, and persistent enough to continue giving presentations to venture capital firms despite being repeatedly declined.

*Self-confidence.* Executive leaders and entrepreneurs need to "know themselves" and have a sense of personal contentment. They are challenged with daily crises and difficult decisions at just about every level, and sometimes they have to operate solely on their own instinctive reactions. They must have confidence that they can bypass whatever impediments might crop up. In addition, they should have enough self-confidence to be comfortable working with and taking advice from people who are smarter than they are.

*Leadership skills.* Successful leaders gain the respect of the people with whom they work. They create an atmosphere of mutual trust, and their employees know they will have their leader's support when they need it.

*The ability to motivate people.* Leaders should be able to analyze and sense what other people are feeling and know how to respond in a way that will continue to motivate them. Motivation is not simple; some people are driven by the fear of losing their jobs, whereas others respond better to cheerleading.

*The ability to establish a positive corporate culture where employees can excel.* Creating a productive environment for employees is very important for the overall success of the company. Successful leaders create a workplace in which employees are allowed to be innovative in their approaches to problems, where they can be challenged and grow, and where they feel like they are making a difference.

*To excel requires a combination of innovative science, attainable business goals, and exquisitely good people skills.*

*A fast and accurate judge of character and the ability to build teams.* It is important to recognize your own business limitations while surrounding yourself with the team best suited to get the job done effectively.

*The ability to handle risk and uncertainty.* Risk takers are attracted to entrepreneurship. Entrepreneurs need to be prepared for the possibility that they won't recoup their personal and financial investments.

*The sense of security that comes from being employed in a large company is very strong. People are reluctant to walk away from a good job, and few prefer the risk of a start-up.*

*Courage.* Entrepreneurship requires a certain amount of fearlessness. You must not be intimidated, especially when working in areas that are new to you. It is important to not be satisfied with the status quo and to expect resistance from the mainstream thinkers and naysayers when you challenge established dogma. Invite creative solutions.

*It can be chaotic and anarchistic to leave mainstream thought and become truly innovative.*

*A willingness to work hard.* Entrepreneurs spend long hours at work and are obsessed about the results. A common misconception is that successful entrepreneurs are just lucky. As Thomas Jefferson said, "I'm a great believer in luck, and I find the harder I work the more I have of it."

***The ability to be outgoing and to gather resources.*** The executive team needs to aggressively network in order to meet potential investors and develop other resources to maximize the company's finances.

***An ability to remain focused on staying the course.*** There will probably be too many product opportunities to pursue at one time, and it is important to realize that you can't do them all. Those who are successful are able to prioritize projects judiciously while being mindful of available resources. That may mean temporarily ignoring the science while remaining focused on product development success.

***Excellent decision-making skills.*** Part of making difficult decisions involves being able to collate a large number of people's thoughts and crystallize them into a single conclusion. Entrepreneurs enjoy solving new, difficult problems. They tend to be so self-confident that they believe they can resolve anything.

*Entrepreneurs have to "see the forest through the trees," but they also have to know every tree, boulder, and area of quicksand!*

***Broad experience within the industry.*** Successfully running biotechnology companies requires extensive familiarity with many disciplines. The ideal background is one of broad experience, with solid foundations in science and business. It is often better to be a generalist and hire experts so that the team's collective knowledge is leveraged and integrated.

*The qualities required to be a successful scientist and the qualities it takes to build a business do not always overlap—it's the rare individual who can do both.*

***Tremendous communication skills.*** The executive team will spend a great deal of time presenting their technology or ideas to potential investors, customers, and employees. You need to be able to articulate the company's goals and technology to various types of audiences.

***You should probably consider a career outside of executive leadership and entrepreneurship if you are...***

- Unable to delegate or want to do everything yourself.
- Intolerant of risk and uncomfortable in an environment with unknowns.
- Expecting to conduct business activities in a resource-rich environment.
- Expecting to be trained on the job.
- Seeking an 8-to-5 job with weekends off or a predictable work environment.
- Expecting to set your own hours.
- Expecting vacations, benefits, and insurance in a start-up.
- Completely focused on technology and ignoring the market or pressing business issues.
- Stubborn and unwilling to listen to others.
- Expecting to delegate all responsibility so that everyone else will do all the hard work.
- Unwilling to share your responsibility with those who have had more business experience.
- Completely demoralized by failure.
- Only out to make a fast buck.

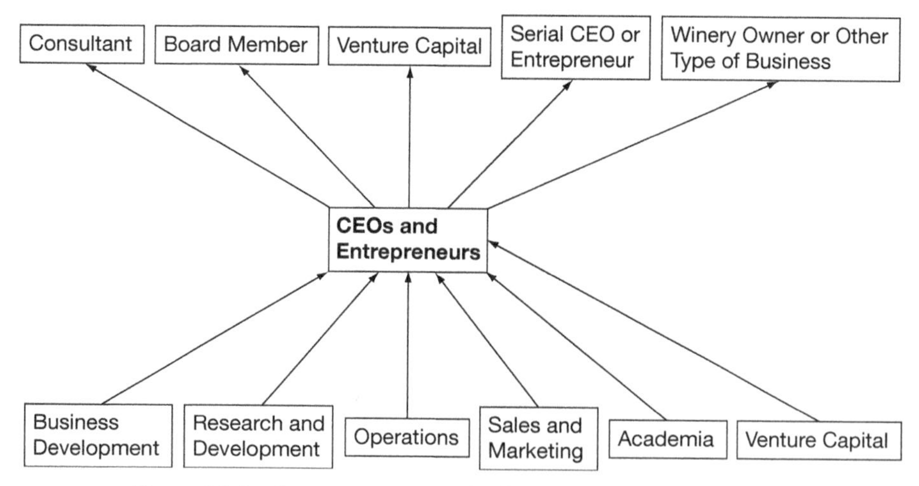


**Figure 22-1.** Common career paths for CEOs and entrepreneurs.

## CAREER POTENTIAL

A common career path for entrepreneurs is to become serial entrepreneurs, who run a company for two or three years and then step aside and begin another. Successful CEOs and entrepreneurs can proceed to manage larger and larger companies. The payoff is in stock options, and if you are good at it, you can become rich. Those who are truly successful eventually graduate to become angel investors, venture capitalists, entrepreneurs in residence (EIRs), and board members of companies (see Fig. 22-1). Some people write books or pass their knowledge on by consulting, whereas others become philanthropists or go on to other ventures, such as opening and running a winery.

### Job Security and Future Trends

*If the team is losing, they usually don't fire the team members—they fire the coach!*

Being a CEO is probably one of the least secure jobs; the CEO is the person who takes the blame when problems arise or when there are disagreements between investors about the company's direction. Raising money for young companies is exceedingly difficult, and most biotechnology companies eventually run out of cash and are sold.

*Working at a start-up is not just a job—entrepreneurs have invested their self worth, pride, and sometimes their homes in the company.*

Regardless of the company's success, there is little permanency in founding CEO positions. Success often brings change, and ironically, a successful CEO who has raised venture capital can be rewarded with the loss of his or her job. Once the company is funded, the board may quickly replace the start-up CEO with an emerging company CEO, who is then supplanted by a successor with operational or IPO experience. It's rare that one person has enough experience to survive the entire process of guiding the company from idea to

IPO. At the same time, there is tremendous market demand for successful CEOs at each successive stage of a company's development. Founders who want to remain active with their start-up companies during later stages typically do so as chief scientific officers or scientific advisory board members.

## LANDING A JOB IN EXECUTIVE LEADERSHIP AND ENTREPRENEURSHIP

### Experience and Educational Requirements

#### *CEOs*

*You are already either a leader or not. These intangible qualities are not acquired from educational training.*

Most CEOs have had previous experience in executive roles and have held meaningful roles within companies as chief operational officers (COOs) or in vice presidential positions (see Fig. 22-1). Other common jobs are chief scientific officer (CSO) and chief financial officer (CFO). If you seek a CEO position, try to obtain as much experience as possible serving on the executive team as a vice president or chief.

Opinions differ about the best educational requirements for a biotechnology CEO. Some people prefer to hire a Ph.D. with operational experience rather than an M.B.A., because the scientist may better understand and convey a company's technology and its significance. The most common choice, however, is an M.B.A. with some kind of medical, science, or math background, because the most important roles of a CEO are to raise money and manage the financial aspects of the business. A law degree helps when negotiating deals and is another advantage.

#### *Entrepreneurs*

*There are no boxes to check to become an entrepreneur.*

Entrepreneurs come from just about any career level and every functional area. All that is required is a willingness to take risks and the ability to corral people to help actualize the vision. The ideal background for an entrepreneur is to have worked on senior levels in multidisciplinary roles for both large and small companies.

Many entrepreneurs initially created programs in major pharmaceutical or biotechnology companies. After the company was downsized or its focus changed, they "spun out" the technology to create their own companies that were focused entirely on their visions and scientific findings.

The other source of entrepreneurs is academia—these founders either remain in university positions or leave to form companies. Professors who start a company "on the side" nearly always fail, because the role of an entrepreneur usually requires more than 100% of one's time.

**How do you really know when to become an entrepreneur?**

When you've identified an exciting business opportunity that promises to be the chance of a lifetime... when you're convinced that the ultimate outcome will be a successful business... when mortgaging your house seems like a reasonable way to fund your start-up... then you're ready to become an entrepreneur!

## Paths to Executive Leadership and Entrepreneurship

*Consider reading "What to look for in a start-up company" in Chapter 3.*

### Executive Leadership

- As the saying goes, "Hitch your sleigh to a rising star": Identify someone who is likely to rise and will take you along. Find a mentor willing to groom you to become a CEO. If possible, choose a successful CEO on whom you can mold yourself. When you encounter problems, you can ask yourself, "How would my mentor have handled the situation?"
- Another saying goes, "Never outshine the master." Your access to the fast track to executive levels is based not only on your ability, but also on how much your boss trusts and depends on you. Do not show up or demean your boss.
- Project management experience provides excellent training for executive leadership. Project managers lead teams and oversee the whole gamut of activity from idea generation to product development. They're responsible not only for the project, but also for the bottom line (see Chapter 9).
- Whenever possible, join companies that give their leaders operating and management experience. Take an active interest in the financial side of the business, such as the budget, operational costs, etc.
- Take career risks. Work outside your domain of expertise so that you can widen your knowledge. Aim for positions where you can orchestrate cross-functional management; these will help you become more familiar with other disciplines.
- Join companies with world-class, renowned investors. One of the surest predictors of a company's success is the stature and reputation of the venture capitalist firm behind it.
- Once you become a CEO, join a CEO forum. These are like "CEOs Anonymous." They provide nonthreatening environments to discuss and solve problems with experienced peers in other industries, and the conversations can remain confidential.

### Entrepreneurship and Executive Leadership

*A company with top-tier investors is like a resume that lists "Harvard University."*

- Unless you are independently wealthy and plan to pay for the company's operational costs, seek venture funding from the top-tiered firms. Most venture firms invest in only two to four start-ups per year out of the many hundreds that are seeking funding, so it is highly competitive—probably more competitive than obtaining NIH grants. The best way to find first-tier venture capital firms is to be introduced to them by your lawyers, accountants, business acquaintances, or consultants.

- If you are an inventor and have an idea that you think is meaningful and has great promise for success, you should really try first to understand it from a business perspective. It is important to consider the market opportunity: Who will use your product? Will the company be profitable?
- If you lack financial prowess, consider taking M.B.A. courses in accounting and finance to increase your business understanding.
- Develop a broad skill set in the processes of running small businesses. You can use the prescribed processes or outsource them; in either case you will be left with more time to spend on other essential issues.
- Become associated with world leaders in your specific area of interest or become an expert yourself. The endorsement of key opinion leaders will put you in a much better position to attract the interest of venture capitalists.
- Serve as a board member for companies and organizations. This will provide exposure to other executives and venture capitalists, and you will learn about the various situations that arise in companies and the best ways to respond to them.
- Develop relationships with venture capitalists. They are highly influential and can help you find a suitable executive position in one of their portfolio companies. As a business acquaintance, you can seek advice and, in particular, ask for their frank guidance regarding the fundability of your percolating business idea. If you are in academia, send your resume to local venture capitalists investing in companies that specialize in the area of your technical background. Sometimes you can provide key assistance conducting due diligence on specific technologies.
- Become an entrepreneur in residence (EIR). EIRs join a venture capitalist's portfolio company at an executive level. It's a terrific way for venture capitalists to ensure that their money is in good hands, and it's a way for entrepreneurs to work in a venture-backed start-up. EIR programs are offered by many venture capital firms.
- Join local "angel" organizations that invest in areas that interest you. Angel groups are loosely organized, local networks of wealthy individuals who invest their own money in enterprising start-ups. Joining one will provide you an opportunity to learn about cutting-edge technologies and managing emerging companies.
- Apply for a Kauffman Fellowship. This is a wonderful way to break into the venture world. Kauffman Fellows are a very select group of people who have been chosen to join specific venture capital firms or to serve as entrepreneurs. Visit www.kauffman.org to learn more.

## RECOMMENDED TRAINING, PROFESSIONAL SOCIETIES, AND RESOURCES

### *Courses and Certificate Programs*

Finance and accounting classes

### *Societies and Resources*

Local CEO forums

### *Books and Magazines*

de Bono E. 1985. *Six thinking hats.* MICA Management Resources, Toronto.

Jaffe D.T. and Levensohn P.N. 2003. *After the term sheet: How venture boards influence the success or failure of technology companies.* November 2003. *A white paper.* At www.dennis-jaffe.com/publications_articles.htm

Lencioni P. 2002. *The five dysfunctions of a team: A leadership fable.* Jossey-Bass, San Francisco.

Werth B. 1994. *The billion dollar molecule: One company's quest for the perfect drug.* Touchstone, New York.

# 23

# Law

## Providing Legal Advice and Protecting Property

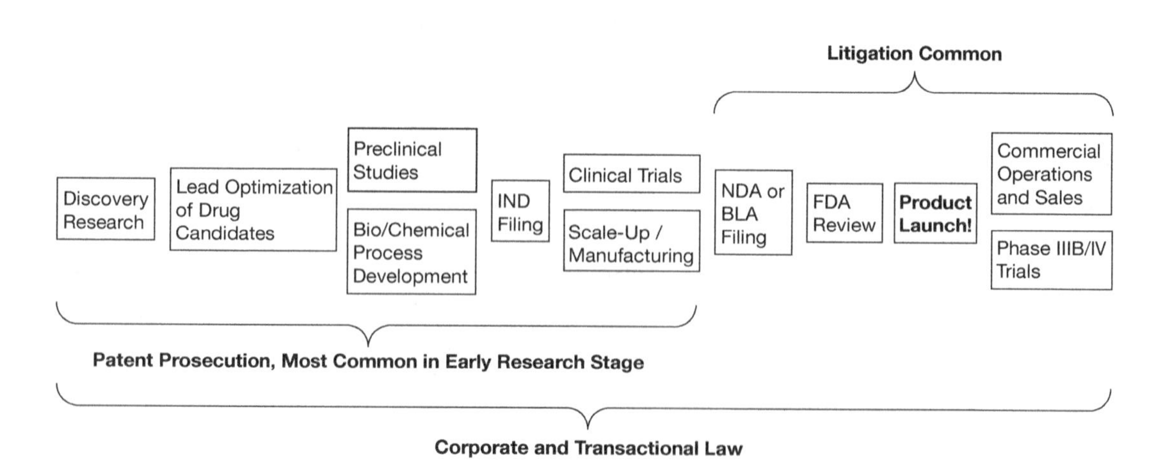


If you have a masterful command of the English language, if you are the type of person who prefers to write about your research findings rather than actually conducting the work, or if you derive pleasure from intellectual discussions and excel at debates, then a career in law may be for you. Be forewarned, however, that this is not a career for the faint of heart or the lackadaisical. Expect hard work and high stress, accompanied by large financial rewards if you are successful.

*Law is not merely a job—it's a profession!*

### THE IMPORTANCE OF LAW IN BIOTECHNOLOGY AND DRUG DEVELOPMENT

Lawyers play an important role in biotechnology. They incorporate companies, provide legal and business counsel, and conduct transactions. They also manage intellectual property (IP), including a company's patents, copyrights, and trademarks.

### *The Importance of Patents*

Making a profit is one of the most important and difficult goals in the life sciences industry. One way to enhance profitability is to possess the right, for a limited time, to exclude others from making and using a product; this right comes from having patent protection. Without such protection, new products would be readily copied. Without the ability to block competition, there would be little incentive for companies to invest in and develop new products. Patent law is instrumental in protecting such investments, and it provides the framework in which the life science industry prospers.

*Patents are the currency of research.*

Companies use dollars to finance research, which eventually results in IP. Patents are the vessels in which companies capture the IP assets generated by research. Unless a company obtains patents to protect its IP, it will not have any formal legal protection to prevent another entity from copying or independently discovering the results.

Patents are also the currency used in the industry to exchange IP assets for cash or other IP assets. When IP assets become valuable (e.g., when they result in a successful new drug), the patents protecting such IP can be leveraged (through license agreements, for example) to generate dollars for the company.

*A big part of a company's IPO is their IP.*

Patents are also used to establish the company's valuation when attracting outside investments. For example, venture capitalists assess a company's patent portfolio when they decide which organizations to invest in. Similarly, when companies go public, a big part of their initial public offering (IPO) is their collection of intellectual property (patents).

## CAREER TRACKS IN LAW

There are four general areas in which biotechnology lawyers flourish: in firms, in biotechnology and pharmaceutical companies, at government and academic research centers, and at the US Patent and Trademark Office (USPTO) (see Table 23-1). Scientists without law degrees can have very rewarding careers as patent agents, consultants, and advisors. Possession of a law degree, however, greatly expands career horizons by allowing one to

**Table 23-1.** Four principal areas of employment for biotechnology lawyers

| Law firm | In house | Government and academic research institutes | USPTO |
|---|---|---|---|
| Partner | General counsel | Patent preparation and prosecution | Patent examiners |
| Of counsel | IP counsel | Technology transfer | |
| Associates | Regulatory affairs | | |
| Patent agents | Patent agents | | |
| | Litigation (in large organizations) | | |

practice in multiple areas of law and to perform a wider variety of functions. Competition is fierce for top law positions, but with a science background, you can leverage your technology expertise into IP prosecution, litigation, or transactional law career tracks, where you may have a clear advantage over others.

### IP Law

Within IP law, there are three main areas in which biotechnology lawyers flourish:

- *Patent attorneys* draft and obtain patents from the USPTO.
- *Patent litigators* serve as advocates for clients in patent disputes and prepare and conduct court cases.
- *Transactional lawyers* draft and negotiate license agreements and business transactions involving IP.

### In Addition, Biotechnology Lawyers Serve Other Special Functions

- *Corporate lawyers* provide legal counsel regarding business and corporation issues. They help with the financing of companies, such as corporate or private venture equity financings or IPOs. They also help to incorporate companies and provide legal assistance for other business needs.
- *General counsel* is the senior-most attorney position in the company, and typically all other lawyers in the company report to him or her. The general counsel provides a broad range of legal counseling for a company, including corporate securities, mergers and acquisitions, transactional law, employment law, IP protection, and more.
- *Government, research organization, and university lawyers* serve at government research labs, universities, and other research centers, either obtaining patent protection for new inventions or assisting the technology transfer office to out-license new inventions.
- *USPTO examiners* work at the USPTO examining patent applications on behalf of the government. Patent examiners often eventually become patent attorneys.
- *Regulatory affairs lawyers* provide regulatory counsel to companies (see Chapter 12).

## LAW ROLES AND RESPONSIBILITIES

*The roles and responsibilities of lawyers in biotechnology vary depending on the area of specialty. The following is a generalized list of possible activities.*

### Drafting Patents and Continuing Applications

Patent attorneys and agents advise companies or inventors about opportunities to obtain patents from the USPTO regarding research and inventions. In addition, they analyze a company's "freedom to operate" by conducting patent searches through a variety of databases to determine whether a company's proposed activities would infringe on any other patent rights.

*In law firms, there are Finders, Minders, and Grinders. Finders find business, Minders mind the law firm, and Grinders do the work.*

Patent attorneys and agents work closely with inventors to help define their inventions and draft patent applications. They offer advice about alternate protection strategies, such as trade secret protection, and help maintain records that might later aid in patent prosecution and litigation.

Once drafted, the patent applications are then submitted to the USPTO, where more than 90% of the time they are rejected! Once they are rejected, the patent attorneys and agents amend the application and negotiate the patent claims with the examiner. This interactive process with the USPTO is called "patent prosecution" and typically must be repeated three or four times until a patent application is either issued as a patent or is abandoned. It currently takes more than 2.5 years, on average, for a utility patent application to go from filing to issuance.

Before a patent is issued, the prosecutor may submit additional applications with related claims to the USPTO. For strategic and technical reasons, it's important to reserve the right to continue claiming various aspects of the invention for a long time.

### *Arguing Appeals and Interferences before the USPTO*

When patent prosecution becomes stalled, it is sometimes necessary to appeal the examiner's decision to the Patent Office's Board of Patent Appeals and Interferences. This involves filing an appeal brief and can also involve arguing the case before a panel of administrative law judges.

This same board also oversees an arcane area of law that is unique to the USPTO: the interference. Interferences are like miniature trials that occur when two patent applications claim the same invention. Because the United States operates as a "first-to-invent" system, interferences are required to determine which of the patent applicants was the first to invent. Interferences operate with a unique set of rules that makes interference practice a specialty.

### *Conducting Due Diligence*

When a prospective investor considers investing in a company, or if there is a potential merger, several different lawyers may be called upon to conduct due diligence on the company's IP. Patent attorneys and agents may be asked to consider the company's patents and their coverage relative to the company's business activities. Patent attorneys and transactional attorneys also study the agreements relating to the company's IP, as well as the IP that the company has acquired through licenses and acquisitions, to understand any legal issues that may arise from these agreements.

### *Offering Patent Portfolio Advice*

The role of patent attorneys and agents is not simply to prepare, file, and prosecute patent applications. Viewed more broadly, patent attorneys and agents devise strategies for protecting products. Taken together, the patent applications that are filed, and more importantly, the patents that are issued, form a patent portfolio that establishes the company's line of defense for their products. Keep in mind that the loss of patent protection in the United States for a blockbuster drug (over $1 billion in sales per year) can mean a loss of over $1 billion in the company's valuation.

### *Structuring and Preparing Legal and Business Agreements*

Transactional lawyers and patent attorneys provide legal advice in preparing business transactions such as confidentiality and collaboration agreements, in-licensing deals, and cooperative research or materials transfer agreements. They translate descriptions of basic technology and fundamental premises into contract language.

### *Litigation*

Litigators become involved when one company is concerned that another company is infringing on its patents, or if it is at risk of being sued for patent infringement. Litigators review laboratory records and company documents and evaluate the underlying scientific issues to construct convincing cases for their clients. They prepare scientific experts and data, and they develop strategies for trials. They also prepare expert witnesses, take court depositions, and write summary judgment motions, trial motions, and appeal briefs. Many patent attorneys also argue their cases before judges, juries, or administrative agencies such as the International Trade Commission, although sometimes they may only assist and advise a general litigator who handles the actual argument.

### *Providing Legal Counsel*

Corporate lawyers assist with incorporating companies, raising money from venture capitalists and corporate partners, arranging commercial transactions, and negotiating mergers and acquisitions (M&As). They help structure business development deals, IPOs, and follow-on public offerings, as well as assisting with other corporate duties such as negotiating employee contracts.

### *Marketing*

Partners are responsible for bringing new business to the law firm. Marketing may include attending or speaking at meetings, writing publications, networking, or even serving as an adjunct professor in a law school. The best way to drum up business is to do outstanding work. (See page 335, "How is success measured?") Partners, therefore, need to set a high standard of quality for workers within the firm.

### *Fulfilling Administrative Responsibilities, Mentoring, and Training*

Most partners oversee personnel, write performance reviews, and are usually involved in hiring decisions. They are also expected to mentor and train associates.

*Doing Pro Bono Work*

All lawyers have an ethical responsibility to do pro bono work on the behalf of companies or individuals who can't afford legal assistance. Although some lawyers merely donate money to organizations that provide legal assistance, many actively seek to personally help others. This work is often in areas outside their specialty. Most law firms are supportive of their attorneys spending some portion of their time on pro bono work, because it makes the firm look good in the community.

## A TYPICAL DAY IN LAW

*A biotechnology lawyer's typical day might include the following activities (this is a generalized list):*

- Drafting and submitting patent applications to the USPTO and negotiating claims (this is how patent-prosecuting attorneys and agents spend most of their time).
- Arguing appeals or interferences before the USPTO.
- Conducting due diligence on patents.
- Performing freedom-to-operate searches and preparing opinions of counsel.
- Preparing for and participating in trials (patent litigators spend much of their time doing this).
- Reviewing legal documents and contracts, and negotiating deals.
- Talking with and advising clients (lots of meetings!).
- Keeping updated on new laws and the latest cases.
- Speaking and networking at industry meetings and conferences.
- Mentoring associates and interviewing candidates.
- Doing pro bono work.

## SALARY AND COMPENSATION

*In general, the pressure is proportional to the compensation.*

Entry-level patent agents with advanced science degrees typically earn incomes more than double, sometimes even triple, that of postdoctoral fellows, even before passing the patent exam. Salary jumps are high, with raises of 8–9% per year. Starting associates (with a law degree) from all law areas earn up to double what patent agents make. This gap can close over time for exceptional patent agents. Only lawyers can become partners in law firms, however, and that is where the big money is. Partners can easily earn two to three times what the most senior associates make. Successful partners can earn over a million dollars per year, depending on their reputation and the number of people working for them. Litigation lawyers usually earn the highest incomes because of the large amounts of time necessary to prepare for trials. In most cases, lawyers in law firms make more money than those working in-house for biotechnology companies. The most money for in-house lawyers is made by senior attorneys (general counsel, chief patent counsel, vice president of IP) who take advantage of stock options when a biotechnology company does well.

*Great work begets more work!*

Salaries vary from firm to firm, depending on factors such as their size, type, and work ethic. In general, large and prominent law firms pay the most money, because associates are expected to work the hardest at these firms.

***How is success measured?***

Ultimately, success is measured in large part by having both an abundance of "happy clients" and an established reputation. If clients are pleased with the service a lawyer or firm has provided, they will become sources of repeat business and new referrals. As the firm's business increases, so does its reputation, which in turn brings in even more business.

A successful lawyer's reputation can also be built on work that has had a significant impact on industry. For example, lawyers can earn a laudable reputation by prosecuting outstanding patents or by being associated with revolutionary inventions that have changed the landscape of biotechnology; successfully defending high-profile cases; or incorporating or helping acclaimed (or wildly successful) biotechnology companies go public.

## PROS AND CONS OF THE JOB

### Positive Aspects of a Career in Law

- One of the greatest appeals to law is the financial reward. Successful partners can earn over a million dollars each year!
- Law is intellectually stimulating. There is steady exposure to the latest cutting-edge technologies, and the continual introduction to new clients and cases ensures that there is never a dull moment.

*Law has Pay Appeal: Successful partners earn over a million dollars each year!*

- A career in law can be highly satisfying. You will be using your skills to solve challenging intellectual problems while helping clients and the biotechnology industry. You will gain a deeper appreciation for the art and elegance of law.
- Law provides a rewarding avenue for using your writing and analytical skills without having to do bench work.
- Coworkers in law firms tend to be exceptionally bright and talented.

- Many lawyers enjoy the nature of law, which often involves verbal sparring—winning a debate with someone and gaining your opponent's respect.
- Law allows you some independence in how you organize your days.
- There are frequent travel opportunities to visit clients.
- Top-tier law firms treat their attorneys luxuriously. They dine at the best restaurants, stay at the best hotels, attend fancy parties, have nice offices and personal secretaries, etc.
- If you are a patent agent, you can raise a family and work from home, part-time, or on an as-needed basis. Your work schedule can be flexible.
- If you work for an IP firm, the environment will be more relaxed than in other types of law firms. There tends to be less hierarchy, people are treated more respectfully, and the atmosphere is usually more informal.
- You can have a long career. Lawyers become increasingly valuable with experience.

## The Potentially Unpleasant Side of Law

- A career in law can be tremendously stressful (see "Greatest Challenges"). The hours are long and there is much work to do. Clients have high expectations and competition for business is extreme. There is great pressure to reach "billable hour" quotas every month, but at the same time, the quality of your work must remain outstanding. Billing rates and quality expectations continue to increase with seniority (an associate's billing rate more than doubles over 8 years). You may constantly be "on call" for clients, which can make any sort of planning for family life activities difficult. For litigators, stress levels can be particularly high during trials.

*There is no "ease" in "legalese!"*

- The path to partner is an uphill battle. It requires constant marketing. Fierce competition from the many other outstanding and bright lawyers in the industry can make it difficult to develop a strong reputation and a cadre of clients. It is especially difficult to make partner when the economy is not doing well.

*It is stressful when there is too much work and equally stressful when there is not enough work.*

- Time is spent according to billable hours, which can restrict your ability to explore your own scientific interests.
- Law firm politics can be complicated. Power struggles sometimes develop between partners competing for clients.
- It may be difficult to work with some partners. Partners do not always train associates systematically, and it can be difficult to find a supportive mentor.

*Clients think that lawyers have big clocks strapped to their backs, and they don't want to pay for the time it takes an attorney to understand scientific nuances.*

- Laws change constantly, and you must continually educate and update yourself.

- Inventors can sometimes be uncooperative or too busy to assist with writing patent applications. Some scientists do not like to interact with lawyers.
- You may incur a heavy debt load by attending law school. Keep in mind, however, that this debt can be quickly paid off afterwards, and if you are employed as a patent agent, many law firms will pay your tuition.
- Sometimes highly aggressive lawyers behave unprofessionally (with overly zealous advocacy), particularly in litigation.

### Specifically for Patent Agents

- Although patent agents perform practically the same work as patent attorneys, they may not be treated or paid as well. Many patent agents ultimately become frustrated with this invisible stratification and go to law school.
- In some law firms, patent agents tend to be pigeonholed into just writing patent applications, and there may not be opportunities to learn new skills.
- It is typically easier to get a lifestyle position in a law firm (particularly a smaller firm) as a patent agent than as an attorney.

## THE GREATEST CHALLENGES ON THE JOB

*Managing Stress*

Why is law so stressful? Demand for work is unpredictable, and clients' needs can change abruptly, so it can be extremely difficult to control your workload. There is a fine line between bringing in enough business for a steady supply of work and bringing in so much that there are not enough people to get it done. Often there is simply too much work to do, and constant deadlines and emergencies add a sense of urgency to the job. There is a strong emphasis on meeting quotas for billing, and clients expect and demand top-quality service. In addition, lawyers bear a heavy burden of responsibility for their clients. A lot of money may be at stake, and mistakes can have huge consequences.

*Keeping the Clients Happy*

Because competition for business is fierce, it is essential for a firm to be service oriented. Clients who were pleased with past work and personal interactions with the lawyers will return for repeat business, so good client management and interpersonal skills are tantamount to success. Successful lawyers are adaptable enough to work with various client and inventor personalities and are quickly and astutely able to recognize and respond to their needs.

Part of keeping customers happy involves managing clients' expectations. Clients often underestimate expenses and the amount of work required. If you want to avoid client dissatisfaction later on, you need to inform clients up front about the turnaround time, the costs, and potential outcomes of their case, positive and negative.

#### *Maintaining a Well-balanced Lifestyle*

With unexpected emergencies and urgent needs from clients, it can be a challenge to meet the demands of work and also maintain a satisfying personal life. As a consequence, overworked lawyers sometimes burn out.

## TO EXCEL IN LAW...

#### *Experience, Knowledge, and Creativity*

Experience and knowledge separate the good from the great. Lawyers with years of experience have such an extensive reservoir of knowledge that they can anticipate issues that do not seem immediately obvious in their legal proceedings. This ability gives them an edge in advising clients and bypassing problems. Great lawyers also tend to be very creative, allowing them to adroitly find innovative solutions to difficult problems.

#### *Good Client and Personnel Management Skills*

Because law is a service industry, lawyers who excel at "client development" tend to be the most successful. Those who have excellent interpersonal skills and are known for being honest, conscientious, and trustworthy develop renowned reputations. When clients feel they have been treated well, they are likely to provide additional business (see "Greatest Challenges").

In addition, prominent lawyers also tend to be highly successful at managing the firm's associate lawyers and staff. They possess leadership skills and can motivate employees by creating a positive work environment.

*When clients feel that they have been treated well and given good counsel, they are likely to return for additional business.*

## Are You a Good Candidate for a Career in the Legal Profession?

#### *People who flourish in law tend to have...*

***Extraordinarily good writing skills.*** You not only need to be a *good* writer, but also a *persuasive* one. For example, patent application claims need to be worded very carefully; seemingly small changes in word choice can have significant consequences. Keep in mind that your writing will not be limited to legal documents, and that much of your work will be conducted through E-mails and letters.

*If you don't like to write, law will be a challenge!*

*Outstanding oral communication skills.* Those who are well-spoken can simplify complicated technical details for clients. They can argue their points convincingly in debates, and they can deftly navigate tricky diplomatic situations with coworkers and clients. In the never-ending hunt for business, good verbal skills increase one's chances of impressing potential new clients.

*Excellent interpersonal skills.* Good lawyers build lasting relationships with their clients. You need to be adaptable enough to establish rapport with clients who have different and sometimes difficult personalities (see Chapter 2).

*Being a chameleon helps— law is a service industry, so the customer is always right!*

*Gregarious, outgoing personalities.* Because this is a very people-oriented vocation, it tends to attract those who are outgoing and social. Because it is so competitive, those who are self-confident and assertive tend to succeed. However, those who are introverted can also be quite successful as patent agents or lawyers not seeking partner status.

*Creative and analytical skills.* Analytical skills are needed to construct solid arguments, to draft creative and strategic claims, and to find innovative solutions to obstacles.

*The ability to pay meticulous attention to technical details.* A small oversight might result in a large negative consequence for your client. Typos are not allowed!

*Excellent time management and prioritization skills.* There are constant deadlines and multiple project assignments. Multitasking and time management are key skills to have or to develop.

*The ability to tolerate stress.* Deadlines are frequent, and the pressure to obtain business and please clients is constant.

*Tenacity and patience.* The process of issuing patents can be a slow and frustrating exercise.

*You need to be able to derive pleasure from moving a file off your desk—there are few immediate accolades in law.*

*Raw intellectual talent.* You must be a fast learner, able to quickly comprehend not just new technologies but also business concepts and to stay updated with the ever-changing laws. To best serve your clients, you need to be business savvy enough to understand your clients' needs and to be able to explain to them the practical realities of their legal options.

*Tireless energy and strong work ethic.* Expect to work hard in law. Because of intense deadlines, you may often have to concentrate for long periods to complete your work while at the same time maintaining high quality standards.

*To be a good lawyer, you need to have unwavering ethics.*

*A well-balanced career and home life.* Most people rely on personal support to balance their lives and to reduce the stress of working in the legal world.

***You should probably consider a career outside law if you are...***

- Unmotivated or unreliable.
- Someone who prefers to work alone and doesn't enjoy interacting with others (patent agents are an exception).
- A linear thinker, someone who needs precise answers before advancing to the next step or who can handle only one project at a time.
- Someone with frequent writing blocks, a person who can think of the words but cannot put them down on paper.
- A person who fears adversarial confrontations (for litigators, in particular).

## LAW CAREER POTENTIAL

*Becoming a partner is like winning a pie-eating contest—where the prize is more pie!*

If you seek a law career, a law degree will greatly expand your career potential. You will earn more and you will be able to explore areas of law that patent agents cannot. Lawyers often work toward becoming a partner in a law firm, but other options include becoming an in-house counsel or working in a company's business development or regulatory affairs departments (see Fig. 23-1).

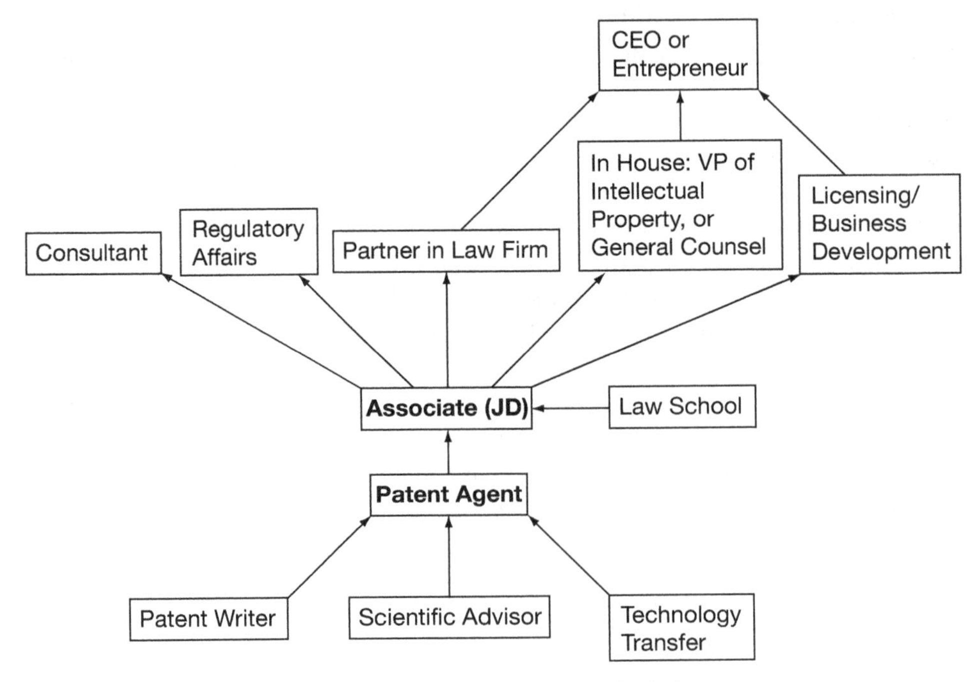


**Figure 23-1.** Common career paths in law.

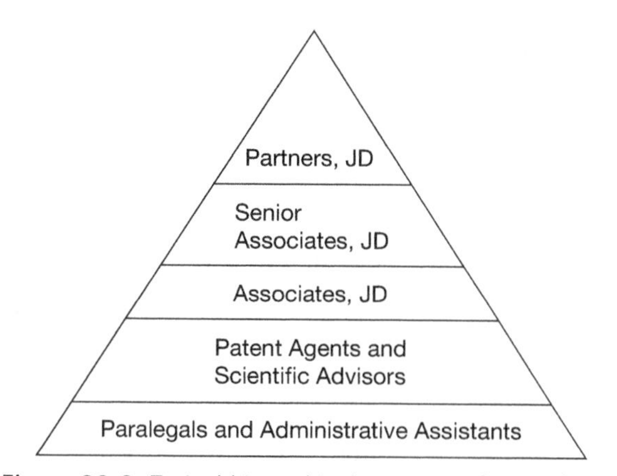


**Figure 23-2.** Typical hierarchical structure of a law firm.

***The path to partner***

The classic career path for lawyers is referred to as the "law firm mill" (see Fig. 23-2). Only 10% of associates become partners, and it takes approximately seven to ten years to make the transition. The likelihood of becoming a partner increases with one's ability to attract clients and establish a good reputation.

Becoming a partner requires a strong commitment, and many people drop out before they reach that level. Some cherish a more relaxed lifestyle—they may want to spend more time with family or avoid the added stress. Others lack the rigorous skill set and fortitude that is required. An alternative to making partner is to become "of counsel," a non-partner track position in which there is no requirement to bring in business. Of counsel provides an additional source of expertise to a law firm. It involves less stress and a somewhat lower (though still healthy) salary. Other possibilities for lawyers who don't want to become partners include going to work in-house for biotechnology companies or working on an as-needed basis for law firms during times of work overflow. Some lawyers are trained in a law firm, join a biotechnology company to gain experience, and then return to a firm to continue their trek toward partner status.

Alternatively, many lawyers become CEOs of start-ups or work as venture capitalists or IP specialists.

A legal background can provide excellent preparation for a CEO position. CEOs need to think analytically and communicate well. They also need to be persuasive and know how to negotiate smoothly...all skills that are honed and valued in a legal career.

## Job Security and Future Trends

The future holds great promise for many significant breakthroughs in biotechnology, and lawyers will be needed to protect and transfer the new technology. Because partners charge higher per-hour fees than patent agents, more patent agents are needed to keep costs lower and clients happier.

***Is it better to work in-house or to join a law firm?***

*Advantages of Working In-House*

- Your efforts are directed toward creating a successful biotechnology company rather than a better revenue center.
- Your lifestyle can be more relaxed: the hours are regular and your workload can be more controllable. (But not always!)
- You can be closer to research. Your work will be related to the scientific focus of the company, and you can participate in research meetings.
- You become part of a team: as such, you share the risks and rewards (e.g., stock options!) of the company. The relationships you form with inventors will be closer than if you work in a law firm.
- You may be exposed to a broader range of company issues, such as dealing with human resource quandaries, regulatory reforms, and Securities and Exchange Commission (SEC) filings.

*Working in a firm is harder, but it teaches you a rigor that pays off.*

*Disadvantages of Working In-House*

- Salaries are typically lower.
- The advancement of your law career may be somewhat restricted: Limited resources mean that there is less training and mentoring in a biotechnology company. You will likely handle fewer cases and they will be less technically diverse, so the breadth of your experiences will be more limited. You will more quickly learn how to do high-quality work in a law firm.
- You are likely to have more administrative and clerical duties when working in-house.
- Relationships with your scientist coworkers can be difficult. Scientists don't always appreciate the value of law, and they sometimes perceive lawyers as people who constrain scientific freedom.
- There may be more risk: Biotechnology companies tend to be more unstable than law firms.

*Your fellow employees in a biotechnology company need to understand what you do, so that you don't seem like an alien from another planet.*

Law tends to be less affected by the biotechnology economy than other service industries. Lawyers play an essential role in incorporating companies, providing legal advice for business transactions, and securing IP protection. These services are needed even in times when the economy is performing poorly. Law firms that work with universities or have a large client base tend to be less affected by fluctuations in the economy, because the need for patents from these groups is steadier, whereas those who work for start-ups will experience a more volatile economic environment.

The future is extremely bright for lawyers interested in litigation. Expectations are that biotechnology lawsuits will become more common and clients will be more insistent on having scientists on their litigation teams. As the biotechnology industry grows, there will be more new technology, more patents, more money to fight over, and larger attorneys' fees.

## LANDING A JOB IN LAW

### Experience and Educational Requirements

In general, a Ph.D. is mandatory for patent agents in the biological and chemical sciences: There are simply too many people with Ph.D.s interested in patent law, and clients are beginning to expect that the person drafting their patent applications will have one. For engineers and physicists, however, a master's or undergraduate degree may suffice.

For partners and associates, a juris doctor (J.D.) degree is required. A Ph.D. is not, but it is certainly helpful, particularly for patent prosecution and when advising or litigating for biotechnology companies. For more business-oriented areas such as corporate or transactional law, an M.B.A. is highly advantageous.

In general, most senior partners today do not have Ph.D.s, because when they were establishing their careers, biotechnology was still an emerging industry. Although a Ph.D. is not presently a requirement for becoming a partner, more of today's new partners have one, and it may be required in the future.

### Paths to Law

*Slackers need not apply!*

- Even if you do not intend to go to law school, talk to law school career counselors. They possess a wealth of advice and information and may provide valuable contacts to appropriate law firms.
- Consider joining a law firm as a scientific advisor or legal analyst. You will be trained as a patent agent and paid well. It will be easier to pass the patent bar exam, and you will be able to "test the waters" before investing time and money in law school.
- Networking may be the best way to secure a scientific advisor or patent agent position in a law firm. This is particularly true for entry-level applicants. When interviewing, try to join a law firm with a large client base and technical diversity to increase your job security.
- Consider becoming a patent agent. The USPTO holds the patent bar exam twice a year. You can take classes on patent law, but the best way to prepare is to work in a law firm for at least a year to learn legalese. The passing rate is usually 30–40%; most applicants pass on the second attempt.
- You can go directly to law school and apply afterward as an entry-level associate in a firm. It is important to try to go to the most prestigious law school possible (*US News and World Report* rates and classifies law schools and IP programs). Although it is common for Ph.D.s to work as patent agents during the day and go to a nearby law school at night, it is far better to attend a top-notch law school full-time to increase the likelihood of securing a more promising future in a prominent law firm. Train for the LSAT exam, because law schools pay close attention to scores.

*The best route to the top law firms is to attend the best law schools and get good grades.*

- It is important to receive good grades in law school. Your summer internship is largely based on the grades you receive in the first year of law school. IP law requires a lot of one-on-one training, so the number of entry-level positions is limited. Firms look at technical backgrounds, the quality of the law school, and grades. All of these factors need to be strong. A Ph.D. from M.I.T. combined with poor grades from a mediocre law school may not be enough to secure a position in a top law firm.

*The extra investment you make by attending law school will more than secure your financial future.*

- If you decide to go to law school, join an IP law society while you are there. It is a congenial place to network and perhaps make lifelong professional partners and friends.
- Before your first final exams, consider taking courses on how to write answers for law school finals, and/or study top-notch answers from previous exams. Law professors expect a very different style of writing from that used in scientific papers. You may know the material cold, but if you answer like a scientist rather than like a lawyer, you may fail.
- While in law school, choose clerkships that broaden your horizons. Try areas such as litigation, corporate law, transactional law, or regulatory affairs. You never can predict which you will prefer, and a summer internship is the best time to explore your options. When selecting law firms to join, opt for the company where you feel the most comfortable and have the sense that you can interact with the lawyers easily and productively.
- Learn how to interview! Companies and law firms look for personable, sociable, confident employees—not modest and timid scientists. Law school career services offer assistance in learning superior interviewing techniques, and there are many books on this topic.
- Take classes in personnel management. This subject is usually not taught in law or graduate school. You will learn how to effectively handle different management styles and personalities.
- Many people recommend that you join a law firm before going in-house. You are more likely to be mentored and trained in a law firm, and you will have greater exposure to a variety of technologies and a higher volume of cases. With a broader skill set, you will be more marketable for future job opportunities.

## RECOMMENDED TRAINING, PROFESSIONAL SOCIETIES, AND RESOURCES

*Courses and Certificate Programs*

Patent classes at local universities and extensions

Classes in personnel management

Business classes

*Societies and Resources*

American Bar Association (ABA; www.abanet.org)

Association of University Technology Managers (AUTM; www.autm.net)

Intellectual Property Law Server (www.intelproplaw.com) provides information about IP law

Law School Admission Council (www.lsat.org) offers a large amount of information about the law school process

Licensing Executives Society International (LESI; www.lesi.org) provides a mini MBA program and has a program that focuses on IP

Local IP societies across the US

Martindale.com (www.martindale.com and www.martindalehubble.com) lists law firms and their specialties and identifies attorneys

Practicing Law Institute (www.mbe.pli.edu) is a nonprofit organization that offers courses, workshops, and a bar review

Patent Resources Group (Kayton courses for the Patent Bar; www.patentresources.com) offers patent law and bar review classes

Robert C. Byrd National Technology Transfer Center (www.nttc.edu)

*US News and World Report* publishes a list of good law firms and classifies law schools and IP programs

United States Patent and Trademark Office (USPTO; www.uspto.gov)

# 24

# Health Care Finance

## Venture Capital, Institutional Investing, Investment Banking, and Equity Research

THE FINANCIAL SIDE OF BIOTECHNOLOGY AND DRUG DEVELOPMENT includes professions such as venture capital, institutional investing, investment banking, and equity research. Professionals in these careers have a unique position in industry: Although they are completely removed from the bench, they indirectly have the potential to dramatically improve people's lives by making it financially possible for companies to develop products or treatments for diseases.

These careers provide great opportunities to delve into the biotechnology industry and learn how the worlds of finance and Wall Street work while you leverage your medical, science, and business knowledge. In addition, these jobs are some of the best-paid in the industry. Consequently, they are the most difficult jobs to obtain.

### THE IMPORTANCE OF HEALTH CARE FINANCE IN BIOTECHNOLOGY AND DRUG DEVELOPMENT

*Venture capitalists convert ideas that emerge from scientific laboratories into companies that can turn those ideas into marketable products.*

Developing biotechnology and pharmaceutical products is very capital intensive. When all the failures along the way are accounted for, a single successful drug can cost hundreds of millions of dollars to develop. Thus, having the capital to develop products, particularly in the early stages of a company, is critical.

Biotechnology and drug development companies go through discrete phases in their progression toward becoming a public entity (see Fig. 24-1). Each phase offers opportunities for the health care financier. Angel investors (high net worth individuals who invest their own money in a company) typically invest in start-up companies; venture capitalists and corporate venture capitalists fund developing private companies; and investment bankers shepherd companies through an initial public offering (IPO) or an acquisition and assist with various other financial transactions. When the company is public, institutional investors purchase stock, whereas sell-side equity research analysts rate the stock and provide analysis (coverage) of the company.

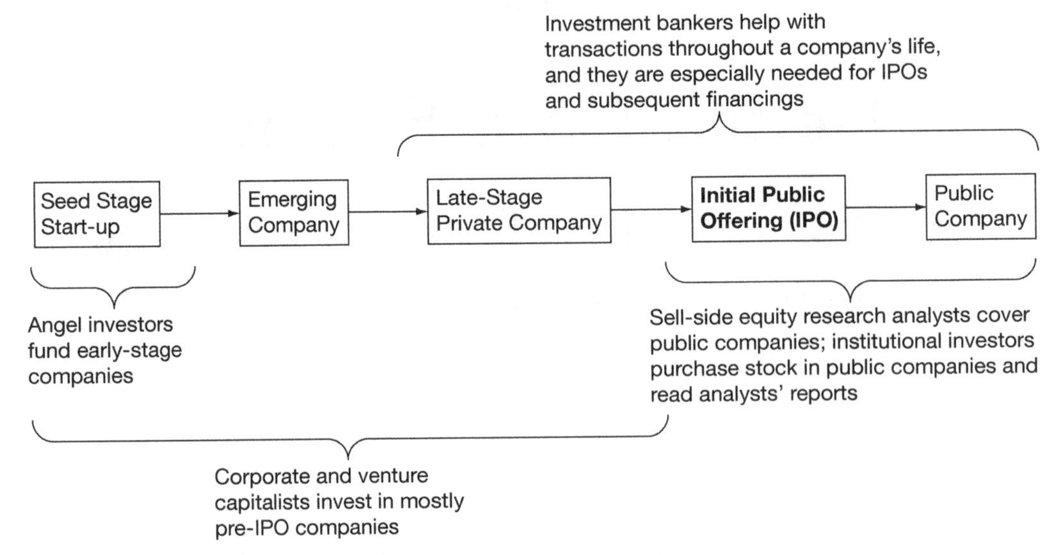


**Figure 24-1.** Private to public company life cycle.

## CAREER TRACKS IN HEALTH CARE FINANCE

### Venture Capitalism: Investing in Private Companies

*The art of venture capital is the ability to make small companies successful and their stock valuable.*

Venture capitalists invest mainly in private companies. They identify people or companies with novel experimental technologies that have promising commercial potential. They then shepherd the companies from the concept stage to a point where public and institutional investors are willing to fund them via acquisitions or IPOs.

Venture capitalists raise large amounts of money—called a fund—from limited partners (such as large pension funds, endowments, foundations, and wealthy individuals) and invest that fund in 10–40 early-stage private companies. They oversee their investments by assisting their portfolio companies, usually by serving on the boards of directors. Typically, when the portfolio companies go public or are acquired, venture capitalists sell their stock and return the initial investment plus significant profits to the limited partners. This process is then repeated with a new round of investors and start-ups.

Venture capitalists specialize in particular investment areas, such as biotechnology, medical devices, health care services, and early- or late-stage companies. Venture capital firms tend to be small, typically employing two to ten partners plus several associates.

There are several types of venture capital. Institutional investors and investment banks sometimes have venture capital branches. There are also corporate venture arms in some large biotechnology and pharmaceutical companies, hospitals, and foundations. These sources of venture capital invest the company's money, whereas venture capitalists invest limited partners' money.

## Institutional Investing: Investing in Public Companies

**Note:** *Institutional investing is also known as "buy-side equity research." To avoid confusion in this chapter, "equity research" will be a term applied only to sell-side equity research.*

*There are two types of equity research, "sell-side" (those who sell stock) and "buy-side" (those who buy stock).*

Institutional investment firms invest money belonging to foundations, endowments, pension funds, funds of funds, corporate investors, and investment banks. There are many types of institutional investment firms—the most common are mutual funds and hedge funds (see below). Some of these funds invest exclusively or partially in health care. Within that group of health care funds, only a few are dedicated to biotechnology; most invest in a wider range of the health care market, including pharmaceutical companies, medical devices, and health care service companies (i.e., HMOs and hospitals).

Like venture capitalists, institutional investors purchase stock and provide capital for companies. However, institutional investors invest primarily in public companies and thus do not assist their portfolio companies by serving on their boards of directors. Often, institutional investment funds are much larger, and unlike venture capital firms, their funds are often public. These public funds are regulated by the U.S. Securities and Exchange Commission (SEC).

### *Institutional Investment Hedge Funds*

*Hedge funders seek high returns on investments because they receive a percentage of the profits; mutual funders seek high returns so that they can advertise the positive results and grow the size of their fund to increase management fees.*

Hedge funds are typically managed by small firms. They make most of their money on the "carry" (i.e., a percentage of the profits). Hedge fund managers tend to be more opportunistic and aggressive, taking advantage of market inefficiencies. These are not public funds and are therefore not regulated by the SEC. There are at least 100 hedge funds investing in the life sciences, and many of these have one or more managers with a substantial focus in the biotechnology sector.

### *Institutional Investment Mutual Funds*

Mutual fund managers earn the lion's share of their money from management fees and are thus incentivized to have large funds. During good years, they advertise their returns, using various stock indexes as benchmarks. Mutual funds are public and are thus regulated by the SEC. Of the more than 500 mutual fund companies, almost all have a health care fund (or several) in their portfolio, and many of these include biotechnology, medical device, pharmaceutical, and other life-science companies.

## Investment Bankers: The Transaction Business

Within investment banking, there are three main finance careers: advisory services, sell-side equity research, and sales and trading.

*Investment bankers are like ultra athletes—they work hard and for long hours in a fiercely competitive environment.*

### *Advisory Services*

Investment bankers provide financial services to help companies raise capital and identify and acquire the assets necessary to fulfill their growth plans. They facilitate mergers and acquisitions, private investments in public entities (PIPES), IPOs, follow-ons, debt offerings, convertible debts, private equity investments, and the sale of public stock.

There are several types of positions within investment banking advisory work. A "relationship manager" is responsible for becoming acquainted with CEOs and executive management teams in order to secure their business. There are also "product specialists," who provide expertise in specific types of transactions, such as mergers and acquisitions or equity capital (i.e., stock deals).

### *Sell-side Equity Researchers, Investment or Research Analysts*

Within investment banks, sell-side equity researchers create and publish informative reports to help their institutional clients. These reports contain due diligence (in-depth background research) on a small number of companies that they cover. The coverage includes analyses of the science, clinical trials, management team, financial models, and valuation, in addition to an analyst's "buy," "sell," or "hold" stock ratings. This information is offered to entice investors to trade through their firm, so that the bank can obtain trading commissions.

### *Sales Traders*

Sales traders interface with buy-side investors (mutual and hedge funds) to promote trading (selling and buying stock) through the bank.

# HEALTH CARE FINANCE ROLES AND RESPONSIBILITIES

## Venture Capital

### *Raising Capital from Limited Partners*

Venture capitalists' primary role is to make money for their investors and their own firms, and in the process they help start-up companies become successful. Raising venture funds is a difficult process for most venture capitalists, and it is not their favorite part of the business. They travel in "road shows" meeting with limited partners to generate interest in their fund. Because the capital is locked up for several years and expectations among investors are high, investments are subject to a rigorous level of due diligence.

*Entrepreneurs often forget that venture capitalists also have to raise money.*

*Portfolio Development: Deal Sourcing and Due Diligence*

Much time is spent sourcing for companies to invest in. Venture capitalists typically make two or three investments per year. They begin by evaluating a large number of companies and then narrow these down to just a few. The sources for potential investments come from a variety of places, such as other venture capitalists who wish to form syndicates, business friends and associates, and lawyers with whom they have worked in the past. Occasionally, the source will be an interesting scientific article, which leads to a conversation with the author about the work and its commercial potential. University technology transfer offices are another source for new ideas and potential investments.

*Before making an investment, one takes a "deep dive" into a company, the market, and the scientific literature.*

A lot of work is involved in making investments. On average, it takes six to eight months from the initial identification of a potential investment to negotiating and signing a deal. During this process, venture capitalists become experts in the company, its technology, and that sector of the market. They conduct extensive "due diligence" to determine the promise of potential investments. This involves analyzing the viability of the technology or science and the validity of the intellectual property, conducting reference checks on the management team, talking to industry experts about the potential product and its market opportunity, and conducting financial analyses to determine the potential returns on investments.

*Managing the Portfolio: Providing Company Support and Serving on Boards*

Venture capitalists do not just invest money—they also contribute their experience, time, and knowledge, and utilize their large network of contacts to assist their portfolio companies. Sometimes serving as board members, venture capitalists provide guidance and strategic input. They play an important role in business development deals, product development strategy, acquisition of additional capital, and recruitment of a management team and additional board members.

*Venture capitalists act as strategic thought partners for their portfolio companies.*

*Keeping Updated*

Venture capitalists keep up to date on the latest industry developments by attending or speaking at conferences and reading scientific journals.

## Institutional Investing

Institutional investing, like venture capital, is on the buy-side; but institutional investors invest mainly in public companies, and as such, their responsibilities differ.

*Communicating with Investors*

Those in hedge funds interact with their limited partner investors once a year. They also communicate pertinent industry news that relates to a portfolio company. Mutual funders send announcements to their public investors.

### *Portfolio Management*

An institutional investor's portfolio management is based on the investment strategies and goals of each fund (such as outperforming the biotechnology index of the New York Stock Exchange). They develop ways of finding investment opportunities in promising undervalued companies and strategically study the market to identify companies that promise success in anticipation of economic shifts.

### *Responding to News*

Institutional investors closely watch the news, particularly for any information related to their portfolio companies. When there is relevant news, they speak to the CEO or investment relation's head to obtain further clarification (see Chapter 21). On the basis of this information, they may adjust their portfolio investment strategy and inform the fund shareholders.

### *Screening Companies as Potential Investments*

Institutional investors evaluate potential investments by screening promising public companies. They often meet with biotechnology management teams who are traveling on "road shows," presenting the virtues of their respective companies in order to raise money (see Chapter 22).

### *Conducting Due Diligence on Prospective Investments*

*Institutional investors kick the company's tires before investing.*

Although institutional investors rely on the research of sell-side equity analysts, they generally also conduct their own due diligence before making investments. They review the scientific integrity of the management team, the clinical or preclinical data, the company's competition, and the validity of their patents, from which they construct a picture of the competitive landscape. They talk to key opinion experts and clinical trial investigators about the technology and market expectations.

### *Investing*

After the information obtained from due diligence is reviewed, investment decisions are made. Making an actual investment is relatively unglamorous: The institutional investors simply direct a trader to buy a certain number of shares of a specific company's stock.

### *Keeping Updated: Monitoring the Global Market Environment*

Institutional investors stay up to date on the general performance of markets. This helps them to take the global economic pulse and to monitor developments within the biotechnology and pharmaceutical sectors. They attend medical conferences and conferences hosted by investment banks to become acquainted with thought leaders in the industry (usually high-profile clinicians or presumed experts in specific technical areas).

## Investment Banking

Investment bankers spend their time either executing deals or looking for new business.

### *Identifying New Business and Building Relationships*

Investment bankers spend much of their time finding new clients and establishing rapport with them. Investment bankers meet company executive teams and display their acumen in finance and biotechnology. They describe the services their firm provides and suggest ways they might be of help, such as by providing financial and strategic evaluations and innovative solutions.

### *Conducting Financial Analyses*

An important responsibility for an investment banker is to conduct financial evaluations to determine the estimated value of companies, their assets, and their strategic financial options. This information is used to compose "pitch books," which are created individually for each client to generate business. Pitch books describe the financial value of the company, its strategic options, and a banker's qualifications for getting the best value on a deal.

### *Executing Transactions*

Investment bankers execute transactions on behalf of their clients. These include mergers and acquisitions, IPOs, business development deals, raising capital, and leverage buy-outs.

### *Holding Investor Conferences*

Investment bankers hold investor conferences in which their public and private clients present their technology and business fundamentals. These can be large events, perhaps attracting the entire investment community.

### *Keeping Updated*

To help their clients solve tough financial problems, investment bankers must have a comprehensive knowledge of the overall market and the unique challenges biotechnology companies face. It is their experience in markets and the breadth and depth of personal relationships that allow bankers to be so successful in conducting transactions on behalf of their clients.

## Sell-side Equity Research

### *Responding to the Daily News*

Each day, equity researchers review the daily news from company press releases. Information that is relevant to a company they cover is transcribed into a short note—an opinion of sorts—which is sent to investors and sales traders.

#### *Researching and Providing Coverage of Companies*

Analysts are typically responsible for providing their clients with stock ratings (i.e., buy, sell, or hold stock) on approximately 20 companies. Analysts' time is spent studying and writing opinions about companies' financial expectations based on clinical trials, potential sales, etc. They spend time reading medical and scientific journals and scanning them for pertinent technical information. Their investment summaries and stock ratings are published and used by the equity sales traders, who provide the information to their clients.

#### *Keeping Updated*

Analysts spend approximately one month a year attending a wide spectrum of investment, medical, and scientific conferences.

## A TYPICAL DAY IN HEALTH CARE FINANCE

### Venture Capital

*A typical day for a venture capitalist might include the following activities:*

- Attending weekly partners' meetings to discuss potential investments.
- Managing portfolio companies. Assisting senior management teams with strategic advice about urgent and important matters such as mergers, business development deals, fund-raising, and recruiting.
- Attending portfolio company board meetings.
- Screening companies. Reading business plan summaries and inviting management teams to present company fundamentals which, if promising, will be followed by due diligence.
- Conducting due diligence calls to industry thought leaders and researchers.
- Helping friends by offering venture capital advice, either business- or career-related.
- Keeping updated. Talking to colleagues, attending conferences, keeping current with financial markets and major transactions, and reading scientific and medical journals.

### Institutional Investing

*Except for the first three points above, a typical day for an institutional investor is similar to that of a venture capitalist. In addition, an institutional investor is busy with:*

- Reading the morning press news. If events are relevant to his investment portfolio, conferring with sell-side equity research analysts to ascertain their interpretation of the data. Checking with the CEOs and heads of investment relations of the newsworthy companies. Working on their investment model and making decisive changes based on the news.

- Making investment decisions and purchasing or selling stock.

## Investment Banking

*A typical day for an investment banker might include the following activities:*

- Teleconferences with team members of the health care group to review the past week's events.
- Visiting with potential and current clients in order to build relationships. Meeting with the chief financial and executive officers of companies.
- Conducting financial evaluations to compute the relative value of companies or assets.
- Executing business transactions and negotiating financial and legal terms.
- Conducting due diligence calls to clinical investigators or intellectual property lawyers.
- Attending or speaking at financial conferences.
- Reading publications and press releases.

## Sell-side Equity Research

*A typical day for an equity researcher might include the following activities:*

- Reviewing relevant press releases in the morning. If there is news that is pertinent to one of the covered companies, analyzing the information and preparing a cohesive statement for the sales force and clients. Calling investors to update them on the news.
- Attending meetings with private and public companies and considering whether to initiate coverage (i.e., company analysis and stock ratings).
- Researching and writing about companies for coverage. Studying details on clinical trials or product launches and speaking with key opinion leaders for their perspectives. Generating reports or notes.
- Building and updating financial models.
- Educating the sales team about various stocks and their significant company issues.
- Preparing for and meeting with clients to talk about investment theses and company stock ratings.
- Attending medical, scientific, and investment conferences.
- Organizing and running investor conferences.
- Reading medical and scientific journals to stay current.
- Taking equity researcher exams to remain certified.

## SALARY AND COMPENSATION

### Venture Capital

If your goal is to become unfathomably wealthy, venture capital should not be your first choice. Successful investment bankers often earn more money than venture capitalists. The few people who become billionaires in this business are the founders and CEOs of highly successful biopharma companies (see Chapter 22 to learn more). That being said, venture capital is still highly lucrative—more than most other careers in biotechnology and drug development.

Salaries in venture capital are derived from fund management fees and the carried interest (the "carry"). A management fee, paid by limited partner investors, covers the salaries of investment professionals and the firm's operating expenses. The carried interest is the net profit of the investments at the end of the fund's life, usually following the sale of stock options from portfolio companies that went public or were acquired. After every invested dollar is returned to their institutional investors, there is an 80:20 split of the remainder, the profit. The 20% that the venture firm retains is the carry, which on a successful deal can be quite substantial. This is split by the senior partners, and in some firms, a small percentage is doled out to junior partners and associates. A senior partner can earn several million dollars or more from the carry, representing a healthy reward for several years of hard work.

*"Carry" is the Holy Grail in venture capital.*

Partners are often required to invest their own money in the fund—usually around one or two percent, which can amount to millions of dollars. Having a personal financial stake in the fund demonstrates that a venture capitalist is committed to the success of these investments—a strong selling point for their limited partner investors.

### Institutional Investing

Compensation for institutional investors is based in part on the performance of stocks they hold. Depending on the economy and an individual's performance, some investors can earn millions of dollars per year. Due to swings in the public market, there is a lot of volatility, but compared to venture capital, institutional investors realize a more immediate return on the investments they manage. Management fees can be quite large for those who work in mutual funds.

### Investment Banking

Investment bankers often earn astronomical bonuses in addition to large base salaries. Managing directors can earn millions of dollars per year—but only a few people reach that level. Compared to venture capital, compensation is more immediate, although the long-term potential may not be as

*For investment bankers, an important question to answer is not which type, but how many, Porsches to buy.*

good. Investment bankers earn substantially more money than people in most other industry positions.

If you are interested in earning very large amounts of money, consider joining one of the larger multinational banks. These banks have more assets and several types of revenue streams, so they can pay employees more than smaller boutique banks and firms.

### Sell-side Equity Research

The days of receiving tens of millions of dollars for saying positive things about bad stocks are over. The coverage generated by analysts is regulated by the SEC, which prevents them from generating that kind of revenue now. Senior biotechnology analysts at top firms can earn handsome base salaries and million-dollar bonuses; however, few individuals reach that level.

*Compensation in equity research is analogous to that in professional sports: The star players earn the exorbitant incomes.*

Investment bankers earn more than their equity research analysts. Salaries for equity researchers vary widely, by a factor of ten or more. Bonuses are based on seniority and on ratings.

***How is success measured?***

***Venture capital***

The success of a venture capitalist is measured in part by how much money he/she is managing, as well as the return generated on investments (i.e., net profit). Building successful companies and taking portfolio companies public or into an acquisition are successes as long as the stock can be sold for a profit. Another measure of success is the ability to add value to portfolio companies by being involved in critical business matters (although such success is hard to quantify). Venture capitalists also measure their own achievements by the enduring success of companies they supported during their formation and early-growth phases, even long after the venture capitalists have sold off the stock. Venture capitalists have a "trophy wall," where they display the framed IPO certificates of their portfolio companies.

***Institutional investing***

In institutional investing, performance is based on annual investment returns and how much money was realized. Potential investors will scrutinize the decision-making patterns of an institutional investor: Were stocks bought low and sold high, or the reverse? Were particular stocks bought or sold at a good time? In addition, an investor's personality and ability to do a good job of representing the firm to clients are important factors in establishing compensation.

For analysts, success is measured by conducting constructive due diligence and obtaining accurate information about companies, including a comprehensive package of data, complete with "the dirt."

***Investment banking***

Success in investment banking is measured by how much money is made for the bank and the potential to make even more money from new business. Money is made from transactions, mostly from follow-on stock offerings and IPOs. Building relationships with companies so that they will select your firm when they need transactions, and having clients return for continued business, are two measures of success.

*Sell-side equity research*
In sell-side equity research, consistently offering accurate and predictive stock-pick advice is important for advancing your career. People who are able to predict stock performances develop a respected reputation, are paid more, and eventually become highly profiled. In addition, if an analyst provides high-quality information that is otherwise unavailable to investors, he will be judged favorably.

Judging equity research performance can be rather subjective, and in some firms, it can devolve into a popularity contest. The compensation of sell-side analysts is subjected to a vote based on performance measurements. Clients and the sales team vote for their favorite research associates and analysts. In addition, an objective measure called StarMine ranks the performance of analysts' stock recommendations. Other metrics include the number of trades in one's name and the quality of the stocks that one covers.

## PROS AND CONS OF THE JOB

### Positive Aspects of a Career in Health Care Finance

*All Health Care Finance Careers*

- The tremendous amount of job variety makes each day different and dynamic. Every day is a learning experience with new intellectual challenges. These careers smoothly blend a combination of business, science, and finance skills.
- There is constant exposure to some of the most exciting cutting-edge technical advances and a lot of stimulating new information to absorb. You are immersed in the fascinating worlds of science and finance. It is one of the few careers where those with a scientific or medical background can enter Wall Street and apply their training.

*Health care finance is an exciting mixture of science, people, and business all rolled into one.*

- You will have an opportunity to deal with extremely bright, famous, and interesting people. They could be Nobel laureates or people who were your heroes, and you may have an opportunity to help them move their ideas forward. It can be very stimulating to be around such dedicated, bright, and highly motivated executives, who are driven to fulfill their dreams.
- Managing millions of dollars and interacting with high-level executives is socially rewarding, because you are having a major impact. It is hard not to feel important.
- Health care finance executives help facilitate the flow of money to, or directly provide funds for, the most promising companies. Investors win, society wins, and biotechnology companies have enough resources to deliver a product to patients.

*A promising company developing cancer therapies would quickly go bankrupt without investors to provide the necessary capital.*

- Health care finance jobs have great cachet. There is an opportunity to interact with a level of industry executives who would otherwise be difficult to meet, such as CEOs, investors, and key opinion leaders.
- Compensation is excellent in health care finance careers. Many people become extraordinarily wealthy. It is an excellent way to acquire solid business skills while being paid handsomely.
- These jobs offer excellent career advancement into other disciplines and provide a great network of contacts to make the transition possible.

### *Specifically for Venture Capital and Institutional Investment Careers*

- These jobs offer autonomy and freedom. You decide what you want to spend your time on and who you want to meet each day. You have the freedom to do your job without significant interference and involvement from others.
- Unlike investment banking, there is no correlation between input hours and productivity, so most people work far fewer hours per week than bankers do. The philosophy is to work smarter, not harder, but it is still hard work.
- For venture capitalists, there is tremendous satisfaction in helping to build and add value to companies. You are making a real difference and having a positive impact on companies and the livelihood of their employees.
- Compared to venture capitalists, general partners in institutional investment firms work more independently. Investors do not require a consortium or consensus to close deals, as is typical of venture capital. In addition, it is not necessary to obtain the cooperation of external limited partner investors.

### *Specifically for Investment Banking*

- It is thrilling to have your hard work pay off by winning a business transaction. Given the highly competitive landscape, such triumphs are personally rewarding.

*If you are an athlete and enjoy competition, every day at an investment bank is a rush.*

- Several years of experience on the sell-side will provide a plethora of career opportunities, including switching to the buy-side world. Working on the sell-side is a convenient way to access all of those worlds.

### *Specifically for Sell-side Equity Research*

- You gain an almost academic thrill from applying your accumulated investor and educational knowledge to create a thesis that later is proved to be correct. For example, when you predict that a particular stock will go up or that a clinical trial will fail, and these things actually do happen, you have validation for your work and ability.

- Your capacity to gain insightful information about companies and to add your educated perspective helps clients make money. It is rewarding to have appreciative clients who are grateful for your quality work.
- Work is not as stressful or intense as it is for investment bankers.
- Equity researchers work closely as members of small teams and enjoy the camaraderie.

## The Potentially Unpleasant Side of Health Care Finance

### *Venture Capital*

- Venture capital is not very team oriented. It does not support the same level of cohesiveness and commonness of purpose as one finds working in a biotechnology company. Some firms are team oriented, whereas others cultivate a "lone wolf" culture. "Friendly rivalry" is common, as coworkers try to steal each others' promising deals. In addition, because most venture firms are small in size, work can be lonely compared with big companies which have many employees in different disciplines.
- The job is somewhat amorphous. You don't work on the same type of tangible things that your portfolio companies do; i.e., you don't develop a product that can be marketed. People who enjoy getting their hands dirty and creating data that can be applied to product development won't get that satisfaction in venture capital.
- It takes a long time to see results or a return on your investments (see Greatest Challenges). It is possible to spend ten years as a venture capitalist and have no successes. As more people enter the field and there is more generic competition from funding sources, it becomes increasingly difficult to distinguish yourself.

*Venture capital investments are long-term propositions; there is seldom immediate gratification from your work.*

- Venture capitalists typically travel one or two times a week, attending portfolio company board meetings and conferences. The amount of travel can be reduced by investing in local companies.
- Venture capital work can be very demanding. You are expected to be fully engaged and very interactive. The sheer amount of personal interfacing can be daunting at times.
- Fund-raising is tedious and tiring. On a standard road show, you may visit ten different cities and give the same presentation 25 times in five days.
- Jobs in venture capital firms are highly coveted and fiercely competitive. The work itself can be highly competitive too, as partners compete for the most promising deals.

### *Institutional Investing*

- There is little time to dwell on any one item for very long.

- Biotechnology investing is volatile. It is depressing when clinical trials fail and the company stock plummets. An institutional investor must learn to accept this aspect of the business.
- Travel can be demanding.

*Investment Banking*

*The only two days when it is acceptable to take off time in investment banking are your own wedding or your own funeral!*

- Expect to work long hours and have no control over your schedule. Nearly all of your waking moments are spent on the job. Junior analysts receive two or three weeks of vacation per year, but they can rarely take the time off. Your vacation plans may be taken away from you at any time (even while traveling), and saying no is unacceptable. Investment banking and an active family life do not mix well. Clients depend on you to accomplish tasks. If you are not there, they will take their business elsewhere.
- Meeting clients and developing prospective business requires a lot of travel. It is important to talk to many people in order to build relationships and present new ideas to clients. It takes time and many meetings before business is acquired. Face-to-face meetings with clients are necessary to build relationships—executives don't make billion-dollar decisions over the telephone.
- Investment banking is stressful. Not only are the hours long, but the inherent unpredictability of the biotechnology industry and the market can wreak havoc on the success of financial transactions. Failure is enormously harmful to your reputation and future business prospects.
- Emergencies are frequent. You need to respond immediately to your client's needs. Solving problems can sometimes take all night.
- Big egos often become obstacles to good business sense, particularly with mergers and acquisitions. This can squash deals. People issues can't easily be solved even if there is financial synergy.
- Some large banks have as many as 200,000 employees. In such an environment, it is difficult to be noticed in order to advance your career. As a result, office politics can be rampant and consumptive in some companies.
- Hiring sprees and layoffs are common, depending on the economic climate. If you seek a position with good job security, investment banking is not a safe choice.

*Sell-side Equity Research*

- Expect to work long hours and weekends. You need to constantly keep track of the news. It is a challenge to meet your work requirements every day and also have a semblance of a personal life. As you gain seniority, work hours will improve modestly, but some people eventually burn out.

- Expect to travel frequently, although not as much as investment bankers.
- There is a great deal of stress associated with giving stock advice to clients. If you have developed a solid rationale for your stock ratings, people will rarely become upset at you when you are wrong. But even with the best analysis, there are factors beyond your control, so it is inevitable that you will make stock predictions that prove incorrect. When people lose money, they can become thoroughly unpleasant. Many have a tendency to yell and berate analysts. They might comment, "You should go back to treating patients, but that isn't a good idea either, because you would probably kill them!" If you are wrong too often, you may damage your reputation and ultimately lose your job!

*Defending a stock that has plummeted simply stinks!*

- Every day, there is tremendous pressure to derive newsworthy information from the media to relay to clients and your sales team.
- Coming to a consensus with investment bankers is not always easy. Investment bankers are driven to conduct as many transactions with as many companies as possible. Equity researchers are required generally to cover the companies that the investment bankers have completed deals with, regardless of the quality of the company. Research is supposed to be conducted independently of investment bankers.
- Personal performance evaluations tend to be rather subjective in equity research. If you receive an unfavorable evaluation, it may directly affect your pay or, even worse, your job.
- Equity researchers are not allowed to personally invest in the companies that they cover, because this represents a conflict of interest. However, they can invest in companies in other industries.

## GREATEST CHALLENGES ON THE JOB

### Venture Capital

#### *Long Time Horizons and an Unpredictable Environment in Which to Prove Yourself*

It takes many years for investments to mature, and during that time, there will be little if any direct feedback on your investment performance. This makes it difficult to develop a solid track record and maintain the confidence that you will do well in this job. For the young and unproven venture capitalist, it can take six to eight years to determine whether you are competent. In addition, your success depends on certain factors that are beyond your control, such as market cycles and whether a clinical trial is approved. You could be an admirable investor, but if the biotechnology sector is being punished for reasons beyond your control, you could fail.

#### *Developing a Reputation in a Competitive Landscape*

When you and your firm develop a reputation for achieving superior returns, it becomes easier to attract limited partner investors to the fund. Therefore, it is a constant battle to develop a good investment reputation and stay ahead of the pack. Venture capitalists tend to flock toward commonly accepted ideologies that are relevant to the current market and economy. Most firms develop similar visions and investment strategies and compete for quality deals (see next point).

### *Too Much Money Chasing Too Few Deals*

There is an excess of investors fishing in a limited pool of available and promising business deals. The basic science that supports biotechnology is funded by competitive grants and fellowships, largely from the NIH, which go to a select group of researchers. Only a few ideas per year percolate out of several hundred medical schools and academic institutions. Due to this limited pool, competition is intense, and it is common for the valuation of deals to be bid up. As the price of a deal goes up, it becomes less attractive for a venture capitalist's risk/return profile. There is always someone willing to pay too much. As a result, a venture capitalist may be without any investments and may be forced to invest in companies with inadequate management teams or technology. Inferior investments carry a greater risk of inadequate returns for the fund. This makes it more difficult to differentiate yourself from other venture capitalists and to ultimately obtain adequate limited partner funding.

### *Financing Early-Stage Companies*

Investing in early-stage deals is financially risky, and as a consequence, fewer young companies are being financed. Currently, venture capitalists prefer to invest in late-stage companies with promising clinical data. This has created a problem in the industry—a large vacuum in financial support exists for early-stage opportunities. Investors can face a situation where after investing in an early-stage company, they will need several financing rounds before the company's products can enter clinical trials. However, because the products are so early, it may be difficult to obtain additional funding. To compound the problem, taking companies public now requires more capital. As fewer early-stage companies are being financed, fewer companies are carrying out clinical trials, and as a result, there is more competition for attractive deals in the clinical stage (see the previous point). This makes venture capital a difficult way to earn a living and in which to perform high-quality work.

## Institutional Investing

### *Just Being Able to Stay Current*

With the overwhelming amount of news and information available today, it is impossible to remain completely current—no one can do it by herself. It is a constant challenge to catch all of the important information.

### *Performing Well in a Constantly Changing Environment*

Because of the ever-changing economic environment, it is difficult to find good thesis ideas that promise sustainable returns. It is disheartening to work through down cycles when stocks are low. It is easy to become emotionally attached to investments and to panic and sell your stock during adverse economic times. The challenge is to be able to think clearly and adhere to your investment principles—business objectivity and perseverance are your salvation during adverse economic times.

### Investment Banking

*Winning*

Winning transaction business is the greatest challenge for investment bankers. Bankers are compensated when they complete financing or mergers and acquisition transactions; but so are their competitors, so the result is an intensely competitive environment.

### Sell-side Equity Research

*Finding Interesting Ideas to Relate on a Daily Basis*

It is difficult to continually present interesting and important information to investors. Sometimes, just to have something to say, you may need to expand on ideas that do not represent your best concepts.

## TO EXCEL IN HEALTH CARE FINANCE...

### Venture Capitalism

*Having Visionary Investment Capabilities*

Those who excel in venture capital can apply years of accumulated knowledge and experience to evaluate a product concept and envision it through to product launch. They understand what will be needed to achieve commercial viability regardless of the current market environment and are able to suggest ways to mitigate potential risks. Successful venture capitalists invest in companies that will deliver what the market needs in three to five years.

*Being a Valuable Contributor to Portfolio Companies*

In venture capital, success can depend on being an effective and valuable contributor and working closely with senior management of your portfolio companies. Venture capitalists who succeed make sure that each of their companies has the most productive management team and solid backup plans in place. They are willing to play an active role in helping the executive team complete major transactions to successfully drive the company forward.

*An Uncanny Ability to Pick the Most Promising Entrepreneurs*

An exceptionally good management team greatly increases the likelihood of success. Prominent venture capitalists innately seem to know how to pick the most promising entrepreneurs. First-rate entrepreneurs gravitate to them as if they were magnets.

## Institutional Investing

*Having Conviction in Your Research and Thesis*

People who excel in institutional investing have great conviction in their investment theses. They can determine the intrinsic value of companies and can see any discrepancy between the stock and the perceived worth of a company. For example, they might invest during seemingly bad times when others think it is the worst time to purchase stock, and might have outstanding returns when everyone else does poorly. The investors who play it safe and just follow the herd simply drive stock prices up.

## Investment Banking

*Salesmanship*

Successful investment bankers have outstanding sales skills and the ability to build relationships and gain trust. Investment banking is an incredibly sophisticated type of sales. You need to stimulate investor appetites if you are going to sell your organization. Bankers who have developed over 20 years' worth of relationships have an advantage— they can consistently call on and gain client trust, and thus win transaction business.

## Sell-side Equity Research

*Constantly Having New Information and Staying in Front of Investors*

Sell-side equity researchers aspire to provide novel and timely information for clients. They provide provocative ideas backed by solid research and evaluation. They also need to be able to explain and disseminate their concepts effectively.

## Are You a Good Candidate for Health Care Finance?

*People who flourish in venture capitalism and institutional investing tend to have...*

*Outstanding interpersonal skills.* You need to be able to work well with others, to be socially adroit, and to be able to communicate effectively. You must possess a good combination of independence and teamwork skills. You need to be comfortable diving into data and also be able to pick up the phone and talk to anyone and everyone. Superior interpersonal skills are called upon for the following reasons:

- Charismatic venture capitalists may have an easier time attracting entrepreneurs seeking funding and executives being recruited by portfolio companies. Exemplary entrepreneurs often pick their favorite venture capitalists to be investors and board members, and will exclude those with reputations for being harsh with entrepreneurial teams.
- Superb social skills are needed to work effectively with management teams. As a board member, you will be called upon to solve problems and make decisions, often using a

consensus-driven process. People skills are needed to manage the egos of CEOs and the management team, as well as to control your own ego. Some venture capitalists develop a God-like complex in this business.

- Strong interpersonal skills are also needed by venture capitalists to attract limited partner investors. Ultimately, you must establish an outstanding reputation built on years of superior investing and trust before limited partners will write a multimillion dollar check. In addition, like entrepreneurs, limited partners selectively choose who they want to work with. Outstanding interpersonal skills enable investors to interact with and gather information from people of diverse backgrounds, training, and ideologies for due diligence.

*A tremendous network.* A large personal network is beneficial for increasing the flow of deals and securing the most promising investments. The larger your network, the more opportunities there will be to identify and contact individuals with the pertinent knowledge base or expertise for due diligence calls.

*Venture capital is a people's business—the more people you know, the better the deal flow.*

*Strong analytical skills.* It is essential to be intellectually flexible and possess strong analytical skills. You need to be mentally comfortable with analyzing the smallest details while simultaneously keeping broader considerations in mind.

*Intellectual courage.* You must be able to handle the risk of venturing into areas where you are not an expert. Venture capitalists deal with a myriad of problems in finance, science, and people management. It is advantageous to have a wide-ranging set of skills.

*Tolerance for high stress and risk.* The outcome of venture investments is often not known for six to eight years. In the interim, there will be the continual stress of not knowing how your investment will perform. For institutional investors, there may be a million reasons why a stock is not successful, including many that an investor could not predict or foresee.

*Strong technical knowledge.* Biotechnology investing requires broad technical proficiency in science and medicine. You may be called upon to understand and be conversant in the languages of biology, clinical trials, and regulatory affairs, to name only a few.

*The ability to focus on more than just the science.* The most common failure of scientific investors is focusing too much on the science and the belief that technical components are the most important considerations. In the majority of cases, the most crucial problems arise within the management team. Solving these people problems has nothing to do with science.

*An ability to multitask.* The ability to work energetically, balance multiple projects simultaneously, and reprioritize at a moment's notice is required.

*The skills to think flexibly.* Biotechnology companies pass through many iterations and permutations of their business plan before they succeed. You need the foresight to recognize that when things go wrong, changes may be called for. Despite the changes, you must be able to continue assisting your portfolio companies or resist the temptation to sell stock during such challenging times.

*Good judgment of people, markets, and science.* It is important to have savvy business intelligence and to appreciate what makes financial sense and where to make money. Having intuition about management teams, being able to read and understand what motivates individuals and to estimate their capabilities are fundamental to making wise investment choices.

*A tolerance for conflict.* Hard decisions will be required, such as firing a founding CEO. You need to be fact-based, diplomatic, empathetic, and maintain a positive attitude during conflicts.

*Diplomacy and excellent negotiation skills.* Excellent negotiation skills are needed for conducting deals with limited partners and making investments. The optimum outcome for a successful deal is for people on both sides to be pleased with the terms. Entrepreneurs are more motivated to be successful when closing a deal on mutually good terms.

*An ability to make objective and intelligent decisions.* This is an area where one should not be emotionally attached to a particular technology or management team, but be able to objectively consider the wisest investment strategies. Institutional investors need to remain calm during adverse economic times and to avoid making panicky decisions, such as selling off stock without thinking through all the consequences.

*Patience and perseverance.* Portfolio companies have their ups and downs, and investments have long time horizons. Patience and perseverance will help you stay with the company or the stock.

*The ability to investigate or delve into details.* When conducting due diligence on companies, institutional investors are aggressive about delving into major problematic elements before making investments. They explore every avenue and investigate innovative sources to obtain insightful information. They are continually on a quest to find the true story behind something. Venture capitalists conduct due diligence on companies as well, but this takes place at an earlier stage of the company, so there may be less company history to review and fewer dark secrets to disclose. Once they invest, it is in their best interest to project their decision in the best light and with confidence.

*Institutional investors are like investigative reporters—they are adept at digging up and uncovering information.*

*People who flourish in investment banking and sell-side equity research tend to have...*

*Excellent communication skills.* Those who are most successful know how to appear thoughtful, yet not condescending. You need to learn how to craft a story, form a recommendation, and fluently express the message.

*An ability to work in a fast-paced environment.* Work is fast paced, and your responses must be quick. This is a time-sensitive career, and a sense of urgency is called for. Things happen in real time and people can lose millions of dollars in seconds and minutes.

*A well-developed ability to multitask.* Time management and efficiency are key. You need to be able to juggle many tasks at the same time. Any single task may be easy enough to accomplish, but it is doing many things at the same time that makes work challenging.

*A willingness to work hard and a solid work ethic.* This is an all-consuming job. Even when at home, investment bankers continue to check the news. Banking is a lifestyle, not just a job. A strong work ethic is mandatory. Investment bankers need to be young and energetic to handle the rigorous workload and long hours.

*A tenacious (yet tempered), aggressive, and competitive drive to win.* Banking is a highly competitive industry. You have to be self-motivated and driven to win. Securing a deal can be fiercely competitive, and successfully completing transactions can be equally challenging. Many ex-athletes work in banks. They enjoy the thrill of the chase and of winning. Their experience in competition provides skills that are directly transferable to investment banking.

*Bankers have an insatiable drive to achieve.*

*A winning sales personality.* Sales skills, such as being a good listener and building rapport with clients, are essential. Sales skills include knowing how to read people's reactions and faces, being perceptive, and knowing when to and when not to ask for business.

*Analytical skills.* A research or analytical background is beneficial in this career. You will need to think logically and know how to exhaustively create, investigate, and test hypotheses. Adeptness at building financial models, researching scientific or clinical trial details, and knowing what is relevant among an extensive amount of data are needed.

*An outgoing personality and excellent interpersonal skills.* It is advantageous to be outgoing, extensively networked, and highly social. Because work occurs in multifaceted teams, at any given time you could be coordinating and interacting with dozens of people, each with a different drive and motivation. Much of the work involves discussing ideas and incorporating concepts with clients. It is important to treat clients well and provide honest and candid advice. In addition, it is important to be politically savvy. Many clients are affluent and may have become egomaniacs after accumulating substantial wealth, so you will need the social fortitude to interact appropriately.

*It is important to be outgoing. You need to be engaging and to develop rapport by demonstrating your financial and industry expertise.*

*Creative problem-solving skills.* New problems are encountered with just about every transaction. You need to remain tenacious, have a firm instinctual feeling about your work, and apply original creativity to solving difficult problems.

*A positive attitude.* Bankers spend a great deal of time at work. When employees become overloaded and overworked, they can easily become surly and irritating. This is such an important point that some hiring managers prefer to hire analysts with exuberant positive attitudes even if they lack some of the trainable technical skills.

*Thick skin.* It is important to be able to handle criticism and to learn from failure. It is accepted that you will lose more deals than you will win, so it is important not to overreact to rejection or become despondent. Senior bankers eventually develop a thick skin and become impervious to rejection.

*A desire and interest to learn about the financial markets.* Working in an investment bank presents an exciting opportunity to learn about how Western capitalism works. The amount of exposure and information that you will acquire about finance in a few years may exceed that gained from working in a biotechnology company.

*An ability to make quick decisions.* Equity researchers need to respond quickly to daily news. If, for example, a clinical trial fails, one needs to decide and announce quickly whether to upgrade or downgrade a stock.

*Excellent writing skills.* Equity researchers need to communicate clearly and concisely. They need to be able to quickly gauge their audience in order to describe the science at an appropriate intellectual level.

***You should probably consider a career outside health care finance if you are...***

- Someone who does not enjoy traveling.
- Unwilling to work hard and for long hours.
- Someone who demands seeing projects all the way to completion. This is one of the few careers that reward those with short attention spans.
- Inexperienced in business.
- Unable to effectively handle problematic people.
- Lacking a broad understanding of what drives business fundamentals and valuations.
- In need of being in complete control over every situation.
- Intolerant of risk.
- Incapable of handling stressful situations, losses, and setbacks.
- Indecisive or unable to make quick decisions.
- Overly shy or timid.
- Only in it for the money.

## HEALTH CARE FINANCE CAREER POTENTIAL

In health care finance, career transitions can include joining or founding companies, becoming an individual or angel investor, and providing capital. Many people join start-ups and become entrepreneurs, either in a business development capacity or in other senior executive leadership roles (see Fig. 24-2).

### *Venture Capital*

Most venture capitalists see their career as an end point, not as a stepping-stone to other occupations. Those who do leave often go into business development or executive leadership positions in young companies.

Venture capital firms tend to be small in size, and job titles across the industry are not standardized. General partners have full liability, manage the carry and the decision mak-

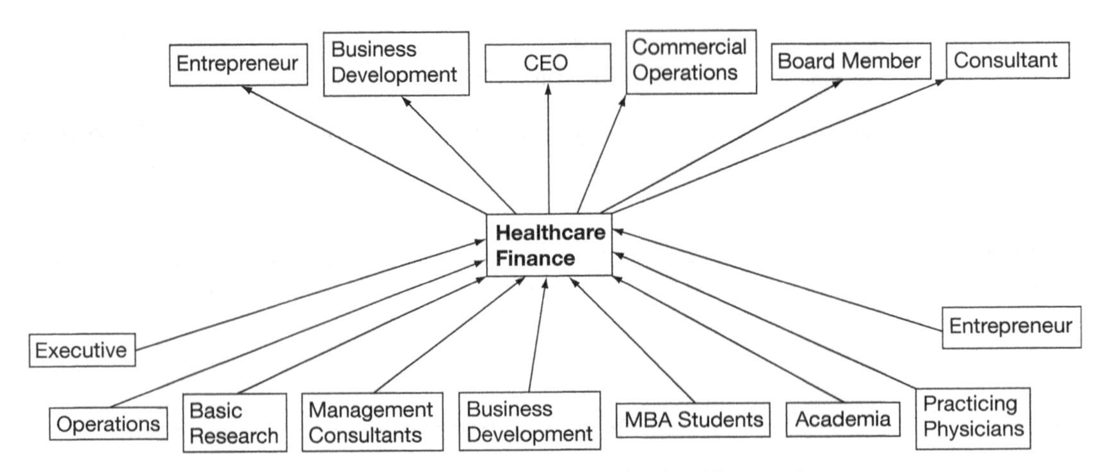


**Figure 24-2.** Common career paths for health care finance.

ing for the firm. They are involved in fund-raising and have the primary responsibility for making and managing investments. Venture partners have the right to co-invest in deals in exchange for a percentage of the carry from those deals. Junior partners have some freedom to operate independently and manage investments. The primary roles of an associate are to assist general partners in doing due diligence and helping their portfolio companies succeed. Most associates will spend a few years in venture capital and expand their experience later by working in portfolio companies, typically in business development or executive management roles. An associate position is not a guaranteed step on the ladder to partnership—associates are often hired with no expectation that they will ever become partners. Analysts are junior staff members who conduct financial models and use other tools taught in business school.

### *Institutional Investing*

There are two basic levels of institutional investors: analysts (buy-side equity researchers) and portfolio managers (senior equity researchers). Both analyze the scientific progress and viability of companies and conduct due diligence. Portfolio managers research and select stocks to invest in, and then manage their investment portfolio. Associates and analysts can eventually become portfolio managers. But the ultimate goal is to raise capital and run your own fund.

### *Investment Banking*

Investment bankers may make the transition into business development roles in companies, or they may move into venture capital or institutional investing.

### *Sell-side Equity Research*

*The number one exit out of sell-side is to transfer to the buy-side.*

Entry-level equity researchers start out as associates and may eventually become senior or lead analysts. Associates and analysts spend their time researching companies and derive investment theses and financial models. Senior analysts are responsible for managing the relationships with their institutional investor clients. Moving into the buy-side world, such as venture or corporate capital, is also a common career path.

## Job Security and Future Trends

Obtaining a job in health care finance is highly competitive. There are many highly qualified people seeking these jobs and relatively few openings. Obtaining your initial position may be the most difficult step.

### *Venture Capital*

Venture capital firms tend to be small organizations composed of two to ten partners. As a consequence, few job opportunities are available, and qualified applicants are plentiful. Firms receive literally hundreds of resumes every month from very talented professionals seeking employment. It is even more difficult to gain a position in the most renowned firms with illustrious reputations and a large amount of capital under management. The total number of active venture capital partners in the life sciences is probably less than 200, assisted by an even smaller number of associates and junior partners.

*The two most common reasons that venture capitalists leave their firm are having (1) a dismal investment track record and, more commonly, (2) personality clashes with the partners.*

Partners wear "golden handcuffs," so they rarely change firms. It takes many years before they can receive their carry, which is a clever mechanism to incentivize the high performers to remain at the firm.

### *Institutional Investing*

Job security in institutional investing is related to the performance of individual funds and is also dependent on the overall biotechnology economy. Funds are rated on their quarterly or annual returns. If the biotechnology universe is down and the markets are weak, investors are less likely to pay institutional investors to manage their money. Hedge funds, however, can prosper in either up or down markets.

### *Investment Banking*

There are only 300–500 investment bankers working in the biotechnology sector. Careers in investment banking experience intense volatility due to whimsical economic forces, resulting in employment expansions or contractions.

### *Sell-side Equity Research*

Job security is based largely on the economy, the stock market, and the profitability of the firm. There are 100–200 sell-side equity researchers who specialize in the biotechnology and pharmaceutical industries.

## LANDING A JOB IN HEALTH CARE FINANCE

### Experience and Educational Requirements

#### *Venture Capital*

To become a venture capitalist, extensive training and a diverse set of skills are required, in fields such as business development, operations, consulting, banking, law, or science. Many venture capitalists have an advanced scientific or medical degree and an M.B.A. Fewer than 15% have only an M.B.A. degree.

Extensive operational experience and an exemplary track record in business are the best preparation for venture capital. To provide the most value to portfolio companies and to serve as a board member, extensive industrial knowledge, a track record of successful achievements, and operational experience are optimal preparation. In addition, when venture capitalists raise capital, their potential investors evaluate the senior partners' track records.

People can gain entry into venture capital work from a variety of different disciplines. Career entrepreneurs or successful executives in a large or small company can become partners. They could have been a founder, CEO, senior vice president of research and development, or in business development at a large company, or serve the same role in several successful start-ups that went public. They tend to have 10–20 years of operating experience, an extensive network of contacts, and deep knowledge about how to run companies.

Entrepreneurs in residence (EIRs) are trusted executives who are temporarily placed in portfolio companies. They also assist venture firms with their due diligence. EIRs usually have had significant industry experience after founding several successful companies, which makes them candidates to become partners in a venture capital firm.

Junior partners have usually served as vice presidents of one or more large organizations, as management consultants, or as venture capital associates.

Associates may have relatively limited industry experience beyond an advanced degree. They often have some kind of technical or clinical background. They could have experience working in business development, operations, management consulting, or in an investment bank. Many have advanced degrees, and half of them have M.B.A. degrees. Associates are rarely hired directly out of academia.

There are many permutations to the scenarios described above. It is common for the more junior staff to rotate out into companies and then rejoin venture firms later after they have gained adequate operational experience.

#### *Institutional Investing*

Increasingly, institutional investors in the health care industry are required to have advanced medical or scientific degrees. An M.B.A. degree is also very valuable. Most institutional investors have some combination of experience in investment banking, sell-side equity research, business development, finance, or operations. The natural progression is to work in sell-side equity research before entering into buy-side institutional investing.

### *Investment Banking*

Approximately 75% of investment bankers have an M.B.A. and no advanced science degree. The remaining have an M.D. or Ph.D. degree, business development or research experience, and/or an M.B.A. degree. Applicants can apply directly from medical school, residency programs, or graduate schools, and start as associates.

### *Sell-side Equity Research*

Previously, analysts were expected to have only M.B.A. degrees, but in recent years there has been a dramatic influx of people with scientific backgrounds, especially for health care coverage. Today, the ideal candidate has a Ph.D. or M.D., an M.B.A. degree, some industry experience, and knowledge of finance.

## Paths to Health Care Finance

- The number of jobs available in health care finance is limited, and many qualified people want them. To get one of these jobs, you need to distinguish yourself from the rest of the qualified pool. One tactic is to meet professionals in the field by networking and attending investment conferences.
- Work with career counselors or contact alumni associations on campus to identify fellow alumni in health care finance. A venture capitalist is more likely to respond to a cold call from an alumnus than from a complete stranger.
- If you have a science background, strengthen your financial interest and acumen. Work in business development for a few years or obtain an M.B.A. degree.
- An M.B.A. degree from one of the top schools will significantly increase your chances of starting a career in health care finance. It will also provide you with high-quality peer and alumni contacts that will serve you well throughout your career. During your M.B.A. program, seek out summer internships at banks and investment firms. Although these summer opportunities are highly competitive, they will give you an advantage when applying for your first job.
- There is a select group of recruiters within the finance world. Look for them on-line. Network and find out who they are. When calling analysts and associates, ask for recruiters who specialize in health care finance.
- Learn all you can about the biotechnology and pharmaceutical industry. Read as much as possible, learn about clinical trials, and find out who the major companies are and the strategies they have adopted for success and growth.

### *For Venture Capitalists*

- One of the best ways to become a venture capitalist is to first become a prestigious Kauffman Fellow. This fellowship is sponsored by a unique nonprofit association that promotes entrepreneurship. As a Kauffman Fellow, you will have an opportunity to work in a venture capital firm or for one of their portfolio companies. Visit www.kauffmanfellows.org to learn more.

- Consider joining the corporate venture capital arm of a large pharmaceutical or biotechnology company. This is probably an easier avenue into the field than applying to venture capital firms. Alternatively, you could work your way into corporate venture capital by initially being on a business development team or, if you are in the discovery or clinical research group, by gaining experience evaluating potential business development opportunities.
- One of the best ways to become acquainted with venture capitalists is to serve on the executive management team in a start-up (preferably a successful one!). There you may have the chance to meet with and pitch your business opportunity to venture capitalists. If funded, you will then have a chance to become well acquainted with your venture capital board members.
- Senior executives should consider working as an entrepreneur in residence (EIR), which many firms offer. EIRs are trusted executives who work temporarily in venture-backed portfolio companies to help build and oversee the company success and investment.
- Venture capitalists often seek outside consultants, who are presumed to be experts, to conduct due diligence on specific technologies or to evaluate business plans. You can submit your resume requesting an opportunity to work as an outside consultant.
- Obtain operational experience. This is considered the optimum venture capitalist background. See Chapters 14 and 22 to learn more.
- Business development and management consulting experience are both very valuable for moving into venture capital.
- Apply for associate positions at firms. As an associate, you may be able to advance slowly into a partnership position.
- When you are interviewing at a firm, be prepared to display broad knowledge of the industry. Be prepared for questions such as: What technical areas should the firm consider for new investments? Which private companies do you think highly of and why? (However, don't mention companies that the firm has already invested in. This merely shows that you have reviewed their Web site.)
- Be sure to conduct some due diligence on the firm in which you are interested, and learn about the partners' reputations. Although some firms provide excellent mentoring of aspiring associates, in other firms, it is a "sink or swim" environment. Because there is no prescribed training to become a venture capitalist, partners develop their own styles.
- Venture partnerships are very much like marriages. Due to the nature of venture capital, small firms are reluctant to hire partners because of the long-term commitments. Analysts and associates are expected to cycle out in two to three years. Internships are an attractive way to determine whether an applicant's personal dynamics work within that firm. If you

are considering a partner-level position, be certain that there is a good fit between you and the firm. After all, you may be working with that team for the next 10–20 years.

- Become an angel investor and join local angel groups to gain more experience with start-ups and investing.

#### For Institutional Investing

- Work on the sell side as an investment banker or equity researcher for at least two years before going to the buy side.

#### Specific for Both Investment Banking and Sell-side Equity Research

- Take finance classes to show your commitment and interest, as well as for honing your financial acumen.
- Contact senior analysts and investment bankers and work your connections. Start with people that you know who work on Wall Street. There are a limited number of analysts. You can call or E-mail each one individually and request an informational interview or inquire about job openings.
- Apply in the spring and summer. People leave their jobs after they have received their salary bonuses, which are awarded in the winter. Musical chairs begins in February.
- For equity research applicants, write a sample research report to showcase your talents and submit it with your resume.

## RECOMMENDED TRAINING, PROFESSIONAL SOCIETIES, AND RESOURCES

#### Courses and Certificate Programs

Certificates in equity research are available if you want to become a certified research analyst

Classes in statistics and finance

Drug development books and classes

#### Societies and Resources

*Finance-specific*

The Kauffman Foundation (www.kauffman.org). This is a unique and fabulous program which provides grants to support entrepreneurship.

National Venture Capital Association (www.nvca.org) is a society for venture capitalists and a good resource when you are reviewing firms.

The Chartered Financial Analyst (CFA) Institute (www.cfainstitute.org). This organization for investment professionals offers a CFA certificate.

United States Securities and Exchange (SEC) EDGAR Database (www.sec.gov). You can search the EDGAR database to learn about companies and corporate governance.

Price Waterhouse Coopers (www.pwcglobal.com) has information on their Web site about companies and trends.

Investment banking salaries are listed annually in the *New York Post.*

*Biotechnology and Pharmaceutical Societies*

Biotechnology Industry Organization (www.bio.org)

The Pharmaceutical Research and Manufacturers of America (PhRMA, www.phrma.org)

#### Books

Jaffe D.T. and Levensohn P.N. 2003. *How venture boards influence the success or the failure of technology companies* (A White Paper). This article about venture capital can be found at www.dennisjaffe.com, under "publications."

Lewis M. 1990. *Liar's poker: Rising through the wreckage on Wall Street.* Penguin Books, New York.

Rolfe J. and Troob P. 2000. *Monkey business: Swinging through the Wall Street jungle.* Warner Business Books, New York.

Vault (www.vault.com). Vault sells a variety of career books and has one about finance.

#### Magazines that Cost Money

*BioCentury* (www.biocentury.com), biotechnology business intelligence

*Bio World* (www.bioworld.com), worldwide biotechnology news and information

*Windhover* journals (www.windhover.com): *Start-Up* and *In Vivo*

*Nature Biotechnology* has sections on investing.

Price Waterhouse Coopers (www.pwcglobal.com) publishes a biotechnology review.

#### Free Electronic Daily Biotechnology News

FierceBiotech (www.FierceBiotech.com)

BioSpace (www.BioSpace.com)

Biotechnology Industry Organization (www.bio.org)

# 25

# Management Consulting

## The Strategy Advisers

*If you have the passion to innovate, drive change, and help companies be more successful, consider management consulting.*

MANAGEMENT CONSULTING CAN BE a wonderful way to put your analytical skills and scientific or medical training to use while developing your business expertise. You will learn how to lead teams and manage people, and you'll be tackling a variety of interesting problems across a wide range of companies. The access you gain to top business professionals will open doors to a broad spectrum of future career opportunities. Perhaps best of all, your efforts could dramatically affect the future of the companies with whom you work!

### THE IMPORTANCE OF MANAGEMENT CONSULTING IN BIOTECHNOLOGY AND DRUG DEVELOPMENT

*Management consultants tackle the problems that keep CEOs awake at night.*

Most high-level executives are busy managing the operational parts of their companies, so they often do not have the time or resources to conduct strategic and other types of analyses. These tasks are frequently outsourced to management consulting firms. Management consultants bring an integrated, cross-functional, cross-industry perspective to their assignments. These firms are great sources of collective intelligence and keen business acumen, which they use when providing tactical assistance to clients or when helping them solve complex business problems.

The assignments that management consultants face vary, ranging from global issues such as the introduction of new business models or technologies, to more fundamental questions about how to anticipate changes in the industry. Typical assignments include those centered on basic business issues, such as developing strategic plans for the future, prioritizing different compounds in development, devising pricing strategies, or developing ways to improve productivity.

## CAREER TRACKS IN MANAGEMENT CONSULTING

### *Management Consulting*

The degree to which each firm specializes in an industry or function varies. Most large firms have practices within specific industry segments (e.g., biotechnology and pharmaceutical versus high tech). They also have practices that are function-specific (e.g., cost reduction, post-merger management, finance).

Among firms that specialize in the health care industry, there tends to be subspecialization for pharmaceutical, biotechnology, insurance, and medical device companies. Also included in this group are firms that work with hospitals and nonprofit organizations interested in global or national health.

There are several major global firms that specialize in the life sciences, and numerous boutiques. Boutiques tend to have more narrowly defined functions and focus on niche issues and areas.

### *Other Types of Consulting*

In the broader scope of the consulting world, there are global and boutique accounting and technology firms and various other consulting companies. These types of consulting firms are very different from management consulting, and the skills they require are also quite different.

## MANAGEMENT CONSULTING ROLES AND RESPONSIBILITIES

### *Understanding the Client's Business Needs*

Generally speaking, in order to serve their clients better, management consultants initially assess their business needs. They conduct interviews with clients and experts to learn more and to develop rapport and trust.

*A successful consulting engagement is a partnership between the client and the firm.*

### *Conducting Primary Research*

Clients often request information that is not currently available or has not been synthesized. They may want to know, for example, the approximate market size for a novel product that will be launched ten years from now, taking into consideration the market's sometimes unpredictable changes over that time period.

Researching in part entails interviewing people in senior positions about the client's products. Thus, for example, if the client is developing a new cancer drug, consultants might interview oncologists, doctors, patient advocacy groups, and renowned experts. In addition to the interviews, the consultants use many other resources, including analyst and market or industry reports, academic articles, newspapers, trade journals, and company data. They also use data from the U.S. Census Bureau, the Food and Drug Administration, and other specialized data services.

#### *Analyzing and Synthesizing Data, Drawing Meaningful Conclusions*

For financial analyses, data are generated and analyzed to detect patterns and trends, and the team may develop financial models to test hypotheses. Team members hold brainstorming sessions to devise novel strategies and solutions for clients. They collect a lot of disparate information and transform it into something digestible.

*Consultants are enlisted to solve complex problems.*

#### *Presenting Results to Clients*

The firm's overall findings and recommendations are presented to the clients as a reasonable, coherent, and logical story. The consultants explain, from their perspective, what the client should do and why. A good presentation can be very interactive, with the client asking tough questions to get comfortable with the recommendations.

## A TYPICAL DAY IN MANAGEMENT CONSULTING

*Depending on the firm, management consultants spend the first one or two years...*

- Conducting interviews with experts or customers to generate data.
- Analyzing data, crunching numbers, and building models.
- Brainstorming at team meetings.
- Generating charts and graphs based on the data and analysis.
- Presenting overall findings and recommendations to clients.

*More senior members and partners spend their time...*

- Supervising teams and overseeing the projects, ensuring that the projects are running smoothly and that team members are working well together.
- Developing relationships with industry executives for potential business, learning about their problems and needs.
- Bringing in additional business projects.
- Mentoring younger consultants.
- Building team morale.
- Recruiting and training new consultants.
- Speaking at meetings and conferences.

## SALARY AND COMPENSATION

Management consulting is one of the most lucrative careers available for those with advanced degrees. Salaries are considerably higher than those in academia. Entry-level compensation tends to be competitive with or slightly more in consulting than in biotechnology companies, and partners are usually exceptionally well paid. Consultants' total compensation (salary plus bonuses), however, is lower than that found in banking or venture capital. If you choose to move on to other careers, your training will pave the way to better-paid positions than you would have had without consulting experience.

***How is success measured?***

In general, success is based on the quality of your contributions toward solving business cases, your ability to analyze data and communicate findings, and how well you managed clients or other team members. The products of your work should be free of errors, creative, timely, and accurate.

For a team, success means having an important and beneficial effect on the client's business. This could mean saving them millions of dollars a year through cost-cutting measures or discovering new business pursuits. One sign of success is having appreciative clients who are happy with your results and would return for repeat business. The results of the team's efforts may also be seen in press releases announcing changes in the client's strategy and resultant increases in prosperity.

## PROS AND CONS OF THE JOB

### Positive Aspects of a Career in Management Consulting

- There is no such thing as "routine" in management consulting—your client's challenges are constantly changing and you are continually learning.
- Consulting is intellectually intense. Expect a tremendous learning curve, as you are continually exposed to new areas. You will have an opportunity to work on interesting topics and projects that you can be passionate about, such as global health issues, cost-cutting measures, product development, trends in economic systems, and much, much more. Your availability and expertise are often taken into account when projects are assigned, and sometimes you can choose your projects.
- Most firms offer training for their employees. They spend a significant amount of time developing their employees, and some offer a mini-M.B.A. program in which the business fundamentals of economics, strategy, and finance are taught. Recruits are sometimes assigned an advisor, who will oversee their development and ensure that they gain the experience needed to excel.

- Most consulting firms are serious about career development. Consultants receive performance reviews after each assignment and a general review every six months. The performance reviews tend to be objective and include ways to improve performance.
- It is rewarding to help clients solve difficult problems and to provide strategic advice so that their businesses can be more successful. Some projects offer the satisfaction of benefiting society. For example, a project could involve determining how to move patients more quickly into the emergency room or how to help children in third-world countries.

*Management consulting is like a terrific postdoctoral fellowship in business fundamentals.*

- One of the advantages of this career is the opportunity to work on innovative strategies with senior and mid-level executives; it is a wonderful way to develop your network. It is not uncommon for consultants to subsequently find employment with their clients.

*Much satisfaction is derived from interacting with clients and shaping the agendas of important companies.*

- The caliber of management consultants is high. They tend to be extremely ambitious, interesting, accomplished, and passionate people.
- This is a team sport. An extensive amount of collegiality is required, and team members tend to be bright and intellectually stimulating people. Other team members help you learn, grow, and develop.
- Management consulting offers a fast-lane career trajectory. There is the potential to achieve partner status, and it also creates opportunities for many other careers in industry.
- If you enjoy traveling, management consulting provides a wonderful opportunity to do so.

### The Potentially Unpleasant Side of Management Consulting

- Expect to work hard and for long hours. 12-hour days are the norm, and you may work 14–16-hour days plus occasional weekends. Management consulting firms are typically hired for limited time periods, during which they must rapidly learn about their client's complicated business.

*If you are unwilling to balance work and life priorities, management consulting may not be for you.*

- Depending on the firm, management consultants spend an average of 40–70% of their time traveling. Local projects require less travel.
- Consulting is typically referred to as an "up or out environment," which reflects its competitive nature. Consulting firms are structured as pyramids: There are 2–3 managers for each partner, 2–3 consultants for each manager, and so forth. If you are not promoted to the next level within a reasonable amount of time (2–3 years), you are asked to leave. The good news is that most firms support you during your transition period and offer coaching sessions, and some even introduce you to potential employers.

- Although many assignments are very exciting, some of them can be truly boring and repetitive.
- Consulting offers you an opportunity to get an overview of an industry or a business system. On the flip side, it is difficult to become an expert in any particular field.

## THE GREATEST CHALLENGES OF THE JOB

### *Being Comfortable with Ambiguity*

There frequently are no "best" answers for problems in management consulting—keep in mind that the cases for which companies use consultants are frequently very difficult. Aggressive deadlines must be met, and there is limited time to explore for more data, so it is important to be comfortable making decisions with incomplete data.

*Management consultants are like a business version of the Marines. They are thrown into very important, often difficult, situations, and frequently have a relatively short time in which to make a big impact.*

## TO EXCEL IN MANAGEMENT CONSULTING...

### *The Ability to Quickly Assess Relevancy, Regardless of the Industry*

Outstanding management consultants are able to enter a new situation and quickly identify the relevant issues and their business implications, regardless of the specific industry. They usually have an extensive knowledge of the industry.

### *Solid Client Management Skills*

Highly successful consultants develop rapport with clients, understand their concerns, have strong problem-solving skills, and are able to anticipate needs and deliver results. They know which analyses to conduct, they focus on the real problems, and they are proactive.

## Are You a Good Candidate for Management Consulting?

*Consulting allows different personality types to be successful. For example, both introverts and extroverts can excel.*

### *People who flourish in management consulting careers tend to have...*

*An aptitude for counseling.* Management consultants enjoy serving as "thought partners" to clients. They empathetically understand their client's problems and find innovative solutions to their business needs. They provide counsel without owning the product.

*A collaborative, team-player attitude.* Projects are worked on in teams, so it is imperative that coworkers are collegial and respectful. In addition, coworkers share a collegial attitude and assist with other projects, even those to which they are not assigned. Firms take great pride in the tremendous amount of collective knowledge they share.

*Extraordinarily good communication skills.* In many consulting assignments, presentations are the only product of your work that the clients see. It is extremely important to make them robust, concise, and defensible. You will have to master PowerPoint and will need to have good written English and presentation skills to be successful.

*Exceptionally good problem-solving and analytical skills.* Usually the problems are vague and present many variations. This work requires the ability to be an analytical and structured thinker, to tackle complicated problems by breaking pieces down into their core components, and to attack each of the pieces individually. This requires the ability to apply the scientific method and deductive reasoning.

*Creative and innovative thinking.* Often, team members need to consider alternative solutions and to be able to integrate ideas from unrelated disciplines.

*Great self-motivation, drive, and intellectual curiosity.* Self-motivation is required for the level of hard work demanded to deeply explore the client's problems and to arrive at innovative solutions.

*Excellent time-management and multitasking skills.* It is important to be diligent and responsible so that projects are completed on time and team members can rely on you. Projects usually have many components, so even if you are working on only one project at a time, you will still need to keep track of several parts at once.

*A tolerance for ambiguity.* Scientists typically seek the absolute truth, but in consulting, you may have to make decisions based on limited information. There will also be situations in which you will need to speak confidently about your ideas, even if you are not an expert on the subject.

*The ability to see the big picture as well as to understand the small details.* It is important to be able to keep the big picture in mind, but it is also important to be able to look for trends and insights that might not be obvious.

*Honesty and integrity.* You need to be strong enough to provide clients with your best answer, even if it is news that the client may not want to hear.

*The ability to balance work and personal life.* You can spend all day working and still never finish, so it is important to balance and prioritize your time.

*Good stamina and stress resistance.* It is not unusual to spend a few all-nighters preparing your final presentation. And sometimes things can go wrong, so the stress level can escalate quickly. You need to be able to stay calm, rational, and positive, even in very dramatic situations.

***You should probably consider a career outside of management consulting if you are...***

- A person who has been working in industry for a long time. You will no longer be the boss but the worker bee, and it can be a difficult transition.
- Impatient; not interested in details.
- One who prefers to work independently or does not want to work under anyone else.
- Not very analytical and lacking an aptitude for math.

## MANAGEMENT CONSULTING CAREER POTENTIAL

In general, there are two career paths for management consultants: One can work toward becoming a partner or enter a different discipline. Consulting is often seen as a short-term career, and due to the wealth of other opportunities available to skilled consultants, the career has a high turnover rate. In addition, firms provide terrific training and preparation for entering other disciplines.

*Becoming a partner takes about the same amount of time as achieving tenure in academia.*

Only a small fraction of consultants become partners. Whether or not you become partner depends partly on your personality fit in the firm as well as your performance. You need to show that you can lead teams, develop strong client relationships, and provide expertise and the raw ability to solve tough problems.

With exposure to the entire industry and the high-level contacts, the alternative career options for those who don't become partners are seemingly endless. The most common paths are to go into functional disciplines such as business development or marketing, venture capital, or senior management (e.g., CEO). Other less common opportunities lie in clinical development, operations, entrepreneurship, and equity research (i.e., finance) (see Fig. 25-1). Very rarely do people return to academia or their previous residencies, but that avenue is also a viable option.

Organizational structures and titles vary among firms. In general, entry-level consultants are responsible for day-to-day analysis and data generation. As consultants gain more experience, they coordinate and oversee projects, run problem-solving sessions, and work closely with clients. As they move up the ranks, they manage client relationships, provide guidance on a higher level, and bring in new business. Eventually, they can obtain partner status, where they accept internal responsibilities for the firm, such as employee development and more.

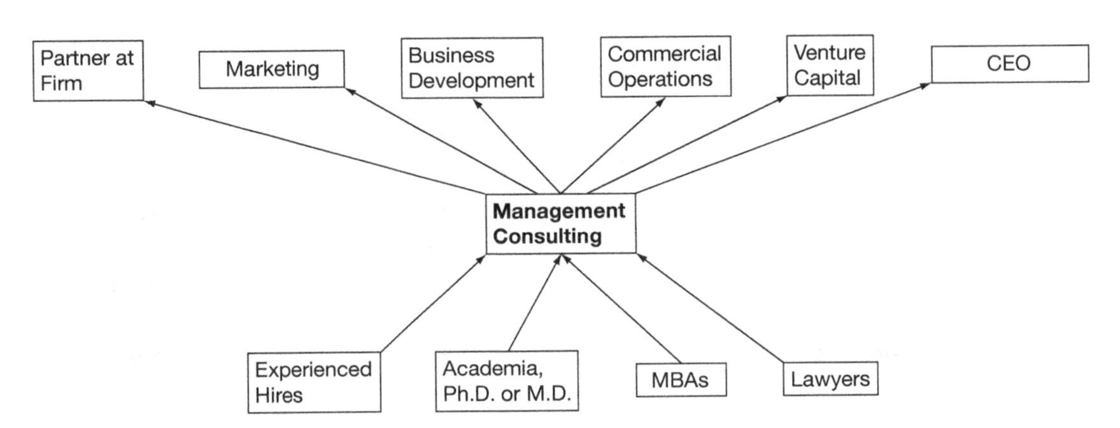


**Figure 25-1.** Common management consulting career paths.

### Job Security and Future Trends

This occupation is not immune to the effects of the economy, but there is a continued need for consultants in both positive and negative economic environments. During economic downturns, companies cannot afford consultants' services, but at the same time, consultants can help companies cut costs. During favorable economic environments, consultants help companies grow.

*Consultants help clients grow when the economy is strong and help them cut costs when the economy weakens.*

## LANDING A JOB IN MANAGEMENT CONSULTING

### Experience and Educational Requirements

Various backgrounds are common in this career. For entry-level positions, an undergraduate degree and sometimes industry-specific knowledge are required. Most people at consultant levels have advanced degrees, and an M.B.A. degree is an advantage. Within health care consulting, companies prefer to hire employees with either Ph.D. or M.D. degrees, although this depends on the firm. Your academic background and your publication record are both taken into consideration. Most first hires arrive straight from residency or graduate school, but specialists ("experienced hires") who can offer specific business expertise are occasionally hired at various levels.

### Paths to Management Consulting

The employment application process for top-level management consulting firms is as competitive as that for top colleges and universities. Here are some ways to improve your chances:

- Apply on-line directly to management consulting firms. The top firms that have a practice in the life sciences are (in alphabetical order): Bain, Boston Consulting Group, L.E.K., McKinsey, and Monitor Group, and there are hundreds of small boutiques that offer specialized services. These companies have a standard application process. It is similar to the college application process, and references, a resume, and a personal statement will be required.
- Apply for positions toward the end of September. Firms have recruiting cycles, and fall is the time to recruit. Offers are generally made at the end of December, and successful hires start work in June or September.
- Attend recruiting events and workshops sponsored by management consulting firms at your local university campus and learn more about their interview process. You can also apply when they are on campus, usually in early September. Find out when they will be visiting by asking your business school office.

- If you are interested in consulting for biotechnology and pharmaceutical companies, apply to locations that are hubs for these industries, such as Boston, San Francisco, San Diego, and New Jersey. If you are consulting in an office in Dallas, for example, where oil and gas companies are more prevalent, your work may be less biotech-related.
- Management consulting firms employ a unique interviewing process based on business cases. To assess your ability to think logically and to find out whether you are comfortable with ambiguity, they will ask about an industry or a topic that you likely know nothing about.
- Start reading business-related news, such as *The Wall Street Journal* and *Business Week,* to gain a deeper appreciation of business activities.
- A few of the consulting firms publish quarterly or yearly journals. Contact each firm and request a copy of its journal so that you can better understand the types of projects that they work on.
- Do not limit yourself to the top firms. There are many boutiques and other types of consulting companies, such as those that specialize in accounting and technology.

***To prepare for interviews:***

It is very important to practice case studies. You should try to review at least 20–30 cases before interviewing.

Visit your business school career center to obtain business cases for practice.

You can also purchase books and guides with practice cases from Wetfeet and Vault.

Visit the management consulting company Web sites and review their cases.

Practice interviewing with people who have been or still are employed at a firm.

Make sure to answer questions with succinct and structured responses.

Be prepared to answer the question "Why do you want to go into consulting?" for each interview.

*The best tip to securing a position in a top firm is to practice, practice, and practice those business case interviews!*

## RECOMMENDED TRAINING, PROFESSIONAL SOCIETIES, AND RESOURCES

WetFeet (www.wetfeet.com) and Vault (www.vault.com) offer career guides with practice cases for management consulting.

*The Wall Street Journal* and *Business Week*

Roberts D.J. 2004. *The modern firm: Organizational design for performance and growth (Clarendon Lectures in Management Studies).* Oxford University Press, New York.

# 26

# Recruiting

## The Business of Matchmaking

ARE YOU A "PEOPLE PERSON" WHO IS ENERGIZED by personal interactions and enjoys networking? Are you a "connector" who enjoys bringing people together? Careers in recruiting (also called "search") offer you the chance to exercise your "people" skills while you make a difference in people's lives and contribute to the scientific community. This occupation is for people with significant industry experience who understand the importance of technical talent and management in companies.

### THE IMPORTANCE OF RECRUITING IN BIOTECHNOLOGY AND DRUG DEVELOPMENT

Recruiters help build teams and serve as hiring consultants. Their strength is in understanding the strategy within an organization and exactly what talent is needed in order to achieve a company's goals. Their expertise and knowledge of the industry and of the leaders in the field provide insight into who is most qualified.

*Competent management is EVERYTHING.*

Venture capitalists invest in promising technologies and, just as importantly, in management teams with a track record of success. The management team can make the difference between a successful company and a debacle, and "human capital" is considered one of a company's greatest assets. As a consequence, organizations pay a premium to secure top talent. It is the recruiter's role to identify, evaluate, and attract that invaluable talent.

*Finding the right candidate can be like finding a needle in a haystack.*

Recruiters provide a valuable service that saves companies time, energy, and money. When human resources (HR) posts a job position on the Internet, they may receive hundreds of resumes, and screening them is an enormous task. A search firm, however, can narrow the field by prequalifying candidates interested in a particular opportunity. In addition, search firms have established extensive networks and can thus increase the probability of finding the most suitable match for the company. Recruiters can expedite the hiring process by serving as third-party arbitrators for both the candidates and companies when negotiating employment contracts. Recruiters also assist growing companies by helping with matters such as organizational development.

### *Making the match*

Some people think that recruiting is easy, but in reality, making the match is difficult! Much luck and experience are needed, as well as the ability to recognize the best-suited candidates. Typical position specifications involve multiple parameters, and it is challenging to find and recruit individuals who match the profile, are interested, and fit into the corporate culture of the company. Candidates are sifted from a very large population of highly qualified individuals. Recruiters sometimes contact hundreds of potential candidates and work many labor-intensive hours before a match is made.

## RECRUITING CAREER TRACKS IN THE LIFE SCIENCES

*In general, there are at least four types of recruiting, but these distinctions can be blurred. For example, some contingency recruiters also do retained work.*

### *In-House Recruiting*

In-house recruiters are employed in the human resources department of biopharma companies and typically work on the lower- and middle-management level hires. They post openings on the company's Web site and on job posting sites, attend job fairs, and work with recruiting agencies. They may participate in the entire process of hiring employees, including screening and interviewing candidates, drafting employment contracts, and handling relocation and other new employee issues. The "hiring managers," the people in companies who request a person to fill a spot, are considered their "customers."

### *Staffing, Temporary, and Direct (Permanent) Hire*

Some firms specialize in "staffing" (hiring staff, usually lower-level employees) and the temporary employment of lower-level technical personnel. Temporary help is needed for employment with a limited time frame, to replace personnel on a leave of absence, or for start-up companies that cannot afford full-time employees. Many staffing companies also provide direct hire or contingency recruiting. There are several large staffing firms that specialize in or have a practice group in the life sciences.

### *Contingency Recruiting*

A client may engage multiple contingency recruiters with the agreement that only the company whose candidate is hired will be paid. Payment is therefore "contingent" on finding the right candidate.

In general, contingency recruiters work in the lower and middle management levels, although some work at the executive level. Some will "market candidates," i.e., help job seekers find positions in companies. There are literally hundreds of contingency firms, and organizational structures vary tremendously. Most are one- to ten-person businesses. There are, in general, two types of positions in staffing firms: recruiters and account managers (sales).

#### *Executive Retained Search*

For these searches, the client "retains" a search firm, i.e., an agency has been hired to identify the appropriate candidate, usually on an exclusive basis. The search firm is paid for its work, *regardless* of how the candidate was identified and whether or not the search was successful.

Typically, these companies work on director, vice president, and executive level searches, such as CXOs (CEO, Chief Operating Officer [COO], Chief Financial Officer [CFO], and Chief Scientific Officer [CSO], etc.). They rely on a large database of contacts and actively recruit the most qualified candidates. Several recruiting roles exist in large retained search firms. Partners bring in business, work on searches, and run the firm. Associates conduct part of the search work along with the partners, and "researchers" generate lists of candidates and identify companies from which to recruit. A select number of worldwide executive retained search firms and many boutiques specialize in the life sciences industry.

## RECRUITING ROLES AND RESPONSIBILITIES

*There are several typical roles and responsibilities for those who work in staffing, contingency, and retained search firms.*

### Finding the Right Clients

#### *Account Management*

Some search professionals spend their time generating business. This is called "business development" in the search industry. It includes client visits to explain the search process and fee structure, demonstrate market knowledge, discuss biographies of coworkers to potential clients, and prepare proposals for potential searches.

### Finding the Right Candidates

#### *Drafting the Position Description or Specification*

*The better you know your client, the better you can make the right match.*

After a business contract for a search has been signed, the first step is to determine the client's needs. Recruiters often visit the client and interview the hiring managers and peers to determine the scope and level of responsibility of the position, the ideal candidate profile, optimal educational requirements, salary level, and other considerations. A position description is then drafted and approved by the client and is used as material for recruiting purposes.

#### *Researching*

Large search firms hire researchers who identify the target companies from which to recruit. This may entail "name generation," where researchers find the names and contact information for possible candidates to approach. Such information is often purchased from vendors.

#### *Sourcing and Contacting Potential Candidates*

Recruiters directly contact prospects by phone or E-mail and try to entice them to consider a particular job opportunity. They also network with sources that can refer qualified people for particular functions. If there is interest and a possible match, the candidate is interviewed for further evaluation.

#### *Interviewing*

Some recruiters spend time either on the phone or in person interviewing prospective candidates. They gauge the qualifications of the candidates to determine whether they have an adequate technical background and whether their personalities will fit into the corporate culture of the client company. Job candidates may also be counseled before interviews with the company.

#### *Presenting a Slate of Candidates and Client Management*

After a slate of candidates has been screened, interviewed, and evaluated, some search firms draft summaries of each candidate and present this list to the hiring manager. The client interviews the candidates and selects finalists. During this time, search professionals also discuss and review the progress and help to arrange interviews.

#### *Reference Checking and Due Diligence*

Recruiters often conduct reference checks to validate the candidate's story and obtain a more objective perspective about the candidate. They may run degree verifications as well.

### Closing the Deal

#### *Negotiating Employment Contracts*

Some recruiters serve as objective, third-party arbitrators to help negotiate contract terms between the candidate and company. Recruiters may also help to smooth the hiring process by introducing real estate agents if relocation is required, conducting informal salary surveys for companies, and more.

#### *Candidate Marketing*

Some search firms market candidates that they deem highly qualified and capable. This means contacting companies and informing them about a terrific person seeking employment and inquiring whether they might need such talent. Most search professionals, however, spend the majority of their time conducting searches to fill positions.

*Networking*

Most recruiters network by attending industry meetings and events. A network of individual contacts and knowledge about people adds tremendous value to their work.

## A TYPICAL DAY IN RECRUITING

*Depending on the type of recruiting position and firm, a typical day might be spent doing some of the following:*

- Visiting prospective customers and describing services.
- Working on a business proposal for a prospective client.
- Interviewing clients to identify the key attributes that are being sought and drafting position specifications.
- Sourcing for candidates: Researching the database and the Internet, and contacting people by E-mail and phone calls.
- Interviewing candidates by videoconference, face-to-face, or by phone, and writing candidate summaries that describe the qualifications and suitability of candidates.
- Discussing the search status on conference calls with clients.
- Setting up client interviews with candidates.
- Checking references.
- Negotiating offers.
- Reading industry news.
- Attending industry conferences to network and generate business.

*One secret in search is to smile when you speak—people can hear the smile on the other end of the phone.*

## SALARY AND COMPENSATION

There is no typical salary for recruiters. In most firms, compensation consists of a base salary plus commissions for successful placements. Commissions are frequently based on a percentage of the first-year income of successful placements. One strategy is to take on a large volume of lower-level positions with smaller commissions, whereas the reverse is to take on a smaller volume of higher-level positions. Executive recruiters conducting CEO and other executive-level searches can earn more money per search, but these searches may require more work and take more time.

*The compensation potential in recruiting is high: You can make as little or as much money as you want, depending on how much effort you are willing to devote.*

Compared to other careers, recruiters tend to earn more than a scientist in industry with the equivalent number of years' experience. In general, they also earn more than the people whom they place. As a rule, recruiters and account managers at staffing firms may earn as much as or more than sales representatives in pharmaceutical companies.

At top executive search firms, the pay depends on the company, the economy, the years of a search professional's experience, and more. Partners in large, top-tier search firms are highly compensated and probably earn the most money compared to other search professionals. Those who own search firms do exceptionally well. In smaller boutiques, the pay varies dramatically. Employees in staffing and contingency firms also do very well.

***How is success measured?***

As this is a highly competitive service industry, happy and loyal customers are the best evidence of success. In the short term, success is measured by the speed and professional execution of a search. Long-term success includes enjoying a continued respected reputation in the industry for quality work and increased brand recognition. In addition, for executive search, success can be measured by the level of the search work (directors, VPs, or CEOs) and the type of clients. As search professionals gain recognition for their work, they move up the chain from recruiting directors to general managers, and on to CEOs and board members.

## PROS AND CONS OF THE JOB

### Positive Aspects to Careers in Recruiting

- It is an immensely valuable service to companies—the success of a company can depend on a good placement. Ultimately, your hire may be instrumental in developing the breakthrough product that helps save people's lives.
- It is very rewarding to help people secure their dream jobs and advance their careers. For candidates who are unemployed or who have limited potential for career advancement in their companies (the proverbial "glass ceiling"), recruiters can provide tremendous new opportunities. Most candidates are grateful for their efforts. In addition, it is gratifying to see your client companies grow and prosper as a result, in part, of your successful placements.

*The thrill of recruiting is finding and making the right match: being the facilitator of a potential long-term relationship.*

- Recruiters meet interesting people from all around the world. The candidates and clients are often accomplished and important, and can include vice presidents of large pharmaceutical and biotechnology companies, venture capitalists, Nobel laureates, and more.
- Recruiters have job flexibility and independence. Particularly true for partners and owners, there is almost complete freedom to manage your schedule and structure your days. Most recruiters do not work on weekends.
- It is a fast-paced, dynamic job, and every day is different. Each search poses unique challenges and every client is different; clients might range from small, venture capital-backed start-ups to large pharmaceutical powerhouses. The opportunity to work on

searches for diverse functions, different therapeutic areas, and geographies also lends variety to the job. Each recruiter will develop an area of specialization over time, and will make use of the efficiencies gained by it.

- Each client represents a chance to learn new technology and discuss the latest trends in research. If you want to, you can be immersed in science and fully apply your technical background.
- You will learn about the inner workings and various vocational areas of the biotechnology industry and how they fit together.
- Recruiters can be well paid, and there is a direct correlation between effort and earnings.
- Each project has a clear beginning and endpoint. When a search is finished, you can move on to new searches.
- Executive search and recruiting becomes easier as you gain more experience. Your network becomes more developed, and consequently, your ability to conduct high-quality work grows.

*The size of your network increases your net worth!*

- Depending on the type of search firm, and particularly for contingency and staffing companies, there might not be much travel required.

### The Potentially Unpleasant Side of Recruiting

- There is a stigma associated with the word "headhunter"; it is a bit like saying you are a "used car salesman."
- Recruiting is labor intensive, and the highly repetitive nature of the work can become tedious.
- Recruiters are under constant pressure to find the best candidates for clients who have urgent business needs that require the right talent. It can feel as if you're on a treadmill.
- Work is fast paced, stressful, and often involves long hours.
- People are unpredictable. Their behavior can disappoint and frustrate, introducing a myriad of uncomfortable situations and events beyond your control. Some act differently from what they claim to be, and they are not always honest. Common examples of stress-inducing situations caused by unpredictable human behavior include the following: the candidate does not show up for his or her first day of work; the "chemistry" is not right for a qualified candidate; the ideal candidate does not want to relocate; the client does not treat candidates respectfully; or the client changes his or her mind about the position's level of responsibility.

*Recruiting is not like FedEx, where you can track packages and know exactly when they arrive—people are unpredictable.*

- It is a highly cyclical business; work depends on the economy. If the biotechnology industry is suffering, the recruiting business will suffer.

- Clients can be unrealistic and often do not appreciate the advice and value that recruiters offer. Most people don't appreciate how hard recruiters work—they think recruiting is a cinch and can be reluctant to pay search fees.
- Although you are providing an important service, the end result is not a quantifiable, tangible product and there is no permanency to your work. Some candidates only stay for a short time or are laid off due to a company downsizing.
- You are not being trained in a marketable occupation with much career development, so it is difficult to parlay your experience to another vocation.
- In many search firms, recruiting is not a team-oriented occupation. You work mostly on your own and are held accountable for your own successes and failures.
- As in sales, frequent rejections to search business proposals and a lack of returned calls from potential candidates can be discouraging.

## THE GREATEST CHALLENGES ON THE JOB

### *Limited Talent Pool*

Search professionals try to identify successful and strong leaders, not just competent people who can fill spots. Such individuals are always in high demand, and they are courted by every agency. This limited pool of talent makes the search process longer and more difficult.

### *Client Management*

Sometimes clients can be challenging to work with. They often do not have a good understanding of what they need, and some change the position specifications during the course of a search. Clients often view their own companies as the best, but if this opinion is not shared by the rest of the world, recruiting talent for them can be difficult.

### *No Barriers to Entry*

There are no barriers to entry into the life science recruiting industry, so people with search experience who lack understanding of the biopharma industry are free to set up business. It is difficult to verify the quality of their work, and there are no industry standards. These unqualified search professionals clamoring for business can create a bad image for the recruiting industry and complicate one's efforts to develop a positive relationship with clients.

### *Developing a Clientele and Reputation*

Recruiting is a highly competitive industry. There are many excellent recruiting firms and recruiters; acquiring new business can be very challenging and takes time.

## TO EXCEL IN RECRUITING...

*Customer Satisfaction and Going the Extra Mile*

People who excel in recruiting are able to provide outstanding customer service and develop mutually respectful relationships with clients and candidates. This ultimately translates into customer loyalty and repeat business. The best relationships are based on quality work and familiarity with the client's needs. In addition, the best recruiters can offer their customers extra benefits that go beyond placing candidates, such as providing highly qualified candidates for positions not yet open, providing consultative knowledge of the industry such as hiring trends and compensation levels, assisting with the organizational development of a company, and providing assistance when negotiating contracts. Executive search professionals can provide start-up companies with referrals to business development leads and industry-focused service professionals, consultants, and venture capitalists. These extra benefits enhance the customer service equation so that recruiters can be seen as business partners and as extensions of human resource departments.

*Experience and Knowledge*

Successful recruiters draw on their extensive industry knowledge, a large network of contacts, and many years of recruiting experience when they confront the many variables that occur during searches. They are able to quickly identify and recruit the most qualified candidates.

*Candidate Qualification Can Be a Tricky Business*

*A recruiter is part matchmaker and part detective.*

One skill to note is the ability to identify qualified and successful candidates based on telephone conversations or personal interviews. This sounds easier than it is. Candidates may not reveal the whole truth at the beginning of a conversation, but through careful listening and by asking the right questions, one can begin to better understand the candidates, their capabilities, and personas. There is a skill, sometimes intuitive, to differentiating between candidates who are truly successful and those who can tell a good story.

### Are You a Good Candidate for Recruiting?

*People who flourish in recruiting tend to have...*

*Being hardworking, smart, and tenacious; having a can-do attitude and superb people skills are the most valued attributes, along with extensive industry knowledge.*

***Boundless energy, drive, and ambition.*** You need to have the personal desire and drive to be successful, which demands hard work and goal-orientation. Recruiters need to have tireless energy to manage the minute details and many tasks. You need to be committed to finding the best people for your client as soon as possible. Proactive client and candidate management skills are a must.

*Superb listening and interpersonal skills.* Interpersonal skills are imperative in developing rapport and trust with clients and candidates (see Chapter 2). In the framework of recruiting, important qualities include:

- Perceptive listening skills, in order to understand the client's needs and to better qualify candidates. It is important to listen carefully during reference checks, not only to what is said, but also to what is NOT said about a candidate.
- Having an inviting personality so that candidates will open up and interview comfortably.
- Being comfortable asking candidates difficult questions.

*Intuition and perceptiveness.* An intense curiosity and intuition about what motivates people is needed. If you sense that a candidate is hiding the truth, referencing may be called for.

*Knowing when to trust a candidate's story requires intuition and perception.*

*Gregarious personalities and excellent networking skills.* Much of a recruiter's value lies in his or her network. This requires being socially confident, able to meet strangers with ease, and being able to make cold calls, perhaps to very important or famous individuals.

*Tenacity and persistence.* You will need to be persistent when contacting the many candidates that are sometimes required for successful searches. Particularly qualified candidates may require extra persistence when you try to contact them.

*Superb multitasking and time management skills.* Recruiters may be working on five to ten searches at a time and may contact anywhere from 30 to 300 people for a single job opening, depending on the difficulty of the search and the level of the position. It is easy to fritter away hours on unproductive calls. To be productive, you need to prioritize carefully and be exacting in how you use your time.

*A sense of urgency.* The client pays a substantial fee for you to quickly solve their business problem, so you need to work diligently until you succeed. It is important to be responsive and to promptly take care of details.

*Strong ethics.* Maintaining candidate and client confidentiality is very important. People's careers and livelihoods are at stake, and it can be very easy to make a slip.

*A selling personality.* You need to sell clients on your skills to conduct the search, sell candidates on opportunities, and sell clients on candidates. Those who can remain upbeat, optimistic, self-confident, and convincing are at an advantage.

*Self-confidence in a corporate environment.* Executive search professionals interact with senior business executives, and it is important to know how to behave appropriately and credibly.

*Thick skin.* Clients can be difficult to work with or disagreeable, and it is important to be diplomatic in tough situations.

***You should probably consider a career outside of recruiting if you are...***

- Too demanding, critical, or have unrealistically high expectations of people.
- Self-centered or self-absorbed.
- Overly affected by the highs and lows in your job.
- Not good at tactical implementation.
- A procrastinator.
- Overly timid and shy.
- More of a talker than a listener.
- Not diplomatic.
- A person who finds it difficult to keep information confidential.

## RECRUITING CAREER POTENTIAL

Recruiting provides experience in developing relationships and in building trust and rapport. These soft skills can be applied to other careers, such as business development, sales, organizational development, human resources, outplacement services, and career counseling (see Fig. 26-1). In general, however, recruiting does not lead to obvious career paths besides climbing the search chain of command and becoming partner in a firm. It is also relatively common for search professionals to work in an established search firm before establishing their own agencies.

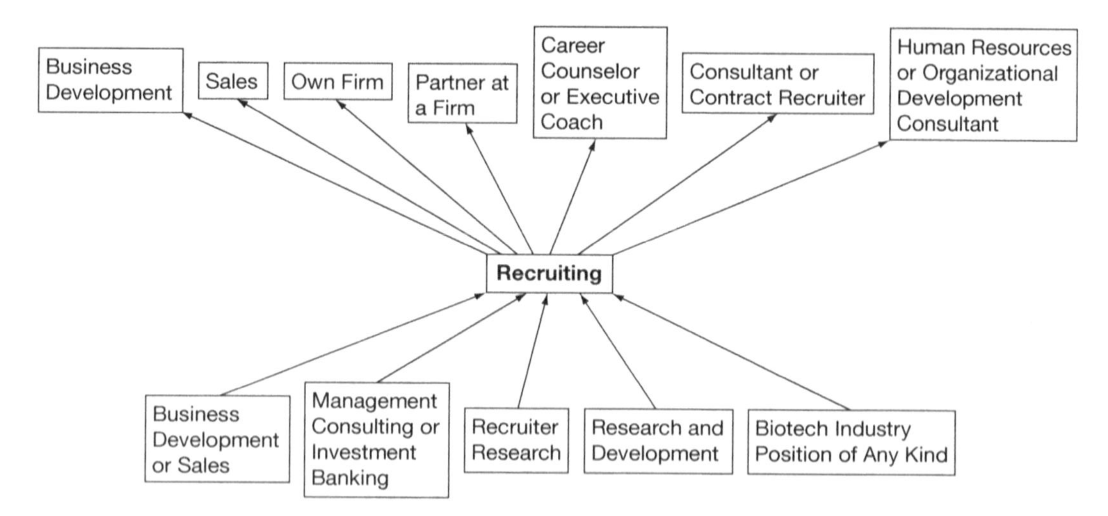


Figure 26-1. Common career paths for recruiters.

### Job Security and Future Trends

*Recruiting is sensitive to the economy.*

The recruiting business directly reflects the biotechnology economy. During boom times, companies are hiring and agencies may have an overabundance of work. During down times, however, business is scarce and search firms spend more time building brand recognition and talking to customers. In-house recruiters suffer the same fate: When the economy is down, there is no need to hire. Instead, they help laid-off employees find new jobs (this is known as "outplacement"). There is some stability in the microeconomics. If one biotechnology sector, such as tools, is slow, one can conduct searches in another, such as drug discovery. If discovery research is being downsized, clinical development might be hiring, etc.

This vocation will not likely be outsourced overseas. To do a good job, you may need to visit the clients and understand the territory. Many firms interview candidates in person. These tasks would be difficult to accomplish from abroad.

## LANDING A JOB IN RECRUITING

### Experience and Educational Requirements

*It is easier to train industry experts to recruit than to train recruiters about the industry.*

For most people, recruiting is a second or third career, and, in general, there are no typical educational requirements. Executive search professionals come from just about every conceivable vocational area (see Fig. 26-1), but those with sales or business development backgrounds and strong interpersonal skills are the most common entrants. An extensive knowledge of the industry and a reasonable network of contacts are required. You should have a deep appreciation of the vertical segment of the biotechnology market for which you are recruiting.

Many search professionals do not have advanced science degrees, but it is a great advantage to have one or else to have significant industry experience. Advanced degrees provide validated credibility with both clients and candidates. You will be better able to convince candidates of the technical merits of an opportunity, and you will have more confidence and authority when contacting leading industry experts. With a strong science background, recruiters can better understand the needs of the client and can appreciate the skills of the candidate.

Recruiters who qualify for the higher-level contingency and executive search firms often possess M.B.A. degrees, because they interact with executives. Those with college degrees who have an interest in executive search firms can initially join as research analysts (researchers). Research analysts move toward research associate positions and eventually become research consultants and recruiters. Alternatively, you can skip the research step by obtaining an M.B.A., or with biotechnology industry experience (of just about any kind), and be accepted directly as a recruiter. The top-tier executive search firms tend to hire VPs and executives with extensive industry experience and a large network of contacts.

### Paths to Recruiting

- The vast majority of recruiters never originally considered a position in a search firm but instead opportunistically happened upon it. Typically, they were called about a specific job opportunity, met a recruiter for an interview, and at some point in the discussion wound up talking about the possibility of a career in recruiting. Thus, one way to enter the field is to interview with recruiters and then inquire about it as a potential vocation.
- Learn about the drug discovery world and the market in which you want to recruit. Become more knowledgeable by familiarizing yourself with the companies, the therapeutic areas that they specialize in, and the key experts.
- If you are considering joining a search firm, it is important to choose one with integrity and the same values that you possess. Inquire from others about which firms have the best reputations and which have high employee turnover rates. Talk to people who have previously worked in the firm and interview their clients.
- Join firms with a life science practice. It is important to build a network of relevant contacts and to learn industry-specific knowledge, which takes years.
- Network, attend meetings, and develop your Rolodex. An extensive network is highly valued in this industry. It's not just the quantity, but also the quality, of your network that is important—your relations with higher-level executives and venture capitalists are advantages.

*Sometimes it's not what you know but who you know that matters.*

- If you are interested in eventually starting your own company, consider first working for an established recruiting firm.
- To identify recruiting firms, Kennedy's International Directory of Recruiters (see www.kennedyinformation.com) lists contact information and Web sites. You can also identify the recruiting companies that specialize in the life sciences industry by conducting a Google search.

## RECOMMENDED PROFESSIONAL SOCIETIES AND RESOURCES

*Societies and Resources*

The Fordyce Letter (www.fordyceletter.com), a monthly newsletter for recruiters

National Association of Executive Recruiters (www.naer.org)

*Books*

*The directory of executive recruiters*, 35th edition. 2006. Kennedy Information, Peterborough, New Hampshire. A large directory reference of executive recruiters.

MacKie R. 2007. *Take this job and sell it! The recruiter's handbook.* QED Press, Fort Bragg, California.

### *Other Books to Consider*

*Harvard Business Review* or other books or magazines that discuss what qualities make a good executive

Books about how to interview people

Books for job candidates about how to interview

# Index

## D

## E



## F

## N



## O



## P

## Q